FACTORIZATION, SINGULAR OPERATORS AND RELATED PROBLEMS

Factorization, Singular Operators and Related Problems

Proceedings of the Conference in Honour of Professor Georgii Litvinchuk

Edited by

Stefan Samko
Universidade do Algarve,
Faro, Portugal

Amarino Lebre
Instituto Superior Técnico,
Lisboa, Portugal

and

António F. dos Santos
Instituto Superior Técnico,
Lisboa, Portugal

KLUWER ACADEMIC PUBLISHERS
DORDRECHT / BOSTON / LONDON

ISBN 978-90-481-6333-5

Published by Kluwer Academic Publishers,
P.O. Box 17, 3300 AA Dordrecht, The Netherlands.

Sold and distributed in North, Central and South America
by Kluwer Academic Publishers,
101 Philip Drive, Norwell, MA 02061, U.S.A.

In all other countries, sold and distributed
by Kluwer Academic Publishers,
P.O. Box 322, 3300 AH Dordrecht, The Netherlands.

Printed on acid-free paper

Contents

Preface

These proceedings comprise a large part of the papers presented at the International Conference *Factorization, Singular Operators and related problems*, which was held from January 28 to February 1, 2002, at the University of Madeira, Funchal, Portugal, to mark Professor Georgii Litvinchuk's 70th birthday.

Experts in a variety of fields came to this conference to pay tribute to the great achievements of Professor Georgii Litvinchuk in the development of various areas of operator theory. The main themes of the conference were focussed around the theory of singular type operators and factorization problems, but other topics such as potential theory and fractional calculus, to name but a couple, were also presented.

The goal of the conference was to bring together mathematicians from various fields within operator theory and function theory in order to highlight recent advances in problems many of which were originally studied by Professor Litvinchuk and his scientific school. A second aim was to stimulate international collaboration even further and promote the interaction of different approaches in current research in these areas.

The Proceedings will be of great interest to researchers in Operator Theory, Real and Complex Analysis, Functional and Harmonic Analysis, Potential Theory, Fractional Calculus and other areas, as well as to graduate students looking for the latest results.

We acknowledge, with gratitude, the financial support of the following organizations:

▷ Fundação para a Ciência e a Tecnologia
  (through Fundo de Apoio à Comunidade Científica);
▷ Centro de Matemática Aplicada (financed by FCT);
▷ Reitoria da Universidade Técnica de Lisboa;
▷ Universidade da Madeira;
▷ Centro de Ciência e Tecnologia da Madeira;
▷ Fundação Luso-Americana para o Desenvolvimento;

and the general support of

- ▷ Universidade da Madeira;
- ▷ Universidade do Algarve;
- ▷ Instituto Superior Técnico.

The Editors would also like to thank Dr. Alexei Karlovich for helping to improve the Latex files for this book.

THE EDITORS

This book is dedicated
to Professor Georgii
Semenovich Litvinchuk.

Foreword

Professor Georgii Semenovich Litvinchuk

This conference was held in honour of Professor Georgii Litvinchuk's outstanding contributions to the theory of singular operators, boundary value problems and factorization problems.

Founder of the modern theory of singular integral equations with shifts, author of several books and the leader of a strong scientific school nowadays spread all over the world, Professor Litvinchuk is known for the invaluable role he played in the development of various fields within the theory of singular operators and factorization problems. He has contributed many pioneering results in the theory of boundary value problems, integral equations of singular type and other fields, which is reflected in the fact that he has written four books.

Professor Georgii Semenovich Litvinchuk celebrated his seventieth birthday on the 27th June, 2001.

He started his scientific career as an assistant professor at the Rostov State University, Rostov-on-Don. He received his PhD degree in 1958 under the supervision of Professor F.D. Gakhov, an outstanding and well-known mathematician who became famous, in particular, due to his brilliant, explicit solution of the Riemann-Hilbert boundary conjugation problem for analytic functions. It was during his professorship at the Rostov State University that Litvinchuk created the theory of singular integral equations with Carleman shift.

But the luckiest and happiest period of Litvinchuk's scientific research started when he accepted a professorship in Odessa, a beautiful city in the southern part of the Soviet Union, now the Ukraine. A young 33 years old associate professor, in 1964 he acquired the Chair of Mathematical Analysis at the Odessa State University. Later he organized and became the Head of the Department of Mathematical Modelling of Economical and Ecological Systems at the Institute of Economics of the Ukrainian Academy of Science.

In 1969 Litvinchuk was awarded the degree of Doctor of Mathematical and Physical Sciences for his thesis "Investigation of boundary value problems with shift for analytic functions and functional singular equations". The same year he was given the status of Full Professor.

His book *"Boundary value problems and singular equations with shift"* (in Russian), published later in 1977 by "Nauka" played a crucial role in the development of the theory of functional singular operators, leading to a substantial increase in the number of research papers in this direction. In 1982 the Chinese translation of this book appeared.

Professor Litvinchuk spent a very fruitful period in Odessa from 1964 to 1992. Very enthusiastic and active, he had an unusual ability to attract young, talented students. It was in Odessa that he created a strong research group, which included many capable and productive students who later became his colleagues and collaborators. He started a seminar on boundary value problems and singular integral equations which he guided for more than 20 years at the Odessa State University.

Factorization problems, naturally related with singular equations, were a subject of special attention of this group. As a result, in 1984 the book

Litvinchuk, G.S. and Spitkovskii, I.M. *Factorization of measurable matrix functions* (in Russian),

appeared, translated then into English:

Litvinchuk, G.S. and Spitkovskii, I.M. *Factorization of measurable matrix functions.* Operator Theory: Advances and Applications, 25. Birkhäuser Verlag, Basel, 1987.

Many of Litvinchuk's students are nowadays university professors spread around all over the world. We mention here only those of them who obtained the second degree in mathematics in the former Soviet Union:
Yuri Karlovich (Universidad Autónoma del Estado de Morelos, México),
Viktor Kravchenko (Universidade do Algarve, Portugal),
Ilya Spitkovsky (College of William and Mary, USA),
Nikolai Vasilevski (CINVESTAV del Instituto Politécnico Nacional, México).
There are many others in Azerbaidjan, Kyrgyzstan, Mexico, Poland, Portugal, Russia, Ukraine, USA, Uzbekistan, indeed an impressive geographical list.

The total number of students who defended their PhD theses under the supervision of Professor Litvinchuk is amazing: there are thirty eight of them!

The range of Litvinchuk's interests in the field of mathematics was really wide. It is reflected, to a certain extent, by the diversity of the talks presented at the conference.

The reader is referred to the interesting paper

B.V. Khvedelidze, L.D. Kudriavtsev, Yu.A. Mitropol'skii, S.M. Nikol'skii and V.S. Valdimirov. "Georgii Semenovich Litvinchuk (on his sixtieth birthday)", *Uspekhi Mat. Nauk* 47:1 (1992), 221-223 (English translation in *Russian Math. Surveys*, 47:1 (1992), 271-274),

relating to the period up until 1992.

In the year 1992, Professor Litvinchuk chose to live in Portugal and pursue his scientific activity there. This new period of his life has now already continued for ten years. He keeps up active research with new Portuguese colleagues at Instituto Superior Técnico (A. Lebre, F. Teixeira) and menaged to complete the preparation of his book with Prof. V. Kravchenko, which was started in Odessa. It appeared in 1994:

Kravchenko, V.G.; Litvinchuk, G.S. *Introduction to the theory of singular integral operators with shift*, Kluwer, 1994.

Since 1994, Professor Litvinchuk has been *Professor Catedrático* at the University of Madeira. He continues to be active, both in teaching and research. His fourth monograph was written during this period:

Litvinchuk, G.S. *Solvability Theory of Boundary value Problems and Singular Integral Equations with Shifts*, Kluwer, 2000.

It was the general impression amongst participants of the conference that Professor Litvinchuk, after decades of an intensive and fruitful scientific career, continues to be active and remains in a very good shape. We share this impression and wish many, many more happy days to Professor Litvinchuk, both in mathematics which he will never be able to leave completely, and in everyday life.

THE EDITORS
December, 2002

Francisco S. Teixeira

A few words about Georgii Litvinchuk (Address at the Opening Session)

I was kindly requested to speak about Professor Georgii Litvinchuk, in particular regarding the very short and specific - but for us very important - time during which his activities were related to the Department of Mathematics of Instituto Superior Técnico.

It all started in December 91, when I received a touching letter from Professor Litvinchuk, in which he stated that he was willing to cooperate with our Department. We were naturally very pleased and honoured by his offer, without however truly realising the significance that this would have for the life of Georgii.

At the time, when he was no longer very young, in 1992, he and his wife started a new life in an unknown country, with an unknown language, social and scientific habits.

It is very hard for me, never having been in Ukraine, to imagine the effort and courage that surely were needed, at the age of 60, in order to carry out such a step.

This fact is all the more noteworthy as he brought with him the ability to charm others, to adapt and, above all, his unusual working capacity, highly contagious for all those around him, and which allowed us to have a glimpse of how someone is able to successfully supervise around 40 Ph D. students, write two books and more than 170 papers, ultimately creating his own internationally renowned scientific school.

It is not up to me to speak here about his scientific work, but it is fundamental to mention that, during the two years he spent with us, he finished his joint book with Viktor Kravchenko and started new research on Singular Integral Operators with shift, which led to several joint papers with his young Portuguese colleagues. Besides all this, including having been away for six months in Australia, he was teaching post-graduate courses and organising a research seminar which are still fondly remembered by all who attended. His stay also enabled us to establish contacts with many of his former colleagues and students, of which I will mention only four that are present here, Viktor Kravchenko, Yuri Karlovich, Ilya Spitkovsky and Nikolai Vasilevski, one of whom eventually joined the University of Algarve, in Portugal.

These two years were also more than enough to create a very strong friendship between us, which endures even when I recall the amazing and extraordinary episodes and tricks I was forced to develop in order to solve the unbelievable "Visa" problems with which our burocratic machine kept harassing Georgii.

I must now remember someone who all of us miss and surely would have liked to be with us today, Professor Erhard Meister, who very wisely concluded a conversation with me in Darmstadt, after I had told him that we would have the possibility of having Georgii Litvinchuk as a Visiting Professor, saying: "How fortunate you are!".

So, just to finish this address, I want to tell Professor José Carmo: "How fortunate of you to have Professor Litvinchuk at your Mathematics Department of the University of Madeira".

CONTRIBUTED PAPERS

SINGULAR INTEGRALS ALONG FLAT CURVES WITH KERNELS IN THE HARDY SPACE $H^1(S^{n-1})$

Ahmad Al-Salman*
Yarmouk University, Irbid-Jordan
alsalman@yu.edu.jo

Hussain Al-Qassem
Yarmouk University, Irbid-Jordan
husseink@yu.edu.jo

To Professor G. S. Litvinchuk on the occasion of his 70th birthday

Abstract We study singular integral operators along subvarieties determined by flat curves and kernels in the Hardy space $H^1(S^{n-1})$. We prove that these operators are bounded on L^p for all $p \in (1, \infty)$. Our results extend previously known results.

Keywords: Singular integrals, rough kernels, maximal functions, and Hardy space.

1. Introduction and statement of results

Let $n \geq 2$ and $\mathbf{S}^{n-1}$ be the unit sphere in $\mathbf{R}^n$ equipped with the induced Lebesgue measure $d\sigma$. Let Ω be a homogeneous function of degree zero on $\mathbf{R}^n$ satisfying $\Omega \in L^1(\mathbf{S}^{n-1})$ and

$$\int_{\mathbf{S}^{n-1}} \Omega(y) d\sigma(y) = 0. \tag{1.1}$$

*Supported by Yarmouk University

For suitable mappings $\Phi : \mathbf{R}^n \to \mathbf{R}^m$ $(m \geq 2)$ and $\Gamma : \mathbf{R}^+ \to \mathbf{R}$, we define the singular integral operator $\mathbf{T}_{\Omega,\Phi,\Gamma}$ on $\mathbf{R}^m$ by

$$\mathbf{T}_{\Omega,\Phi,\Gamma} f(x) = \text{p.v.} \int_{\mathbf{R}^n} \frac{\Omega(y')}{|y|^n} f\left(x - \Gamma(|y|)\Phi(y')\right) dy, \tag{1.2}$$

where $y' = y/|y|$ for $y \neq 0$ and $f \in \mathcal{S}(\mathbf{R}^m)$.

For the sake of simplicity, if $m = n$, and $\Phi(y') = y'$ we denote $\mathbf{T}_{\Omega,\Phi,\Gamma} = \mathbf{T}_{\Omega,\Gamma}$. Also, if $\Gamma(t) = t$, $\mathbf{T}_{\Omega,\Gamma}$ is the classical Calderón-Zygmund singular integral operator which will be denoted by $\mathbf{T}_\Omega$.

In their fundamental work ([4]), Calderón and Zygmund showed that $\mathbf{T}_\Omega$ is bounded on L^p for all $p \in (1, \infty)$ provided that $\Omega \in L\log^+ L(\mathbf{S}^{n-1})$. Moreover, Calderón and Zygmund also showed that the condition $\Omega \in L\log^+ L(\mathbf{S}^{n-1})$ is nearly optimal for the L^p boundedness to hold. Subsequently, it was proved independently by Connett ([6]) and Ricci-Weiss ([14]) that T_Ω is bounded on L^p for all $p \in (1, \infty)$ if $\Omega \in H^1(\mathbf{S}^{n-1})$ (which contains $L\log^+ L(\mathbf{S}^{n-1})$ properly). The definition of $H^1(\mathbf{S}^{n-1})$ will be reviewed in Section 2.

One notices that the operator $\mathbf{T}_\Omega$ has singularity along the diagonal $\{x = y\}$. Recently, many problems in analysis have led to study singular integrals with singularity along more general sets, some in the form of $\{x = \Lambda(y)\}$ (see for example, [15], [16]). Subsequently, the study of singular integrals with singularity along sets of the form $\{x = \Gamma(|y|)y'\}$ has attracted the attention of many authors. We shall content ourselves here with recalling only the following pertinent results:

Theorem A ([1]) *Suppose that $\Gamma(t)$ is a real-valued polynomial with $\Gamma(0) = 0$. Then the operator $\mathbf{T}_{\Omega,\Gamma}$ is bounded on $L^p(\mathbf{R}^n)$ for every $1 < p < \infty$ provided that $\Omega \in \mathbf{H}^1(\mathbf{S}^{n-1})$ (for $p = 2$ it was proved by Fan and Pan in [11]).*

Theorem B ([9]) *Suppose that $\Gamma(t)$ is a real-valued function defined on $\mathbf{R}_+$ that satisfies:*

$$|\Gamma(t)| \leq C_1 |t|^h, \quad \left|\Gamma''(t)\right| \leq C_2 |t|^{h-2}, \tag{1.3}$$

$$C_4 |t|^{h-1} \leq |\Gamma'(t)| \leq C_3 |t|^{h-1} \tag{1.4}$$

for some $h \neq 0$ and $t \in (0, \infty)$, where C_1, C_2, C_3 and C_4 are positive constants independent of t. Then the operator $\mathbf{T}_{\Omega,\Gamma}$ is bounded on $L^p(\mathbf{R}^n)$ for every $1 < p < \infty$ under the condition $\Omega \in \mathbf{H}^1(\mathbf{S}^{n-1})$.

Theorem C ([3]) *Suppose Γ is convex increasing with $\Gamma(0) = 0$. If $\Omega \in \mathbf{H}^1(\mathbf{S}^{n-1})$, then $\mathbf{T}_{\Omega,\Gamma}$ is bounded on $L^p(\mathbf{R}^n)$ for $1 < p < \infty$.*

The main purpose of this paper is to study the L^p boundedness of the more general class of operators $\mathbf{T}_{\Omega,\Phi,\Gamma}$. In fact, we have the following results.

Theorem D *Suppose that* $\Phi \mid_{\mathbf{S}^{n-1}}$ *is real analytic and* $\Gamma : \mathbf{R}^+ \to \mathbf{R}$ *is a generalized polynomial of the form* $\Gamma(t) = \mu_1 t^{d_1} + \cdots + \mu_l t^{d_l}$ *for some* $l \in \mathbf{N}$, *distinct real numbers* $d_1, \ldots, d_l$, *and real numbers* $\mu_1, \ldots, \mu_l$. *Then the operator* $\mathbf{T}_{\Omega,\Phi,\Gamma}$ *is bounded on* $L^p(\mathbf{R}^m)$ *for every* $1 < p < \infty$ *provided that* $\Omega \in \mathbf{H}^1(\mathbf{S}^{n-1})$. *Moreover, the bound for the operator norm* $\|\mathbf{T}_{\Omega,\Phi,\Gamma}\|_{p,p}$ *is independent of the coefficients* $\mu_1, \ldots, \mu_l$.

Theorem E *Suppose that* $\Gamma(t)$ *is a real-valued function which satisfies* (1.3)-(1.4), *and* $\Phi \mid_{\mathbf{S}^{n-1}}$ *is real analytic. If* $\Omega \in \mathbf{H}^1\left(\mathbf{S}^{n-1}\right)$, *then* $\mathbf{T}_{\Omega,\Phi,\Gamma}$ *is bounded on* $L^p(\mathbf{R}^m)$ *for* $1 < p < \infty$.

Theorem F *Suppose* Γ *is convex increasing with* $\Gamma(0) = 0$ *and* $\Phi \mid_{\mathbf{S}^{n-1}}$ *is real analytic. If* $\Omega \in \mathbf{H}^1\left(\mathbf{S}^{n-1}\right)$, *then* $\mathbf{T}_{\Omega,\Phi,\Gamma}$ *is bounded on* $L^p(\mathbf{R}^m)$ *for* $1 < p < \infty$.

Throughout this paper, the letter C will stand for a positive constant that may vary at each occurrence but it is independent of the essential variables. Finally, the authors would like to thank the referee for his helpful comments.

2. The Hardy space $\mathbf{H}^1(\mathbf{S}^{n-1})$

In this section we recall the definition of the Hardy space $\mathbf{H}^1(\mathbf{S}^{n-1})$ and some of its important properties.

Definition 2.1 The Hardy space $\mathbf{H}^1(\mathbf{S}^{n-1})$ is defined as

$$\mathbf{H}^1(\mathbf{S}^{n-1}) = \{f \in L^1(\mathbf{S}^{n-1})^{n-1}) : \|P^+(f)\|_{L^1(\mathbf{S}^{n-1})} < \infty\}$$

with $\|f\|_{\mathbf{H}^1(\mathbf{S}^{n-1})} = \|P^+(f)\|_{L^1(\mathbf{S}^{n-1})}$.

Here

$$P^+(f)(x) = \sup_{0 \le r < 1} \left| \int_{\mathbf{S}^{n-1}} P_{rx}(y) f(y) d\sigma(y) \right|,$$

and P_{rx} is the Poisson kernel on $\mathbf{S}^{n-1}$ given by

$$P_{rx}(y) = \frac{1 - r^2}{|rx - y|^n}$$

for $r \in [0, 1)$ and $x, y \in \mathbf{S}^{n-1}$.

A very useful characterization of the space $\mathbf{H}^1(\mathbf{S}^{n-1})$ is its atomic decomposition.

Definition 2.2 (a) A function $b(\cdot)$ on $\mathbf{S}^{n-1}$ is a regular atom if there exist $\zeta \in \mathbf{S}^{n-1}$ and $\rho \in (0, 2]$ such that

$$\operatorname{supp}(b) \subset \mathbf{S}^{n-1} \cap \mathbf{B}(\zeta, \rho), \text{ where } \mathbf{B}(\zeta, \rho) = \{y \in \mathbf{R}^n : |y - \zeta| < \rho\}; \tag{2.1}$$

$$\|b\|_\infty \leq \rho^{-n+1}; \tag{2.2}$$

$$\int_{\mathbf{S}^{n-1}} b(y')d\sigma(y') = 0. \tag{2.3}$$

(b) A function $b(\cdot)$ on $\mathbf{S}^{n-1}$is an exceptional atom if $b(\cdot) \in L^\infty(\mathbf{S}^{n-1})$ and $\|b\|_\infty \leq 1$.

The atomic decomposition of $\mathbf{H}^1(\mathbf{S}^{n-1})$ is given in the following proposition which can be found in ([5]).

Proposition 2.3 *For any $f \in \mathbf{H}^1(\mathbf{S}^{n-1})$ there are complex numbers λ_j and atoms (regular or exceptional) b_j such that*

$$f = \sum_{j=1}^{\infty} \lambda_j b_j \text{ and } \sum_j |\lambda_j| \approx \|\Omega\|_{\mathbf{H}^1(\mathbf{S}^{n-1})}.$$

3. Preparation

We start this section by recalling the following version of Lemma 5.2 in ([8]):

Lemma 3.1 *Let $l, m \in \mathbf{N}$and $\{\sigma_{s,k} : 0 \leq s \leq l,\ k \in \mathbf{Z}\}$ be a sequence of measures on $\mathbf{R}^m$ with $\sigma_{0,k} = 0$ for every $k \in \mathbf{Z}$. Let $\{\alpha_{sj} : 1 \leq s \leq l$ and $1 \leq j \leq 2\} \subseteq \mathbf{R}^+$, $\{a_k\}$ be a lacunary sequence with $\inf_k \frac{a_{k+1}}{a_k} = a > 1$, $\{M_s : 1 \leq s \leq l\} \subset \mathbf{N}$, and let $L_s : \mathbf{R}^m \to \mathbf{R}^{M_s}$ be linear transformations for $1 \leq s \leq l$. Suppose that:*

(i) *$\|\sigma_{s,k}\| \leq 1$ for $k \in \mathbf{Z}$ and $1 \leq s \leq l$;*

(ii) *$|\hat{\sigma}_{s,k}(\xi)| \leq C\,|a_k L_s(\xi)|^{-\alpha_{s2}}$ for $\xi \in \mathbf{R}^m, k \in \mathbf{Z}$ and $1 \leq s \leq l$;*

(iii) *$|\hat{\sigma}_{s,k}(\xi) - \hat{\sigma}_{s-1,k}(\xi)| \leq C\,|a_{k+1} L_s(\xi)|^{\alpha_{s1}}$ for $\xi \in \mathbf{R}^m, k \in \mathbf{Z}$ and $1 \leq s \leq l$;*

(iv) *For some $q > 1$ there exists $A_q > 0$ such that*

$$\left\|\sup_{k\in\mathbf{Z}} |\sigma_{s,k}| * f\right\|_{L^q(\mathbf{R}^m)} \leq A_q \|f\|_{L^q(\mathbf{R}^m)} \tag{3.1}$$

for all $f \in L^q(\mathbf{R}^m)$ and $1 \leq s \leq l$.

Then for any $p \in (\frac{2q}{q+1}, \frac{2q}{q-1})$, there exists a positive constant C_p such that the inequality

$$\left\|\sum_{k\in\mathbf{Z}} \sigma_{l,k} * f\right\|_{L^p(\mathbf{R}^m)} \leq C_p \|f\|_{L^p(\mathbf{R}^m)} \tag{3.2}$$

holds for all $f \in L^p(\mathbf{R}^m)$. The constant C_p is independent of the linear transformations $\{L_s\}_{s=1}^{l}$.

We recall the following lemma due to Van der Corput ([16], p. 332):

Lemma 3.2 *Suppose ϕ and $\psi$are real-valued and smooth in (a,b), and that $\left|\phi^{(k)}(t)\right| \geq 1$ for all $t \in (a,b)$. Then*

$$\left|\int_a^b e^{-i\lambda\phi(t)}\psi(t)dt\right| \leq C_k\left|\lambda\right|^{-\frac{1}{k}}\left[\left|\psi(b)\right| + \int_a^b \left|\psi^{'}(t)\right| dt\right],$$

holds when:

(i) $k \geq 2$, *or*

(ii) $k = 1$ *and* $\phi^{'}$ *is monotonic.*

The bound C_k is independent of a, b, ϕ, and λ.

The following lemma can be found in ([3]):

Lemma 3.3 *Let $\Omega \in L^1(\mathbf{S}^{n-1})$ and $\Psi : \mathbf{R}^n \to \mathbf{R}^m$ be such that $\Psi(ty') = R(t)\Phi(y')$ where $R(t)$ is a convex increasing function with $R(0) = 0$ and Φ is real analytic on $\mathbf{S}^{n-1}$. Then*

$$Mf(x) = \sup_{k\in\mathbf{Z}}\left|\int_{2^k\leq|y|<2^{k+1}} f(x+\Psi(y))\frac{|\Omega(y')|}{|y|^n}dy\right|$$

is bounded on L^p for all $1 < p < \infty$. Moreover

$$\|Mf\|_p \leq C_p\|\Omega\|_1\|f\|_p.$$

Our oscillatory estimates are based on the following two lemmas:

Lemma 3.4 ([2]) *Let $n, m \in \mathbf{N}$ and $F : \mathbf{S}^{n-1} \times S^{m-1} \to \mathbf{R}$ be such that for each $\eta \in S^{m-1}$, $F(\cdot,\eta)$is a nonconstant real-valued analytic function on $\mathbf{S}^{n-1}$. Then there exist positive constants δ and A such that for each $\eta \in S^{m-1}$, there exist open subsets $U_1, \ldots, U_{l(\eta)}$ of $\mathbf{S}^{n-1}$ which cover $\mathbf{S}^{n-1}$ such that*

$$\sup_{y\in U}\int_U |F(x,\eta) - F(y,\eta)|^{-\delta} d\sigma(x) \leq A \tag{3.3}$$

for $U \in \left\{U_1, \ldots, U_{l(\eta)}\right\}$.

Lemma 3.5 *Suppose $\rho \in (0, \frac{1}{\sqrt{n}})$, $n \geq 2$, $m \geq 2$,$\mathbf{e} = (0,0,\ldots,1)$, and $\mathbf{B}_\rho = \{x^{'} = (x_1^{'}, \ldots, x_n^{'}) \in \mathbf{S}^{n-1} : |x^{'} - \mathbf{e}| < \rho\}$. For a multi-index $\alpha = (\alpha_1, \ldots, \alpha_{n-1}) \in (\mathbf{N} \cup \{0\})^{n-1}$and $u = (u_1, \ldots, u_{n-1}) \in \mathbf{R}^{n-1}$ let $u^\alpha = u_1^{\alpha_1} \ldots u_{n-1}^{\alpha_{n-1}}$. Let $P(u) = (P_1(u), \ldots, P_m(u)) : \mathbf{R}^{n-1} \to \mathbf{R}^m$ be a polynomial* *mapping with degree* d *(*$d = \max_{1\le j\le m} P_j$*). Then there exists a linear transformation* $L_d : \mathbf{R}^m \to \mathbf{R}^{m_d}$ *such that*

$$\int_{\mathbf{B}_\rho} \frac{d\sigma(y')}{|\xi \cdot P(\pi_{n-1}(y'))|^{\epsilon}} \le C\rho^{n-1}\left|\rho^d L_d(\xi)\right|^{-\epsilon},$$

for every $\epsilon \in (0, \frac{1}{2d})$ *with constant* C *independent of* ρ *and* $\xi \in \mathbf{R}^m\backslash\{0\}$. *Here* m_d *is the number of distinct multi-indices* $\alpha \in (\mathbf{N}\cup\{0\})^{n-1}$ *with* $|\alpha| = \sum_{j=1}^{n-1}\alpha_j = d$ *and* $\pi_{n-1}(y) = \pi_{n-1}(y_1,\dots,y_n) = (y_1,\dots,y_{n-1})$.

Proof. Let $P_j(u) = \sum_{|\alpha|\le d} a_{\alpha j} u^{\alpha}$ and $\xi = (\xi_1,\dots,\xi_m)$. For each multi-index $\alpha \in (\mathbf{N}\cup\{0\})^{n-1}$ with $|\alpha| = d$, let $L_\alpha(\xi) = \sum_{j=1}^m \xi_j a_{\alpha j}$. Then $\xi\cdot P(u) = \sum_{|\alpha|=d} L_\alpha(\xi)u^{\alpha} + R(\xi,u)$ where $R(\xi,u)$ is a real valued polynomial in the u variable with degree less than d in each coordinate $u_1,\dots,u_{n-1}$. Let $L_d : \mathbf{R}^m \to \mathbf{R}^{m_d}$ be the linear transformation defined by $L_d(\xi) = (L_{\alpha_1}(\xi),\dots,L_{\alpha_{m_d}}(\xi))$, where $\alpha_1,\dots,\alpha_{m_d}$ are the distinct multi-indices $\alpha\in(\mathbf{N}\cup\{0\})^{n-1}$ with $|\alpha| = d$. Thus by a proper substitution, it is easy to see that

$$\begin{aligned}
&\int_{\mathbf{B}_\rho} \frac{d\sigma(y')}{|\xi \cdot P(\pi_{n-1}(y'))|^{\epsilon}} \\
&\le \sqrt{\frac{n}{n-1}}\int_{-\rho}^{\rho}\int_{-\rho}^{\rho}\cdots\int_{-\rho}^{\rho}\frac{du_1\cdots du_{n-1}}{\left|\sum_{|\alpha|=d}L_\alpha(\xi)u^\alpha + R(\xi,u)\right|^{\epsilon}} \\
&\le C\rho^{n-1}\left|\rho^d L_d(\xi)\right|^{-\epsilon},
\end{aligned}$$

where the last inequality follows by a repeated use of Ricci-Stein's result (p. 182, [13]). □

4. Proof of main results

Proof of Theorem F. Let $\Omega \in \mathbf{H}^1(\mathbf{S}^{n-1})$ and satisfies (1.1). Then we can choose regular atoms $\{b_j\}$ such that $\Omega = \sum_j \lambda_j b_j$ and $\sum_j |\lambda_j| \approx \|\Omega\|_{\mathbf{H}^1(\mathbf{S}^{n-1})}$. Hence $\mathbf{T}_{\Omega,\Phi,\Gamma}$ can be decomposed as

$$\mathbf{T}_{\Omega,\Phi,\Gamma}(f)(x) = \sum_j \lambda_j \mathbf{T}_{b_j,\Phi,\Gamma}(f)(x). \tag{4.1}$$

Therefore, to establish the L^p boundedness of $\mathbf{T}_{\Omega,\Phi,\Gamma}$, it suffices to show that

$$\left\|\mathbf{T}_{b_j,\Phi,\Gamma}(f)\right\|_p \le C\|f\|_p \tag{4.2}$$

for all $1 < p < \infty$ and for any atom $b_j = b$, where C independent of the atom $b(\cdot)$.

Let b, ζ and ρ be as in Definition 2.2. By using a proper rotation if necessary, we may assume that $\zeta = (0, \ldots, 1)$. For $k \in \mathbf{Z}$, let $a_k = \Gamma(2^k)$ and $\sigma_{b,\Phi,\Gamma,k}$ be the measure defined by

$$\int_{\mathbf{R}^m} f d\sigma_{b,\Phi,\Gamma,k} = \int_{2^k \leq |y| < 2^{k+1}} f(\Gamma(|y|)\Phi(y')) \frac{b(y')}{|y|^n} dy\,. \tag{4.3}$$

Then

$$\mathbf{T}_{b,\Phi,\Gamma} f(x) = \sum_k \sigma_{b,\Phi,\Gamma,k} * f(x). \tag{4.4}$$

Let $\xi = (\xi_1, \ldots, \xi_m) \in \mathbf{R}^m$ be arbitrary but fixed. Then we claim that

$$|\hat{\sigma}_{b,\Phi,\Gamma,k}(\xi)| \leq C \left| a_k \rho^{\frac{2(n-1)}{\delta}} \xi \right|^{-\frac{\delta}{2}} \tag{4.5}$$

for some $\delta > 0$. Let $F(y', \xi') = \xi' \cdot \Phi(y')$ where $\xi' = \frac{\xi}{|\xi|}$. Now if $F(y', \xi')$ is constant in y', by the mean zero property (2.3), the inequality (4.5) holds trivially. So assume $F(y', \xi')$ is a nonconstant function in y'. Let A, δ, $U_1, \ldots, U_{l(\eta)}$ be as in Lemma 3.4. Construct in the usual way a smooth partition of unity

$$\sum_{U \in \{U_1, \ldots, U_{l(\eta)}\}} P_U(y) \equiv 1 \text{ for } y \in \mathbf{S}^{n-1}$$

with $\operatorname{supp}(P_U) \subseteq U$. Then

$$\hat{\sigma}_{b,\Phi,\Gamma,k}(\xi) = \sum_{U \in \{U_1, \ldots, U_{l(\eta)}\}} H_{U,k}(\xi), \tag{4.6}$$

where

$$H_{U,k}(\xi) = \int_1^2 \int_{\mathbf{S}^{n-1}} b(y) e^{-i\Gamma(2^k t)|\xi| F(y', \xi')} P_U(y)\, d\sigma(y) \frac{dt}{t}.$$

Now, to apply Lemma 3.4, we measure $\left|H_{U,k}(\xi)\right|$ as follows:

$$\begin{aligned} &\left|H_{U,k}(\xi)\right|^2 \\ &\leq \textstyle\int_{\mathbf{S}^{n-1} \times \mathbf{S}^{n-1}} |b(y')|\, |b(x')|\, |\mathbf{I}_k(y', z')|\, P_U(y') P_U(x') d\sigma(x') d\sigma(y'), \end{aligned} \tag{4.7}$$

where

$$\mathbf{I}_k(y', x') = \int_1^2 e^{-i\Gamma(2^k t)|\xi|(F(y', \xi') - F(x', \xi'))} \frac{dt}{t}. \tag{4.8}$$

Now we claim that

$$|\mathbf{I}_k(y', x')| \leq C \left| a_k |\xi| \left(F(y', \xi') - F(x', \xi')\right)\right|^{-1}. \tag{4.9}$$

To see (4.9), we apply Lemma 3.2 with $\phi(t) = \frac{\Gamma(2^k t)}{a_k}$, $\lambda = a_k\,|\xi|\,(F(y',\xi') - F(x',\xi'))$, $a = 1$, and $b = 2$. To this end, notice that since the convexity of Γ implies that ϕ' is increasing, we only need to show that $|\phi'(t)| \geq 1$ for all $t \in (1,2)$. The latter can be seen as follows:

Since Γ is convex increasing, it must satisfy $r\Gamma'(r) \geq \Gamma(r)$ for all $r \geq 0$. Therefore,

$$|\phi'(t)| = \frac{2^k\Gamma'(2^k t)}{a_k} \geq \frac{\Gamma(2^k t)}{a_k}.$$

The last inequality with the fact that $\Gamma(2^k t) \geq \Gamma(2^k) = a_k$ for all $t \in (1,2)$ (since Γ is increasing), we obtain $|\phi'(t)| \geq 1$ for all $t \in (1,2)$.

Now by interpolation between (4.9) and the trivial bound $|\mathbf{I}_k(y',x')| \leq \ln 2$, we have

$$|\mathbf{I}_k(y',x')| \leq C\,|a_k\,|\xi|\,(F(y',\xi') - F(x',\xi'))|^{-\delta}\,. \tag{4.10}$$

Thus

$$\left|H_{U,k}(\xi)\right| \leq C\,|a_k\,|\xi||^{-\frac{\delta}{2}}\,\|b\|_\infty\,(J(\xi',\delta))^{\frac{1}{2}}, \tag{4.11}$$

where

$$J(\xi',\delta) = \int_{\mathbf{S}^{n-1}\times\mathbf{S}^{n-1}} P_U(y')P_U(x')\,|F(y',\xi') - F(x',\xi')|^{-\delta}\,d\sigma(y')d\sigma(x').$$

Therefore, by (2.2), (4.11) and Lemma 3.4 we get (4.5).

Now we need to consider two cases:

Case 1. $0 < \rho < \frac{1}{\sqrt{n}}$.

In this case we argue as in ([8]). Since $\Phi\mid_{\mathbf{S}^{n-1}}$ is real analytic there exists a polynomial mapping $\mathcal{P} : \mathbf{R}^{n-1} \to \mathbf{R}^m$, $\mathcal{P}(u) = (P_1(u), \ldots, P_m(u))$ with degree at most $d = [\frac{2(n-1)}{\beta}] + 1$ such that

$$|\Phi(y') - \mathcal{P}(\pi_{n-1}(y'))| \leq C\rho^d \tag{4.12}$$

whenever $y' \in \mathbf{B}(\zeta,\rho) \cap \mathbf{S}^{n-1}$. For $1 \leq j \leq m$, let $P_j(u) = \sum_{|\alpha|\leq d} a_{\alpha,j}u^{\alpha}$, $u \in \mathbf{R}^{n-1}$. For $1 \leq s \leq d$, let $m(s)$ be the number of distinct multi-indices $\alpha \in (\mathbf{N}\cup\{0\})^{n-1}$ with $|\alpha| = s$. Let $\alpha_{s1}, \ldots, \alpha_{sm(s)}$ be the distinct multi-indices $\alpha \in (\mathbf{N}\cup\{0\})^{n-1}$ with $|\alpha| = s$.

For $1 \leq s \leq d$, define the linear transformations $L_s : \mathbf{R}^m \to \mathbf{R}^{m(s)}$ by

$$L_{d+1}(\xi) = \rho^{\frac{2(n-1)}{\delta}}\xi$$

and

$$L_s(\xi) = (L_{s1}(\xi), \ldots, L_{sm(s)}(\xi)),$$

where

$$L_{sj}(\xi) = \rho^s \sum_{i=1}^{m} \xi_i a_{\alpha_{sj},i}.$$

For $0 \le s \le d+1$, let $\mathcal{P}_s = (P_1^s(u), \dots, P_m^s(u))$ and $\mathcal{P}_{d+1} = \Phi$, where $P_j^s(u) = \sum_{|\alpha| \le s} a_{\alpha,j} u^{\alpha}$.

For $0 \le s \le d$, let $\sigma_{b,s,\Gamma,k}$ be the measure defined as in (4.3) with Φ replaced by $\mathcal{P}_s \circ \pi_{n-1}$. Now by Lemma 3.2 and Lemma 3.5, there exist positive constants $\varepsilon_1, \dots, \varepsilon_d$ such that

$$|\hat{\sigma}_{b,s,\Gamma,k}(\xi)| \le C \, |a_k L_s(\xi)|^{-\epsilon_s} \tag{4.13}$$

for all $1 \le s \le d$. On the other hand, by (4.12), the definition of $\sigma_{b,s,\Gamma,k}$, and the properties of b, we get

$$|\hat{\sigma}_{b,s,\Gamma,k}(\xi) - \hat{\sigma}_{b,s-1,\Gamma,k}(\xi)| \le C \, |a_{k+1} L_s(\xi)| \tag{4.14}$$

for all $1 \le s \le d+1$.

Now for $1 \le s \le d+1$, let

$$(\sigma_{b,s,\Gamma})^*(f)(x) = \sup_{k \in \mathbf{Z}} \left| |\sigma_{b,s,\Gamma,k}| * f(x) \right|. \tag{4.15}$$

By Lemma 3.3, we have

$$\|(\sigma_{b,s,\Gamma})^*(f)\|_{L^p(\mathbf{R}^m)} \le A_p \, \|f\|_{L^p(\mathbf{R}^m)} \tag{4.16}$$

for all $f \in L^p(\mathbf{R}^m)$ and $p \in (1, \infty)$ with a constant A_p independent of the atom b. Finally it is clear that

$$\|\sigma_{b,s,\Gamma,k}\| \le C \tag{4.17}$$

and

$$\sigma_{b,0,\Gamma,k} = 0 \tag{4.18}$$

for all $1 \le s \le d+1$.

Now, we are in a position to apply Lemma 3.1 to obtain

$$\|\mathbf{T}_{b,\Phi,\Gamma}\,(f)\|_p \le C \, \|f\|_p \tag{4.19}$$

for all $1 < p < \infty$, where C is a constant independent of the atom $b(\cdot)$. To see (4.19), we argue as follows:

Given an arbitrary $1 < q < \infty$. Then by (4.16), we have

$$\|(\sigma_{b,s,\Gamma})^*(f)\|_{L^q(\mathbf{R}^m)} \le A_p \, \|f\|_{L^q(\mathbf{R}^m)} \tag{4.20}$$

for all $f \in L^q(\mathbf{R}^m)$. Therefore, by (4.5), (4.13)-(4.14), (4.17)-(4.18), and (4.20), it is easy to see that the measures $\{\sigma_{b,s,\Gamma,k} : 0 \leq s \leq l,\ k \in \mathbf{Z}\}$ satisfy the assumptions of Lemma 3.1. Hence, by Lemma 3.1, we have

$$\|\mathbf{T}_{b,\Phi,\Gamma}(f)\|_p \leq C_p \|f\|_p \tag{4.21}$$

for all $p \in (\frac{2q}{q+1}, \frac{2q}{q-1})$. Since (4.21) holds for all $p \in (\frac{2q}{q+1}, \frac{2q}{q-1})$ and any $1 < q < \infty$, we obtain (4.19).

Case 2. $\rho \geq \frac{1}{\sqrt{n}}$.

The argument in this case is very much simplified, because ρ is harmless. By noticing that $\|b\|_1 \leq 1$, (2.3), and since Γ is increasing, we get

$$|\hat{\sigma}_{b,\Phi,\Gamma,k}(\xi)| \leq C\{|a_{k+1}|\xi||\} \tag{4.22}$$

with constant C independent of b. Next, since $\rho \geq \frac{1}{\sqrt{n}}$, (4.5) implies that

$$|\hat{\sigma}_{b,\Phi,\Gamma,k}(\xi)| \leq C\,|a_k|\xi||^{-\delta} \tag{4.23}$$

for some $\delta > 0$. Also, it is easy to see that

$$\|\sigma_{b,\Phi,\Gamma,k}\| \leq C. \tag{4.24}$$

By Lemma 3.3, we have

$$\|(\sigma_{b,\Phi,\Gamma})^*(f)\|_{L^p(\mathbf{R}^m)} \leq A_p \|f\|_{L^p(\mathbf{R}^m)} \tag{4.25}$$

for all $f \in L^p(\mathbf{R}^m)$ and $p \in (1, \infty)$ with a constant A_p independent of the atom b. Now, set $\sigma_{b,1,\Gamma,k} = \sigma_{b,\Phi,\Gamma,k}$ and $\sigma_{b,0,\Gamma,k} = 0$. Thus, by (4.22)-(4.25), we see that the measures $\{\sigma_{b,s,\Gamma,k} : 0 \leq s \leq 1,\ k \in \mathbf{Z}\}$ satisfy the assumptions of Lemma 3.1. Hence by Lemma 3.1 and a similar argument as in the last part of Case 1, we get

$$\left\|\mathbf{T}_{b_j,\Phi,\Gamma}(f)\right\|_p \leq C\,\|f\|_p$$

for all $1 < p < \infty$, where the constant C is independent of the atom $b(\cdot)$. This ends the proof of Theorem F. □

Sketch of the proof of Theorem E. Suppose that $\Gamma(t)$ is a real-valued function defined on $\mathbf{R}_+$ satisfying (1.3)-(1.4). Let $\Omega \in \mathbf{H}^1(\mathbf{S}^{n-1})$ that satisfies (1.1). Let $\{\lambda_j\}$ and $\{b_j\}$ be as in the proof of Theorem F. By the same argument as in the proof of Theorem F, it suffices to show that

$$\left\|\mathbf{T}_{b_j,\Phi,\Gamma}(f)\right\|_p \leq C\,\|f\|_p$$

for all $1 < p < \infty$ and for any regular atom $b_j = b$, where C is independent of the atom $b(\cdot)$.

Let b, ζ, and ρ be as in the proof of Theorem F and let h be as in Theorem B. We will only sketch the proof for the case where ρ is relatively small.

For $k \in \mathbf{Z}$, let $a_k = 2^{kh}$. Also let d, $\sigma_{b,\Phi,\Gamma,k}$, L_{d+1}, L_s, and $(\sigma_{b,s,\Gamma})^*$ be as in the proof of Theorem F. To apply the same argument as in the proof of Theorem F, we only need to observe the following:

$$|\hat{\sigma}_{b,\Phi,\Gamma,k}(\xi)| \leq C \left| a_k \rho^{\frac{2(n-1)}{\delta}} \xi \right|^{-\frac{\delta}{2}} \tag{4.5'}$$

for some $\delta > 0$;

$$|\hat{\sigma}_{b,s,\Gamma,k}(\xi)| \leq C \left| a_k L_s(\xi) \right|^{-\epsilon_s} \tag{4.13'}$$

for all $1 \leq s \leq d$ and some $\epsilon_s > 0$;

$$|\hat{\sigma}_{b,s,\Gamma,k}(\xi) - \hat{\sigma}_{b,s-1,\Gamma,k}(\xi)| \leq C \left| a_{k+1} L_s(\xi) \right| \tag{4.14'}$$

for all $1 \leq s \leq d+1$;

$$\|(\sigma_{b,s,\Gamma})^*(f)\|_{L^p(\mathbf{R}^m)} \leq A_p \|f\|_{L^p(\mathbf{R}^m)} \tag{4.16'}$$

for all $f \in L^p(\mathbf{R}^m)$ and $p \in (1, \infty)$ with a constant A_p independent of the atom b;

$$\|\sigma_{b,\Phi,\Gamma,k}\| \leq C \text{ for all } 1 \leq s \leq d+1; \tag{4.17'}$$

for all $1 \leq s \leq d+1$;

$$\sigma_{b,0,\Gamma,k} = 0. \tag{4.18'}$$

The estimate (4.14′) follows by the fact that $|\Gamma(t)| \leq C_1 |t|^h$ and following exactly the same argument as that led to (4.14). The same arguments that yield to the estimates (4.5) and (4.13) can be used to obtain (4.5′) and (4.13′), respectively. The only difference here is that in obtaining (4.5) and (4.13), we used Lemma 3.2, but this time we use the following estimates which can be proved by an integration by parts and the conditions (1.3)-(1.4):

$$\left| \int_1^2 e^{i\lambda\Gamma(2^k t)} \frac{dt}{t} \right| \leq C \left| \lambda 2^{kh} \right|^{-\varepsilon}$$

for all real λ and some $\varepsilon > 0$.

The estimate (4.16′) can be obtained by a bootstrapping argument as in the proof of Lemma 3.3 in ([3]). Finally, the proofs of (4.17′) and (4.18′) are clear. This ends the sketch of the proof of Theorem E. □

Finally, a proof of Theorem D can be obtained by using an induction argument on the number of terms of Γ and the arguments in the proof of Theorem F. The model case of Theorem D which occurs in the first step of the induction argument is when $\Gamma(t) = \mu t^h$ for some non zero real number h. This case as one may observe follows by exactly the same argument as in the proof of Theorem E. The problem here is that the L^p bounds have to be independent of μ. But as long as the L^p bounds are independent of the linear transformations L_s (see Lemma 3.1), the parameter μ can be absorbed in the linear transformations. Hence, L^p bounds independent of μ are available.

References

[1] Al-Hasan, A. J. and Fan, D., L^p-boundedness of a singular integral operator, *Canadian Math. Bulletin*, **41** (1998), no. 4, 404–412.

[2] Al-Qassem, H., Al-Salman, A., and Pan, Y., Singular integrals defined by homogeneous mappings with rough kernels, preprint.

[3] Al-Salman, A., *L^p Estimates of Singular Integral Operators of Convolution Type with Rough Kernels*, Ph. D. Thesis in Mathematics, University of Pittsburgh, U.S.A., 1999.

[4] Calderón, A.-P. and Zygmund, A., On singular integrals, *Amer. J. Math. Soc.*, **78** (1956), 289–309.

[5] Coifman, R. and Weiss, G., Extension of Hardy spaces and their use in analysis, *Bull. Amer. Soc.*, **83** (1977), 569–645.

[6] Connett, W. C., Singular integrals near L^1. In: *Harmonic analysis in Euclidean spaces* (Proc. Sympos. Pure Math., Williams Coll., Williamstown, Mass., 1978), Part 1, 163–165. AMS, Providence, R.I., 1979.

[7] Duoandikoetxea, J. and Rubio de Francia, J. L., Maximal and singular integral operators via Fourier transform estimates, *Invent. Math.*, **84** (1986), 541–561.

[8] Fan, D., Guo, K., and Pan, Y., L^p estimates for singular integrals associated to homogeneous surfaces, *J. Reine Angew. Math.*, **542** (2002), 1–22.

[9] Fan, D. and Pan, Y., A singular integral operator with rough kernel, *Proc. Amer. Math. Soc.*, **125** (1997), 3695–3703.

[10] Fan, D. and Pan, Y., Singular integral operators with rough kernels supported by subvarieties, *Amer. J. Math.*, **119** (1997), 799–839.

[11] Fan, D. and Pan, Y., L^2-boundedness of a singular integral operator, *Publ. Mat.*, **41** (1997), 317–333.

[12] Kim, W., Wainger, S., Wright, J., and Ziesler, S., Singular integrals and maximal functions associated to surfaces of revolution, *Bull. London Math. Soc.*, **28** (1996), 291–296.

[13] Ricci, F. and Stein, E. M., Harmonic analysis on nilpotent groups and singular integrals I: Oscillatory integrals, *J. Func. Anal.*, **73** (1987), 179–194.

[14] Ricci, F. and Weiss, G., A characterization of $H^1(\Sigma_{n-1})$. In: *Harmonic analysis in Euclidean spaces* (Proc. Sympos. Pure Math., Williams Coll., Williamstown, Mass., 1978), Part 1, 289–294. AMS, Providence, R.I., 1979.

[15] Stein, E. M., Problems in harmonic analysis related to curvature and oscillatory integrals. In: *Proceedings of the International Congress of Mathematicians*, Vol. 1, 2 (Berkeley, Calif., 1986), 196–221. AMS, Providence, R.I., 1987.

[16] Stein, E. M., *Harmonic Analysis: Real-Variable Methods, Orthogonality, and Oscillatory Integrals*, Princeton Mathematical Series, **43**. Princeton University Press, Princeton, NJ, 1993.

[17] Stein, E. M. and Wainger, S., Problems in harmonic analysis related to curvature, *Bull. Amer. Math. Soc.*, **84** (1978), 1239–1295.

ON FUNCTIONAL EQUATIONS WITH OPERATOR COEFFICIENTS

Anatolij Antonevich
Belarusian State University, Belarus
and
University in Bielastok, Poland
antonevich@bsu.by

To Professor G. S. Litvinchuk on the occasion of his 70th birthday

—

Abstract The paper contains a short review of results on functional equations. This investigation is based on considerations of C^*–algebras generated by functional operators. As an example, the construction of symbolic calculus for non-local functional pseudodifferential operators is considered.

Keywords: Weighted shift operator, linear extension, hyperbolicity, spectrum, essential spectrum, Fredholm operator, index of an operator.

1. Introduction

The main problems we are dealing with in the present report originate from the study of operators and corresponding functional equations of the form

$$b := \sum_k a_k u(\alpha_k(x)) = f(x), \ x \in X, \tag{1.1}$$

where $\alpha_k : X \to X$ are certain transformations of the domain X and a_k are operators from some operator algebra A, the latter being as a rule well investigated. If X is a manifold and $a_k = a_k(x, D)$ are pseudodifferential operators (PDO) on X, the operator (1.1) is a non-local functional pseudodifferential operator (FPDO). If $a_k = a_k(x)$ are operators of multiplication by functions or matrix functions, the operator of the form (1.1) is a "pure" functional operator. An operator of the form

$$bu = a(x)u(\alpha(x))$$

is called a weighted shift operator. These equations and operators arise in connection with various problems.

The first part of the report contains a short review of the results on the subject given in various publications, for example, in [1-3]. Section 6 contains new results.

We present a solution to the following problems:

1. General approach to investigate the operators (1.1) based on the theory of C^*-algebras.
2. Construction of a symbolic calculus for non-local functional pseudodifferential operators (FPDO).
3. A description of the spectrum of the weighted shift operator and invertibility conditions for a "pure" functional operator and, in particular, for the constructed symbols.
4. Calculation of the index for non-local functional pseudodifferential operators.
5. Description of the essential spectra of a model weighted shift operator.

The origin of this work is related to the investigations of Prof. G. Litvinchuk fulfilled in the 60s. In his investigations, singular integral operators with Carleman shift were considered. It is a class of operators of the form (1.1) in the case when X is a contour in the complex plane, a_k are singular integral operators and the mappings α_k are the iterations of a periodic mapping α_1. A periodic (or Carleman type) mapping is a mapping such that some its iteration is identity: $\alpha_n(x) \equiv x$. Prof. Litvinchuk constructed an auxiliary matrix singular integral operator $\tilde{b}$ (without shifts) and proved that operators b and $\tilde{b}$ are Fredholm only simultaneously. In particular, the well-known symbol of $\tilde{b}$ which is a couple of continuous matrix functions on X, may be considered as a symbol of b and it has all the properties of the classical symbol of a singular integral operator.

An index formula of the form

$$\operatorname{ind} b = \frac{1}{n} \operatorname{ind} \tilde{b}.$$

was obtained by G. Litvinchuk as well (see [6]).

Let us consider more general problem: to investigate the operators of the form (1.1) in the case where a_k are pseudodifferential operators on a multidimensional manifold. We present some examples of results in this direction.

2. C^*-algebras, generated by dynamical systems. The isomorphism theorem

The main idea of this investigation is based on the consideration of the algebra generated by operators of the form (1.1). It is the so-called algebra, generated by dynamical systems.

A C^*–algebra B is called an *algebra generated by dynamical systems with a transformations group* G *and an algebra of coefficients* A (denote $B = C^*(A, G, T_g)$), if a C^*–subalgebra A in B is indicated, a unitary representation: $G \ni g \mapsto T_g \in B$ of the group G in B is given and the following conditions hold

i) the mapping $\tilde{T}_g : A \ni a \mapsto T_g a T_g^{-1}$ is an automorphism of A for every $g \in G$;

ii) the set of all finite sums

$$\sum a_g T_g,\ a_g \in A,$$

is dense in B with respect to the norm defined.

Example 1. Let M be a compact smooth n-dimensional manifold and let A be the algebra of PDO of order zero acting in the space $L_2(M)$. Let $\{\alpha_g\}$ be a given action of a group G of diffeomorphisms on M. Then the operators T_g defined in the space $L_2(M)$ by the formula

$$(T_g u)(x) = \rho(x)^{1/2} u(\alpha_g^{-1}), g \in G, \tag{2.1}$$

where $\rho(x) = J(\alpha_g^{-1}(x))$ is the modulus of the Jacobian of diffeomorphism α_g^{-1} at the point x, form a unitary representation of G. The C^*–algebra B, generated by A and $\{T_g, g \in G\}$, is an algebra of FPDO and it is an algebra of the type $C^*(A, G, T_g)$. The property i) is the well known "invariance of PDO with respect to change of variables".

Algebras of the type $C^*(A, G, T_g)$ play an important role in the general theory of C^*–algebras, but the problems we deal with, were not well studied in the classical theory.

To construct a symbolic calculus for some class of operators means to write out objects in terms of which the Fredholmness condition of the operators could be verified. Let K be the ideal of all compact operators and B/K the quotient algebra. The element $[b] = \{b + K\} \in B/K$ has all the desired properties of the symbol. For example, b is a Fredholm operator if and only if $[b]$ is invertible as an element of the algebra B/K. In order to construct a symbol it is necessary to construct explicitly an algebra isomorphic to B/K.

In this way a natural isomorphism problem arises which could be posed in C^*–algebras terms as follows:

Let us consider two algebras $B_1 = C^*(A_1, G, T_g)$ and $B_2 = C^*(A_2, G, V_g)$, and suppose there exists an $*$–isomorphism

$$\varphi : A_1 \to A_2$$

such that

$$\varphi(T_g a T_g^{-1}) = V_g \varphi(a) V_g^{-1} \ \ \forall\, g \in G, a \in A. \tag{2.2}$$

The question is: does these conditions imply that the mapping

$$\sum a_g T_g \mapsto \sum \varphi(a_g) V_g$$

(defined on the set of finite sums) could be extended up to a $*-$isomorphism between B_1 and B_2?

In general, the answer is negative. The isomorphism theorem includes additional conditions when the answer is positive and describes the most general situation.

Let $PrimA$ be the space of primitive ideals of algebra A. Then every automorphism $\tilde{T}_g$ generates a homeomorphism τ_g of the space $PrimA$. We say that group G *acts on* A *topologically freely by automorphisms* $\tilde{T}_g$ if for any finite set $\{g_1, \ldots, g_p\} \subset G$ and any open set $U \subset Prim\ A$ there exists a point $x \in U$ such that all points $\tau_{g_i}(x), \quad i = 1, \ldots, p$, are distinct.

Lemma 2.1 *If the group* G *acts on* A *topologically freely then the inequality*

$$\|\sum a_g T_g\| \geq \|a_0\| \tag{2.3}$$

holds for finite sums.

If we consider the algebra B as a quotient algebra modulo A, the inequality (2.3) means strong independence of T_g as elements of this algebra module.

Theorem 2.2 (Isomorphism theorem) *Let* $B_1 = C^*(A_1, G, T_g)$, $B_2 = C^*(A_2, G, V_g)$ *and let* (2.2) *hold. Then the mapping*

$$\sum a_g T_g \mapsto \sum \varphi(a_g) V_g$$

can be extended to a $*-$*isomorphism from* B_1 *to* B_2 *if and only if*

1) *the group* G *is amenable, and*

2) *inequality* (2.3) *holds in* B_1 *and in* B_2.

3. The symbolic calculus for FPDO

Let A be an algebra of PDO on a compact smooth n-dimensional manifold M and B be the algebra of FPDO generated by action of the group G introduced in Example 1.

It is well known that the principal symbol $symb(a)$ of a PDO $a \in A$ is a continuous function on the space S^*M of all unit vectors of the cotangent bundle $T^*(M)$ of M. This means that $PrimA = S^*M$.

The diffeomorphism α_g of M generates a homeomorphism β_g of S^*M defined by

$$\beta_g(\eta) = \frac{\partial\alpha_g(\eta)}{\|\partial\alpha_g(\eta)\|}, \quad \eta \in S^*M,$$

where $\partial\alpha_g : T^*(M) \to T^*(M)$ is the codifferential of α_g. These homeomorphisms give an action of G on S^*M. Moreover, S^*M is a smooth manifold and the mappings β_g are diffeomorphisms.

We can define a unitary representation $\{U_g\}$ of G in the space $L_2(S^*M)$ by a formula analogous to (2.1), namely

$$(U_\phi)(\eta) = J(\beta_g^{-1}(\eta))^{1/2}\phi(\beta_g^{-1}(\eta)).$$

Definition By the symbol of a FPDO $b = \sum a_g T_g$, $a_g \in A$, in the space $L_2(M)$ we call the operator $symb(b)$ in the space $L_2(S^*M)$ given by the formula

$$symb(b) = \sum symb(a_g)U_g,$$

where $symb(a_g)$ is treated as the operator of multiplication by a continuous function.

Let B_s be C^*- algebra generated by symbols, that is, by $C(S^*M)$ and U_g.

Theorem 3.1 (Symbolic calculus for FPDO) *Let G be a discrete amenable group and let it act topologically freely on M by diffeomorphisms α_g. The mapping*

$$Symb : b \mapsto symb(b)$$

can by extended to a $*-$epimorphism

$$B \to B_s$$

such that its kernel is the ideal K of compact operators in $L_2(M)$.

In particular, an operator $b \in B$ is Fredholm if and only if the operator $symb(b)$ is invertible (as an operator in the space $L_2(S^*M)$).

4. The invertibility conditions for functional operators. The hyperbolic approach

The symbolical calculus for FPDO reduces the investigation of its Fredholm properties to an investigation of invertibility of "pure" functional operators, that is, operators of the form (1.1) whose coefficients are functions.

One approach to solve this problem is the so-called hyperbolic approach (in the case $G = Z$). This approach looks as follows (in a particular case).

Let X be a compact topological space with measure, $\alpha : X \to X$ be an invertible mapping and T a unitary operator in the space $L_2(X, C^m)$ of vector-functions of the dimension m, generated by α. Let us consider a two-term functional operator of the form $b = I + aT$, where a is a continuous matrix-function on X. By E we denote the product $X \times \mathbf{C}^m$ and consider E as a vector bundle over the space X. The fiber E_x over a point $x \in X$ is the set $\{x\} \times \mathbf{C}^m$. Let $\beta : E \to E$ be the mapping defined by the formula

$$\beta(x, \xi) = (\alpha(x), a(x)\xi), \ (x, \xi) \in E_x.$$

The mapping β maps linearly the fiber E_x into the fiber $E_{\alpha(x)}$ and it is the so-called *linear extension* of α (see [5,7]).

A linear extension is called *hyperbolic* if there exist two (continuous) subbundles E^s and E^u of E, and constants $C > 0$ and γ, $0 < \gamma < 1$, such that

$$E = E^s \oplus E^u$$

and

$$\begin{aligned} \|\beta^n(\eta)\| &\leq C\gamma^n \|\eta\|, \quad \forall \eta = (x, \xi) \in E^s, n = 1, 2, \ldots; \\ \|\beta^n(\eta)\| &\geq C\gamma^{-n} \|\eta\|, \quad \forall \eta = (x, \xi) \in E^u, n = 1, 2, \ldots, \end{aligned}$$

where

$$||\eta|| = ||\xi||_{C^m}.$$

Theorem 4.1 (Hyperbolicity theorem) *If the mapping α gives rise to a topologically free action of the group $G = \mathbf{Z}$ on X, then the operator $b = I + aT$ is invertible if and only if the corresponding linear extension β is hyperbolic.*

Corollary 4.2 *If the space X is $\alpha-$connected, the spectrum $\sigma(aT)$ of the weighted shift operator aT in the space of vector-functions of the dimension m has the form*

$$\sigma(aT) = \cup_1^p K_j, \ K_j = \{\lambda \in \mathbf{C} : r_j \leq |\lambda| \leq R_j\}, \ \textit{where } p \leq m,$$

i.e. the spectrum is a union of annuli centered at the origin, the number of these annuli being less or equal to m.

In particular, if $m = 1$, the spectrum $\sigma(aT)$ consists of one annulus: $\sigma(aT) = \{\lambda \in \mathbf{C} : r \leq |\lambda| \leq R\}$.

Theorem 4.3 (Spectral radius theorem) *If $m = 1$, then the following formula is valid for the spectral radius $R = r(aT)$*

$$\ln r(aT) = \max_{\mu \in Mes_\alpha} \int_X \ln |a(x)| d\mu(x),$$

where Mes_α is the set of all probabilistic measures on X ergodic and invariant with respect to α.

There exist a lot of examples where the set Mes_α was well described and the conditions of invertibility of two-term functional operators and the conditions of Fredholmness of FPDO were obtained in explicit form [1-3].

5. The case of a finite group. Formulas for the index

Let us consider the algebras B of FPDO (Example 1) in the case of a finite group G of the order N. In this case

$$b = \sum_{g \in G} a_g T_g, \ a_g \in A.$$

We consider the space $\tilde{L}_2(M)$ of N-dimensional vector functions on M whose components are functions in $L_2(M)$ indexed by the elements of G and define an operator $\tilde{b}$ by means of the operator matrix

$$[\tilde{b}]_{g,h} = T_g a_{g^{-1}h} T_g^{-1}.$$

Then $\tilde{b}$ is a pseudodifferential operator without shifts.

Theorem 5.1 *If the finite group G acts topologically freely on M, the following propositions are equivalent:*

i) b *is Fredholm operator;*

ii) $\tilde{b}$ *is Fredholm operator;*

iii) det $symb(\tilde{b})(x,\xi) \neq 0 \ \ \forall \ (x,\xi) \in S^*M$, *i.e* $\tilde{b}$ *is an elliptic PDO.*

In the space $\tilde{L}_2(M)$ consider the unitary representation $\tilde{T}_g$ of G by the operator matrices $[\tilde{T}_g]_{h,l} = T_g$, if $l = g^{-1}h$, and $[\tilde{T}_g]_{h,l} = 0$, if $l \neq g^{-1}h$. Then $\tilde{b}\tilde{T}_g = \tilde{T}_g\tilde{b}$ and the *Lefschetz numbers* $L(\tilde{b}, g)$ are well defined:

$$L(\tilde{b}, g) = Tr\tilde{T}_g|_{\ker \tilde{b}} - TrT_g|_{\operatorname{coker} \tilde{b}},$$

where Tr denotes the trace of an operator. We remind the reader that the formula for the Lefschetz numbers of elliptic PDO was obtained by Athyah and Singer.

Theorem 5.2 *If a finite group G acts topologically freely on M and the operator b is Fredholm, then*

$$\operatorname{ind} b = \frac{1}{N}\,[\operatorname{ind} \tilde{b} + \sum_{g \neq e} L(\tilde{b}, g)].$$

Theorem 5.3 *If the action of finite group G on M is such that all the mappings $\alpha_g : M \to M$ have only finitely many fixed points, then*

$$\operatorname{ind} b = \frac{1}{N}\,\tilde{b}.$$

The results of this section represent a generalization of G. Litvinchuk's results for the case of multi-dimensional manifolds.

6. The essential spectra of weighted shift operators

In the previous investigation, the main attention was paid to the description of the spectrum of weighted shift operators. But the structure of the operator $\lambda I - b$ for spectral values of λ in the spaces of vector functions is still unknown.

We shall use the following classification of the spectral values. Let b be a linear bounded operator in a Banach space. Let us consider the following subsets of $\mathbf{C}$:

$$\begin{aligned}
\sigma_G(b) &= \{\lambda \in \mathbf{C} : \overline{\mathrm{im}\,(\lambda I - b)} \neq \mathrm{im}\,(\lambda - b)\}, \\
\sigma_k(b) &= \{\lambda \in \mathbf{C} : (\lambda I - b \ \text{ is not semi-Fredholm}\}, \\
\sigma_{F_\pm}(b) &= \{\lambda \in \mathbf{C} : \lambda I - b \ \text{ is not } F_\pm - \text{operator}\}, \\
\sigma_F(b) &= \{\lambda \in \mathbf{C} : \lambda I - b \ \text{ is not Fredholm}\}, \\
\sigma(b) &= \{\lambda \in \mathbf{C} : \lambda I - b \ \text{ is not invertible}\}.
\end{aligned}$$

The set $\sigma_G(b)$ is called the essential Goldberg spectrum, the set $\sigma_k(b)$ is called the essential Kato spectrum, $\sigma_{F_\pm}(b)$ - the $F_\pm$ spectrum, $\sigma_F(b)$ - the Fredholm spectrum and $\sigma(b)$ is the (ordinary) spectrum of b. Up to now only in some typical cases of weighted shift operators b , the operator $\lambda I - b$ was known to be non Fredholm for all spectral values λ, i.e., $\sigma_F(b) = \sigma(b)$.

We consider now a model example when it is possible to obtain a description of all these essential spectra of weighted shift operators. In particular, from those results it will follow that the essential spectra may be all different. The method used in this section is a development of the hyperbolic approach.

Let X be the segment [0,1] and $\alpha : X \to X$ be a diffeomorphism having only two fixed points: 0 and 1. Let us consider the weighted shift operator b in the space $L_p(X, C^m)$ of the form

$$bu = a(x)[\alpha'(x)]^{-1/p} u(\alpha^{-1}(x)),$$

where the matrix function $a : X \to End(C^m)$ is continuous and non degenerate at every point.

In the scalar case ($m = 1$) the spectrum $\sigma(b)$ is an annulus and for interior points λ of $\sigma(b)$ the operator $\lambda I - b$ is unilaterally invertible (see [6]). But in the case $m > 1$ the essential spectra were unknown even for this example.

Let us consider the product $E = X \times C^m$ and the linear extension $\beta : E \to E$, corresponding to the operator b:

$$\beta(x, \xi) = (\alpha(x), a^{-1}(x)\xi), \quad x \in X, \quad \xi \in = \mathbf{C}^m. \tag{6.1}$$

We have a definition of hyperbolicity for a linear extension and by the hyperbolicity theorem it is equivalent to the invertibility of the operator $I - b$. The invertibility of an operator b is equivalent the following collection of properties:

$\ker b = 0, \ker b^* = 0$, $\operatorname{im} b$ is closed. In accordance with this, we represent the hyperbolicity property as a collection of some others properties.

For the sake of definiteness we shall assume that $\alpha(x) < x$ for $0 < x < 1$.

Definition A subbundle $E^{s,0}$ on [0,1) is called *locally stable* at the attracting point 0 for β if it is an invariant subbundle with respect to β and there exists a constant γ, $0 < \gamma < 1$, such that for every $\delta > 0$ there exists a constant $C > 0$ and the following inequality holds

$$||\beta_n(x,\xi)|| \leq C\gamma^n ||\xi||, \quad x \in [0, 1-\delta], \quad (x,\xi) \in E^{s,0}, \quad n = 1, 2, ...$$

A subbundle $E^{u,0}$ on [0,1) is called *locally unstable* at 0 if it is invariant with respect to β and there exists a constant γ, $0 < \gamma < 1$ such that for every $\delta > 0$ there exists a constant $C > 0$ and the following inequality is fulfilled

$$||\beta_n(x,\xi)|| \geq C\gamma^{-n} ||\xi||, \quad x \in [0, 1-\delta], \quad = \quad (x,\xi) \in E^{u,0}, \quad n = 1, 2, ...$$

The point 1 is an attracting point for the mapping α^{-1} and we define subbundles $E^{s,1}$ locally stable at the point 1 with respect to β^{-1} and the subbundle $E^{u,1}$ unstable with respect to β^{-1}.

Definition A linear extension β is said to be *locally hyperbolic at the point* 0 (point 1) if $E = E^{s,0} \oplus E^{u,0}$ ($E = E^{s,1} \oplus E^{u,1}$, respectively) on a neighborhood of 0 (of 1, respectively).

Note that the maximal stable subbundle is defined in a unique way while there exist many unstable subbundles, namely any subbundle V, which is complementary to $E^{S,0}$ (that is, $E^{S,0} \oplus V = E$) and invariant with respect to β, is unstable.

Let $\lambda_j(0)$ be the eigenvalues of $a(0)$ and χ_0 be the number of eigenvalues (according to multiplicity) satisfying the condition $|\lambda_j(0)| > 1$.

Lemma 6.1 *The following conditions for the linear extension β are equivalent:*

1) *β is locally hyperbolic at the point 0 ;*

2) *$|\lambda_j(0)| \neq 1$ for all j;*

3) *the matrix-function a admits a local factorization with shift, i.e. there exists a non degenerate continuous matrix-function $S_0(x)$ on $[0,1)$ such that*

$$S_0(x)a(x)S_0^{-1}(\alpha(x)) = d(x), \tag{6.2}$$

where the matrix $d(x)$ is block-diagonal:

$$d(x) = \operatorname{diag}\{d_u(x), d_s(x)\} \tag{6.3}$$

and the modulus of all the eigenvalues of the matrix $d_u(0)$ (which is of dimension χ_0) is greater than 1, and the modulus of all the eigenvalues of $d_s(0)$ is less than 1.

Let $\lambda_j(1)$ be the eigenvalues of the matrix $a(1)$ and χ_1 the number of eigenvalues (according to multiplicity) satisfying the condition $|\lambda_j(1)| < 1$. If $|\lambda_j(1)| \neq 1$ for every j, then β is locally hyperbolic at the point 1, the stable subbundle $E^{s,1}$ is determined and the matrix $a(x)$ can be reduced on (0,1] to an analogous block-diagonal form (with the help of some other matrix $S_1(x)$, the dimensions of the blocks may differ from the corresponding dimensions at the point 0).

Our main result shows that properties of the operator $I-b$, such as to have a closed image, to be F_+ - operator, F_- - operator or to be an invertible operator depend on the mutual arrangement of the stable subbundles $E^{s,0}$ and $E^{s,1}$. A similar result in the case of the space of continuous function was obtained in [4].

Let us fix any point t, $0 < t < 1$ and the segment $\Theta = [\alpha(t), t]$.

A mutual arrangement of $E^{s,0}$ and $E^{s,1}$ may be described by the function

$$d(\tau) = \dim(E^{s,0}_\tau \cap E^{s,1}_\tau), \quad \tau \in \Theta.$$

Theorem 6.2 *Let $X = [0,1]$, the matrix a and the diffeomorphism α satisfy the assumption mentioned above and let b be an operator of the form (6.1) in the vector functions space $L_p(X, C^m)$. Then*

I. *For the image of the operator $I - b$ to be closed ,it is necessary that the linear extension β be locally hyperbolic at the points 0 and 1.*

II. *If β is locally hyperbolic at the points 0 and 1 then the following statements describe all the possible situations:*

   IIa. *The image of the operator $I - b$ is closed if and only if $d_\tau = const$;*

   IIb. *The operator $I - b$ is a F_+-operator if and only if $d_\tau \equiv 0$ and in this case there exists a bounded left inverse operator for $I - b$.*

   IIc. *The operator $I - b$ is a F_--operator if and only if $d_\tau = \chi_0 + \chi_1 - m$ and in this case there exists a bounded right inverse operator for $I - b$.*

III. *The operator $I - b$ is Fredholm if and only if $d_\tau \equiv 0, \chi_0 + \chi_1 = m$ and in this case $I - b$ is invertible.*

We can apply this theorem to operators of the form $\lambda I - b$ and transform its statements into the spectral language.

The operator b is invertible and $\lambda = 0$ is a regular value. If $\lambda \neq 0$ then to the operator $\lambda I - b = \lambda(I - \frac{1}{\lambda}b)$ there corresponds the linear extension

$$\beta_\lambda(x, \xi) = (\alpha(x), \lambda a^{-1}(x)\xi).$$

According to the first statement of Theorem 6.2, the resonance circles $|\lambda| = |\lambda_j(0)|$ and $|\lambda| = |\lambda_i(1)|$ belong to the essential spectrum $\sigma_G(b)$. The complex plane is decomposed by these circles into a finite number of open annuli K_j, one internal disc and the exterior of the maximal circle. A part of these annuli belongs to the spectrum and it may be that some annuli K_j belong to the resolvent set of b. For $\lambda \in K_j$ the linear extension β_λ is locally hyperbolic, there exist the stable subbundles $E_\lambda^{s,0}$ and $E_\lambda^{s,1}$ and the corresponding function $d_\lambda(\tau)$ is determined.

An annulus K_j is called *extreme*, if one of the following inequalities holds for $\lambda \in K_j$:

$$|\lambda| < |\lambda_j(0)| \ \forall j; \ |\lambda| > | = \lambda_j(0)| \ \forall j; \ |\lambda| < |\lambda_i(1)| \ \forall = i; \ |\lambda| > |\lambda_i(1)| \ \forall i.$$

Theorem 6.3 *For all the λ in a fixed annulus K_j the stable subbundles are the same, the function $d_\lambda(\tau)$ depends only on the j and all these λ belong to the same type of essential spectra. This type is described by the corresponding function $d_j(t)$.*

For every λ in the extreme annuli the operator $\lambda I - b$ is semi-Fredholm and unilaterally invertible.

References

[1] Antonevich, A., *Linear Functional Equations: Operator Approach.* Universitetskoe, Minsk, 1988 (in Russian). English translation: Operator Theory: Advances and Applications, **83**. Birkhäuser, Basel, 1996.

[2] Antonevich, A. and Belitskii, G., The essential spectra of a weighted shift operator in the spaces of continuous and differentiable functions. In: *Spectral and evolutionary problems*, Vol. 8 (Sevastopol, 1997), 3–12. Tavria Publ., Simferopol', 1998.

[3] Antonevich, A., Belousov, M., and Lebedev, A., *Functional Differential Equations: 2. C*-applications. Parts 1, 2*, Pitman Monographs and Surveys in Pure and Applied Mathematics **94, 95**. Longman, Harlow, 1998.

[4] Antonevich, A. and Lebedev, A., *Functional Differential Equations: I. C*-theory*, Pitman Monographs and Surveys in Pure and Applied Mathematics, **70**. Longman Scientific & Technical, Harlow, 1994.

[5] Bronshtein, I., *Nonautonomous Dynamical Systems.* Shtiintsa, Kishinev, 1984 (in Russian).

[6] Kravchenko, V. G. and Litvinchuk, G. S., *Introduction to the Theory of Singular Integral Operators with Shift*, Mathematics and its Applications, **289**. Kluwer Academic Publishers, Dordrecht, 1994.

[7] Nitecki, Z., *Differentiable dynamics. An introduction to the orbit structure of diffeomorphisms.* The M.I.T. Press, Cambridge, Mass.-London, 1971.

ELLIPTIC SYSTEMS WITH ALMOST REGULAR COEFFICIENTS: SINGULAR WEIGHT INTEGRAL OPERATORS

Stanislav Antontsev*
Departamento de Matemática, Universidade da Beira Interior, Covilhã, Portugal
anton@ubi.pt

This paper is dedicated to Prof. G. S. Litvinchuk on the occasion of his 70th birthday

Abstract We consider the linear elliptic system of two first-order equations

$$\partial_{\overline{z}} w + \mu_1(z)\partial_z w + \mu_2(z)\overline{\partial_z w} = A(z)w + B(z)\overline{w} + F(z),$$

where $w(z, \overline{z}) = u + iv$ is an unknown complex-valued function, and the related integral operators and boundary value problems. We assume that $A, B, F \in L_p(\Omega)$, $p \leq 2$, in contrast to the regular Vekua's theory where $p > 2$. We prove that in this case the solutions of the system still preserve the properties, which correspond to the regular case with respect to: the structure of zeros, Liouville's theorem, solvability of Riemann-Hilbert boundary value problems etc.

Keywords: Elliptic system, generalized analytic functions, Riemann-Hilbert problems, integral operators.

1. Introduction

We consider the most general linear elliptic system of the first-order equations written as a single complex-valued equation

$$\partial_{\overline{z}} w + \mu_1(z)\partial_z w + \mu_2(z)\overline{\partial_z w} = A(z)w + B(z)\overline{w} + F(z), \tag{1.1}$$

$$2\partial_{\overline{z}} = (\partial_x + i\partial_y), \; z = x + iy,$$

*Work partially supported by Project FCT-POCTI/34471/MAT/2000

in the unit circle $\Omega = \{z : |z| < 1\}$, where $w(z, \overline{z}) = u + iv$ is a complex-valued function. System (1.1) is assumed to be uniformly elliptic i.e.

$$|\mu_1(z)| + |\mu_2(z)| \leq \mu_0 < 1, \; z \in \Omega. \tag{1.2}$$

In the works [19] and [6], there was created the complete theory of generalized analytic functions, which are the solutions of system (1.1) with $\mu_1 = \mu_2 = F = 0$, *i.e.*

$$\partial_{\overline{z}} w = A(z)w + B(z)\overline{w}, \tag{1.3}$$

under the assumption $\Lambda(z) \equiv (A, B) \in L_p(\Omega), \; p > 2$. The theory is called the regular Vekua's theory. Vekua's theory for solutions of (1.3) is closely related to the theory of holomorphic functions, but it is not valid if $\Lambda(z) \in L_p(\Omega), \; 1 < p \leq 2$. In the monograph [5] Vekua's theory is generalized to equations (1.3) with coefficients belonging to Nikolskij-Besov spaces $B^{\alpha}_{p,1}$, see [4]. Various singular cases, in which the coefficients A, B do not belong to $L_p(\Omega), \; p > 2$, but have analytic singularities, were studied in numerous papers, see for example [14].

In the present paper we suppose that $\Lambda(z) \equiv (A, B)$ belongs to some space $L^{\alpha}_2(\Omega)$, $(L_2(\Omega) \subset L^{\alpha}_2(\Omega))$, which does not coincide with the spaces considered before. We prove that in this case the solutions of system (1.1) still preserve the properties, which correspond to the regular case with respect to: the structure of zeros, Liouville's theorem, solvability of Riemann-Hilbert boundary value problems etc.

We consider also related integral operators and boundary value problems. In this connection it is necessary to study boundary value problems in some functional spaces similar to Hardy spaces, see [10], for analytic functions. These results are applied to the study of some integral weight operators, which arise in the theory of Riemann-Hilbert boundary value problems for system (1.1) and smoothness properties of quasi-conformal mappings.

The present paper generalizes and evolves some approaches developed in [2], [3], [1].

2. Definitions. Embedding theorems

First we consider the system for a generalized analytic function w

$$w_{\overline{z}} = A(z)w + B(z)\overline{w}, \; (\partial_{\overline{z}} w = w_{\overline{z}}), \tag{2.1}$$

in the regular case, when $A, B \in L_p(\Omega), \; p > 2$. It is well known [19] that any solution of this system can be presented in the form

$$w(z) = \Phi(z)e^{\omega(z)}, \tag{2.2}$$

where $\Phi(z)$ is an analytic function in Ω $(\Phi_{\overline{z}}(z) = 0)$ and

$$\omega(z) = -\frac{1}{\pi}\int_{\Omega}\frac{\varphi(t)}{t-z}d\Omega_t \equiv (T\varphi)(z), \quad \varphi(z) = A(z) + B(z)\frac{\overline{w}}{w}. \tag{2.3}$$

In (2.3) and everywhere below $d\Omega_t$ with $t = x + iy \in \Omega$ stands for $d\Omega_t = dxdy$.

According to [19]

$$||\omega||_{C^\alpha(\overline{\Omega})} \leq C(p)||\varphi||_{L_p(\Omega)}, \qquad \alpha = \frac{p-2}{p} > 0.$$

In the case $p = 2$, representation (2.2) still holds but the function $\omega(z)$ is unbounded in general.

For the choice of $\alpha = \frac{p-2}{p}$ considered in [5] , $L_q(G) \subset B^\alpha_{p,1}(G) \subset L_2(G)$. However, the operator T is bounded in these spaces $B^\alpha_{p,1}$ and consequently the function $\omega(z)$ in (2.2) is bounded.

Our aim is to study properties of the function $e^{\omega(z)}$ in the case where the function φ belongs to some space $L^\alpha_2(\Omega) \supset L_2(\Omega)$ and the operator T is unbounded in this space.

Let us formulate properties of the function $\omega(z)$ in a form suitable for further considerations.

Definition 2.1 *We say that a function φ measurable on Ω belongs to the space $L^\alpha_2(\Omega)$, $\alpha \geq 0$, if it admits the estimate*

$$|\varphi(z)| \leq \varphi_2(z) \ln^\alpha(1 + \varphi_1(z)),$$

where $0 \leq \varphi_i(z) \in L_i(\Omega)$, $i = 1, 2$.

Definition 2.2 *We say that a function φ measurable on Ω belongs to the space $V^\alpha_2(\Omega)$, if its derivatives $\varphi_{\overline{z}}, \varphi_z$ belong to $L^\alpha_2(\Omega)$.*

It is evident that $L_2(\Omega) \subset L^\alpha_2(\Omega)$, $\alpha > 0$ and $L_2(\Omega) = L^0_2(\Omega)$ as well as $V^0_2(\Omega) \cap L_2(\Omega) = W^1_2(\Omega)$.

Theorem 2.3 *Let $\varphi \in L^\alpha_2(\Omega)$. Then $\omega(z) = (T\varphi)(z)$ belongs to the Orlicz's spaces $L^*_M(\Omega)$, $L^*_M(\Gamma_r)$ defined by an entire function $M(t)$ of order $\rho < 2/(1+2\alpha)$ and any finite type $\sigma > 0$. Moreover, for any $r \in (0,1)$ the following estimate holds*

$$||\omega||_{L^*_M(\Omega)} \leq C, \quad ||\omega||_{L^*_M(\Gamma_r)} \leq C, \tag{2.4}$$

where $\Gamma_r = \{z : |z| = r\}$, $r \in]0,1[$ and $C = C(\alpha, \sigma, \rho, ||\varphi_1||_{L_1(\Omega)}, \varphi_2||_{L_2(\Omega)})$.

Remark 2.4 In (2.4) we use the Orlicz space norm

$$||u||_M \equiv ||u||_{L^*_M(\Omega)} = \inf\left\{k > 0 : \int_\Omega M\left(\frac{u(x)}{k}\right) dx \leq 1\right\};$$

for details on Orlicz spaces we refer to [12]. Note that inequalities (2.4) imply that

$$\int_\Omega e^{\pm\sigma|\omega|^\rho} d\Omega \leq C, \quad \int_{\Gamma_r} e^{\pm\sigma|\omega|^\rho} d\Gamma \leq C. \tag{2.5}$$

In this connection, the order ρ may be greater than one for any $\sigma > 0$, if $\alpha < 1/2$, since $\rho < 2/(1+2\alpha)$.

Proof. First of all, we note that for $\alpha = 0$ the first assertion of the theorem is a consequence of a general theorem of the paper [16] about the embedding $W_p^l(\Omega) \subset L_M^*(\Omega)$, $pl = n$, where $L_M^*(\Omega)$ is the Orlicz's space corresponding to an entire Young function of order $\rho = p/(p-1)$ and any finite type. It is easy to see that the first of inequalities (2.4) is a consequence of the second one. Now we shall prove the second inequality of the theorem with $\alpha \neq 0$. According to [19] it is easy to obtain for any $k > 2$ the following estimate

$$|(T\varphi)(z)|^k \leq \frac{1}{\pi}||\varphi_2||_2^{k-2}||\ln(1+\varphi_1)||_{8\alpha k}^{k\alpha}\int_\Omega \frac{|\varphi(t)|^2}{|t-z|^{\frac{1}{2}}}d\Omega_t\left(\int_\Omega |t-z|^{\lambda-2}d\Omega_t\right)^{\frac{1}{4k\lambda}}$$

with $\lambda = \frac{2}{4k-1}$ and $\|\cdot\|_q = \|\cdot\|_{L_q(\Omega)}$. Hence we conclude that

$$I^k \equiv \int_{\Gamma_r} |T(\varphi)|^k\,|dz| \leq (||\varphi_2||_2 I_1 I_2)^k\, I_3,$$

where

$$I_1 = \frac{1}{\pi}||\ln(1+\varphi_1)||_{8\alpha k}^{\alpha},\ I_2 = \sup_{z\in\Omega}\left(\int_\Omega |t-z|^{\lambda-2}d\Omega_t\right)^{\frac{1}{4k\lambda}},$$

$$I_3 = \sup_{t\in\Omega}\left(\int_{\Gamma_r} |t-z|^{-\frac{1}{2}}|dz|\right).$$

Using the inequality

$$\ln(1+t) \leq \frac{1}{\varepsilon}t^\varepsilon,\ 0 \leq t < \infty,\ 0 < \varepsilon \leq 1,$$

with $\varepsilon = 1/(8k\alpha)$, we obtain the following estimate

$$I_1 \leq \frac{1}{\pi}(8k\alpha)^\alpha\,||\varphi_1||_1^{\frac{1}{8k}} \leq C_1 k^\alpha.$$

Introducing the polar coordinates $re^{i\theta} = t - z$, we can evaluate I_2 in the following way

$$I_2 \leq \left(\int_0^2\int_0^{2\pi} r^{-1+\lambda}drd\theta\right)^{\frac{1}{4k\lambda}} \leq 2^{\frac{1}{4k}}(\pi(4k-1))^{\frac{4k-1}{8k}} \leq C_2 k^{\frac{1}{2}}.$$

Using the formulas $re^{i\theta} = z$, $Re^{i\gamma} = t$, I_3 may be estimated analogously:

$$\begin{aligned} I_3 &\leq \sup_{r,R\leq 1}\left(\frac{r}{(R^2+r^2)^{\frac{1}{4}}}\int_0^{2\pi}\frac{d\theta}{(1-\lambda_0\cos\theta)^{\frac{1}{4}}}\right) \\ &\leq \int_0^1 \frac{dt}{(1-t)^{\frac{1}{4}}} \leq C_3,\ (\lambda_0 = \frac{2rR}{R^2+r^2}). \end{aligned}$$

Finally, we obtain

$$I^k \leq C_3 \left(||\varphi_2||_2 C_1 C_2\right)^k k^{k(\alpha+\frac{1}{2})} \leq C_3 C_4^k k^{k(\alpha+\frac{1}{2})}. \tag{2.6}$$

Now, let $M(t) = \sum_2^\infty a_k t^k$ be an entire Young function of order ρ and some type σ. Then according to [13] the coefficients a_k satisfy the condition

$$\overline{\lim}_{k\to\infty} \left(k^{\frac{1}{\rho}} \sqrt[k]{|a_k|}\right) = (e\rho\sigma)^{\frac{1}{\rho}}.$$

Therefore, for sufficiently large k

$$|a_k| \leq \left((e\rho\sigma)^{\frac{1}{\rho}} + \varepsilon\right)^k k^{-\frac{k}{\rho}}, \tag{2.7}$$

where $\varepsilon > 0$ is any arbitrary small number. Let us consider now the integral

$$I \equiv \int_{\Gamma_r} M\left(|T(\varphi)|\right)|dz| = \int_{\Gamma_r} \sum_2^\infty a_k \left|T\varphi\right|^k |dz|.$$

According to (2.6) and (2.7)

$$I \leq \sum_2^\infty a_k I^k \leq C_3 \sum_2^\infty \left(C_4\left((e\rho\sigma)^{\frac{1}{\rho}} + \varepsilon\right)\right)^k k^{k(-\frac{1}{\rho}+\alpha+\frac{1}{2})}. \tag{2.8}$$

Therefore, the series in the right hand side of (2.8) converges if $\rho = 2/(1+2\alpha)$, and σ satisfies the inequality

$$C_4 (e\rho\sigma)^{\frac{1}{\rho}} < 1.$$

Thus, for

$$\rho < 2/(1+2\alpha) \tag{2.9}$$

the series (2.8) converges for any $\sigma > 0$. This finishes the proof. □

Remark 2.5 It is evident that the assertions of the theorem are also valid for the operators

$$(T_\varkappa\varphi)(z) = (T\varphi)(z) + z^{2\varkappa}\overline{(T\varphi)\left(\frac{1}{\bar{z}}\right)}, \qquad \varkappa \geq 0, \tag{2.10}$$

$$(T_{-\varkappa}\varphi)(z) = (T\varphi)(z) + \frac{1}{z}\overline{T\left(t^{2|\varkappa|+1}\varphi(t)\right)\left(\frac{1}{\bar{z}}\right)}, \qquad \varkappa < 0, \tag{2.11}$$

where $\varkappa$ is an integer, see [19]. In this connection, the order ρ may be greater than one for any $\sigma > 0$, if $\alpha < 1/2$, according to (2.9).

Theorem 2.6 *Let* $\varphi(z) \in V_2^\alpha(\Omega)$, $\psi(z) \in L_2^\alpha(\Omega)$, $0 \le \alpha < 1/2$. *Let* $\chi(z) \in W_p^1(\Omega)$, $p > 2$ *be a homeomorphism of the domain* Ω *onto itself, such that*

$$\chi_{\bar z} - \mu(z)\chi_z = 0, \ z \in \Omega,$$

$$\chi(0) = 0, \ \chi(1) = 1, \ |\mu(z)| \le \mu_0 < 1.$$

Then

$$\varphi(\chi) \equiv \varphi(z(\chi)) \in V_2^\alpha(\Omega), \ \psi^*(\chi) \equiv \frac{\psi(z(\chi))}{\overline{\chi_{\bar z}}(1 - |\mu|^2)} \in L_2^\alpha(\Omega).$$

Proof. Let us show that the derivatives φ_χ, $\varphi_{\overline{\chi}}$ admit the estimate of Definition 2.1. According to the conditions of the theorem we have

$$|\varphi_\chi| = |\varphi_z z_\chi + \varphi_{\bar z}\bar z_\chi| \le 2|z_\chi|\varphi_2(z)\ln^\alpha(1 + \varphi_1(z)), \ \varphi_i \ge 0,$$

$$|\varphi_{\overline{\chi}}| = \left|\varphi_z z_{\overline{\chi}} + \varphi_{\bar z}\bar z_{\overline{\chi}}\right| \le 2|z_\chi|\varphi_2(z)\ln^\alpha(1 + \varphi_1(z)), \ \varphi_i \ge 0.$$

We introduce the functions

$$\widetilde{\varphi}_1(\chi) = |\varphi_1(z(\chi))|^{\frac{p-2}{p}}, \ \widetilde{\varphi}_2(\chi) = |z_\chi(\chi)||\varphi_2(z(\chi))|$$

and show that $\widetilde{\varphi}_i(\chi) \in L_i(\Omega)$, $i = 1, 2$. In fact,

$$\begin{aligned}\int_\Omega \widetilde{\varphi}_1(\chi)d\Omega_\chi &= \int_\Omega |\varphi_1(z)|^{\frac{p-2}{p}}(|\chi_z|^2 - |\chi_{\bar z}|^2)d\Omega_z \\ &\le \left(\int_\Omega |\varphi_1(z)|d\Omega_z\right)^{\frac{p-2}{p}}\left(\int_\Omega |\chi_z|^p d\Omega_z\right)^{\frac{2}{p}}.\end{aligned}$$

Analogously,

$$\int_\Omega \widetilde{\varphi}_2^2(\chi)d\Omega_\chi = \int_\Omega |\varphi_2(z)|^2|z_\chi|^2(|\chi_z|^2 - |\chi_{\bar z}|^2)d\Omega_z \le C(\mu_0)\int_\Omega |\varphi_2(z)|^2 d\Omega_z.$$

Using the inequality

$$\ln(1 + t^{\frac{p}{p-2}}) \le \frac{p}{p-2}\ln(1 + t), \ 0 \le t < \infty, \ p > 2,$$

we obtain finally

$$|\varphi_\chi|, \ |\varphi_{\overline{\chi}}| \le C\widetilde{\varphi}_2\ln^\alpha(1 + \widetilde{\varphi}_1)$$

with $C = C(p, \alpha, \mu_0)$ and $\widetilde{\varphi}_i(\chi) \in L_i(\Omega)$, $i = 1, 2$. The second assertion of the theorem is an immediate consequence of the first one, because

$$|\psi^*(\chi)| \equiv \left|\frac{\psi(z(\chi))}{\overline{\chi_{\bar z}}(1 - |\mu|^2)}\right| = |\psi(z(\chi))z_\chi| = \widetilde{\varphi}_2(\chi).$$

□

According to Theorem 2.6, the space $V_2^\alpha(\Omega)$ is invariant with respect to quasi-conformal mappings.

3. Integral equations. Weight integral operators

3.1 Two-dimensional integral equations

Let us consider an integral equation

$$\varphi(z) = (P\varphi)(z) + F(z), \qquad z \in \Omega, \tag{3.1}$$

where

$$(P\varphi)(z) \equiv A(z)(T\varphi)(z) + B(z)\overline{(T\varphi)(z)}$$

and the operator T is defined by (2.3).

Theorem 3.1 *Let $A, B \in L_2^\alpha(\Omega)$, $0 \le \alpha < 1/2$. Then the integral equation (3.1) has a unique solution $\varphi(z) \in L_p(\Omega)$, $1 < p < 2$, for any function $F(z) \in L_q(\Omega)$, $p < q$. Moreover, the following estimate holds*

$$\|\varphi\|_{p,\,\Omega} \le C\,\|F\|_{p+\varepsilon,\,\Omega}\,, \tag{3.2}$$

where $\varepsilon = \varepsilon(\alpha) > 0$ is a sufficiently small number such that $\varepsilon(0) = 0$ and $C = C(\alpha,\, p,\, \|A\|_{L_2^\alpha(\Omega)},\, \|B\|_{L_2^\alpha(\Omega)})$.

Proof. First we consider the case $\alpha = 0$. In this case the operator P is bounded in the function space $L_q(\Omega)$. In fact, according to the well-known embedding theorem [19]

$$\|T\varphi\|_{m,\,\Omega} \le C\,\|\varphi\|_{p,\,\Omega}\,, \quad m \le \frac{2p}{2-p}, \quad 1 < p < 2, \tag{3.3}$$

so that by the Hölder inequality we obtain

$$\|P\varphi\|_{p,\,\Omega} \le C(\|A\|_{2,\Omega} + \|B\|_{2,\Omega})\,\|\varphi\|_{p,\,\Omega}\,.$$

Note that in the considered case it is not possible to prove compactness of the operator P in contrast to the regular case. Now let us show that the operator $I - P$ (I is the identical operator), can be represented as a sum of a boundedly invertible operator and a compact one, or which is the same, as a composition of a boundedly invertible operator with an operator of the form $I - T$, where T is a compact operator. This means that equation (3.1) is equivalent to another one with a compact operator.

First we approximate the functions $A(z)$ and $B(z)$ by continuous functions $A_\delta(z)$ and $B_\delta(z)$ so that

$$\left(\|A - A_\delta\|_{2,\Omega} + \|B - B_\delta\|_{2,\Omega}\right) C \le \nu(\delta) < 1 \tag{3.4}$$

which is possible since continuous functions are dense in $L_2(\Omega)$. Now we can rewrite equation (3.1) in the following equivalent form

$$\varphi - \Lambda_\delta\varphi = P_\delta\varphi + F,$$

where

$$\Lambda_\delta = (P - P_\delta) \quad \text{and} \quad P_\delta \equiv A_\delta T + B_\delta \overline{T}.$$

It is well known that the operator P_δ is compact. On the other hand, according to (3.4)

$$||\Lambda_\delta||_p \leq \nu(\delta) < 1.$$

This means that the operator $I - \Lambda_\delta$ is boundedly invertible and

$$||(I - \Lambda_\delta)^{-1}|| \leq (1 - \nu(\delta))^{-1}.$$

Then equation (3.1) can be rewritten as

$$\varphi = \Lambda_\delta^{-1} P_\delta \varphi + \Lambda_\delta^{-1} F, \tag{3.5}$$

where the operator $\Lambda_\delta^{-1} P_\delta$ is compact. Thus, for equation (3.1) to be uniquely solvable, it is sufficient to prove that the corresponding homogeneous equation (with $F \equiv 0$) has only the trivial solution in the space $L_p(\Omega)$, $1 < p < 2$. Assume in contradiction, that there exists a non trivial solution $\varphi_0 \in L_p(\Omega)$. Then the function $w(z) = T\varphi_0 \in W_p^1(E) \cap L_{\frac{p}{p-2}}(\Gamma_r)$ $(E = \{z : |z| < \infty\})$ satisfies the equation

$$w_{\overline{z}} = \widetilde{A}(z) w,$$

where

$$\widetilde{A}(z) = A + B\frac{\overline{w}}{w} \in L_2^\alpha(\Omega), \quad (\widetilde{A}(z) = 0,\ z \notin \Omega).$$

Then, according to (2.2),

$$w(z) = \Phi(z) e^{T\widetilde{A}}.$$

Here, $\Phi(z) = w(z) e^{-T\widetilde{A}}$ is an analytic function in the complex plane E which belongs to Hardy's space $H_q(\Gamma_r)$, $1 < q < \frac{p}{2-p}$, such that

$$\Phi^+(e^{i\gamma}) = \Phi^-(e^{i\gamma}),\ \gamma \in [0,\ 2\pi],\ \Phi^-(\infty) = 0.$$

According to Liouville's theorem for analytic functions in Hardy's spaces, we obtain that $\Phi(z) \equiv 0$, respectively $w(z) \equiv 0$, $\varphi_0 \equiv 0$. Thus equations (3.1), (3.5) are uniquely solvable for any function F and estimate (3.2) (with $\alpha = 0$) is a direct consequence of Banach's theorem for the inverse operators. The case $\alpha = 0$ is proved.

Let $0 < \alpha < \frac{1}{2}$. In this case it is not possible to prove that the operator P in (3.1) is bounded. For simplicity we assume that $B(z) \equiv 0$. Then it is easy to verify that the functions $w(z) = T\varphi = e^{TA}T(Fe^{-TA})$ and φ are solutions to the equations

$$w_{\overline{z}} = A(z)w \quad \text{and} \qquad \varphi = AT\varphi + F,$$

respectively. Thus, the existence of a unique solution of the integral equation (3.1) is proved. Then inequality (3.2) is a consequence of Theorem 2.3, Hölder's inequality and (3.3):

$$\begin{aligned}||w_{\overline{z}}||_{p,\Omega} = ||w_{\overline{z}}||_{p,\Omega} &= ||Ae^{TA}T(Fe^{-TA}) + F||_{p,\Omega} \\ &\leq ||Ae^{TA}T(Fe^{-TA})||_{p,\Omega} + ||F||_{p,\Omega} \\ &\leq ||A||_{2,\Omega}||e^{TA}T(Fe^{-TA})||_{\frac{2p}{p-2},\Omega} + ||F||_{p,\Omega} \\ &\leq C||A||_{2,\Omega}||e^{TA}||_{\frac{p}{\delta},\Omega}||e^{-TA}||_{\frac{p}{\delta},\Omega}||F||_{p+\varepsilon,\Omega} + ||F||_{p,\Omega} \\ &\leq \widetilde{C}||F||_{p+\varepsilon,\Omega}.\end{aligned}$$

Here $\varepsilon = \varepsilon(\alpha) > 0$ is an arbitrary sufficiently small number and $\delta = \varepsilon p/(\varepsilon + p)(p+1)$. Theorem 3.1 is proved. □

We introduce the singular operator

$$S(\varphi|\, z) = \frac{\partial T\,(\varphi|\, z)}{\partial z} = -\frac{1}{\pi}\int_{\Omega}\frac{\varphi(t)}{(t-z)^2}d\Omega_t. \tag{3.6}$$

It is well known [19] that

$$\|S(\varphi)\|_{p,\,\Omega} \leq \lambda(p)\,\|\varphi\|_{p,\,\Omega}\,,\ 1 < p < \infty, \tag{3.7}$$

where $\lambda(p)$ is a convex function such that $\lambda(2) = 1$.

Let us consider the integral equation

$$\varphi(z) + \mu_1(z)S\varphi + \mu_2(z)\overline{S\varphi} = P\varphi + F(z), \qquad z \in \Omega, \tag{3.8}$$

where

$$|\mu_1(z)| + |\mu_2(z)| \leq \mu_0 < 1,\ z \in \Omega. \tag{3.9}$$

Theorem 3.2 *Let $A, B \in L_2(\Omega)$. Then there exists a number $p_0(\mu_0) = (2-\varepsilon)$, ($\varepsilon > 0$ is a sufficiently small number), such that the integral equation (3.8) has a unique solution $\varphi(z) \in L_p(\Omega)$, $p_0 < p < 2$, for any function $F(z) \in L_p(\Omega)$. Moreover, the following estimate holds*

$$\|\varphi\|_{L_p(\Omega)} \leq C\,\|F\|_{L_p(\Omega)} \tag{3.10}$$

where $C = C(\alpha,\ \rho,\ \|A\|_{L_2(\Omega)}, \|B\|_{L_2(\Omega)}, \mu_0)$.

Proof. According to (3.7) and (3.9), it is possible to choose a number $p_0 < 2$ sufficiently close to 2, such that

$$\left\|\mu_1 S + \mu_2 \overline{S}\right\|_{p_0,\, \Omega} \leq \mu_0 \lambda(p_0) = K < 1.$$

Now, we can rewrite equation (3.8) in the following equivalent form

$$\varphi - \Lambda_\delta \varphi = P_\delta \varphi + F,$$

where

$$\Lambda_\delta = \mu_1 S + \mu_2 \overline{S} - (P - P_\delta), \ P_\delta \equiv A_\delta T + B_\delta \overline{T}.$$

We choose the parameter δ in such a way that

$$\left\|\mu_1 S + \mu_2 \overline{S} - (P - P_\delta)\right\| \leq \mu_0 \lambda(p_0) + \nu(\delta) = K_1 < 1.$$

Thus by Banach's theorem the operator

$$\left(I + \mu_1 S + \mu_2 \overline{S} - (P - P_\delta)\right)^{-1}$$

is bounded and the equation (3.8) can be represented in an equivalent form [19] as

$$\varphi = \left(I + \mu_1 S + \mu_2 \overline{S}\right)^{-1} (P\varphi + F) \equiv P^*(\varphi) + F^*(z),$$

for $p \in]p_0,\, 2[$.

Similarly to the proof of Theorem 3.1 we show that the operator P^* is compact. Thus, for equation (3.8) to be uniquely solvable, it is sufficient to prove that the corresponding homogeneous equation (with $F \equiv 0$) has only the trivial solution in the space $L_p(\Omega)$, $p \in]p_0,\, 2[$. Assume to the contrary that there exists a non trivial solution $\varphi_0 \in L_p(\Omega)$. Then the function $w(z) = T\varphi_0 \in W_p^1(E) \cap L_{\frac{p}{p-2}}(\Gamma_r)$ $(E = \{z : |z| < \infty\})$ satisfies the equation

$$\partial_{\overline{z}} w + \widetilde{\mu}(z) \partial_z w = \widetilde{A}(z) w, \tag{3.11}$$

where

$$|\widetilde{\mu}(z)| = \left|\mu_1 + \mu_2 \frac{\partial_{\overline{z}} \overline{w}}{\partial_z w}\right| \leq \mu_0 < 1, \ \widetilde{A}(z) = A + B\frac{\overline{w}}{w} \in L_2(\Omega), \ z \in \Omega,$$

$$\widetilde{\mu}(z) = \widetilde{A}(z) = 0, \ z \notin \Omega.$$

We introduce the function $\chi(z) = w(z) e^{T\widetilde{A}} = e^{T\widetilde{A}} T\varphi_0$. According to the properties of the operator T, the function $\chi(z)$ belongs to $W_{p^*}^1(E) \cap L_{\frac{p^*}{p^*-2}}(\Gamma_r)$, where $p^* = p - \varepsilon \in]p_0,\ 2[$, with an arbitrary sufficiently small $\varepsilon > 0$, and satisfies the relations

$$\partial_{\overline{z}}\chi + \widetilde{\mu}(z)\partial_z\chi = 0,\ z \in \Omega,$$

$$\partial_{\overline{z}}\chi = 0,\ z \notin \Omega,\ \chi(\infty) = 0.$$

But for any function $\chi(z) \in W^1_{p^*}(E)$ the following representation holds [19]

$$\chi(z) = T(\partial_{\overline{z}}\chi) + \Phi(z),$$

where $\Phi(z) = (\chi(z) - T(\partial_{\overline{z}}\chi)) \in W^1_{p^*}(E) \cap L_{\frac{p^*}{p^*-2}}(\Gamma_r)$ is an analytic function, $\Phi(\infty) = 0$. Thus, the analytic function $\Phi(z)$ is identically zero according to Liouville's theorem. Therefore, the derivative $\partial_{\overline{z}}\chi$ satisfies the integral equation

$$\partial_{\overline{z}}\chi + \widetilde{\mu}(z)S(\partial_{\overline{z}}\chi) = 0,\ z \in \Omega,$$

which has only the trivial solution $\partial_{\overline{z}}\chi \equiv 0$ in the space $L_{p^*}(\Omega)$. This proves the theorem. □

Remark 3.3 It is evident that the assertions of Theorems 2.3 and 3.2 are also valid if the operator T in (3.1), (3.8) is replaced by the operators $T_{\pm\varkappa}$ defined in Remark 2.5.

3.2 Two-dimensional weight integral operators

Theorem 3.4 *Let the operators T and S be defined by (2.3), (3.6) and $\omega \equiv T(A)$, $A \in L_2(\Omega)$. Then the following estimates hold*

$$\left\|e^{\omega}T(\varphi e^{-\omega})\right\|_{\frac{2p}{2-p},\,\Omega} \leq C\,\|\varphi\|_{p,\,\Omega}\,,\quad \left\|e^{\omega}S(\varphi e^{-\omega})\right\|_{p,\,\Omega} \leq C\,\|\varphi\|_{p,\,\Omega}\,, \tag{3.12}$$

where $1 < p < 2$ and $C = C(p,\ \|A\|_{2,\,\Omega})$.

Proof. Let us prove the first inequality. Consider the functions φ and $w = e^{TA}T(\varphi e^{-TA})$. It is easy to verify that the function w satisfies the equation

$$w_{\overline{z}} = Aw + \varphi.$$

On the other hand, the function w can be represented in the form $w = T\psi$, where ψ solves the equation

$$\psi = AT\psi + \varphi.$$

Then, estimate (3.2) for ψ holds. Therefore, taking into account (3.3), we obtain that

$$\|w\|_{\frac{2p}{2-p},\,\Omega} = \left\|e^{\omega}T(\varphi e^{-\omega})\right\|_{\frac{2p}{2-p},\,\Omega} = \|T(\psi)\|_{\frac{2p}{2-p},\,\Omega} \leq C\,\|\psi\|_{p,\,\Omega} \leq C^2\,\|\varphi\|_{p,\,\Omega}\,.$$

To prove the second inequality in (3.12), we use the relation

$$e^{\omega} S(\varphi e^{-\omega}) = S(\psi) - wS(A)$$

and (3.7). As a result, we have

$$\left\| e^{\omega} S(\varphi e^{-\omega}) \right\|_{p,\,\Omega} \le \lambda(p) \left\| \psi \right\|_{p,\,\Omega} + \left\| A \right\|_{2,\,\Omega} \left\| T(\psi) \right\|_{\frac{2p}{2-p},\,\Omega} \le C \left\| \varphi \right\|_{p,\,\Omega} .$$

□

3.3 One-dimensional singular weight integral operator

Now we prove the boundedness of a one-dimensional singular operator with the weight e^{ω} in the spaces $L_p(\Gamma_r)$, $1 < p < \infty$. We will use the following statement for analytic functions in Hardy space $H_p, 1 < p < \infty$, see [1], [10].

Lemma 3.5 *Let $\Phi(z) \in H_p(\Omega)$, $1 < p < \infty$, be an analytic function. Then*

$$\left\| \Phi \right\|_{2p,\,\Omega} \le C \left\| \Phi \right\|_{p,\,\Gamma},$$

where

$$C^{\frac{2}{p}} = \frac{2\pi^2 + 1}{4\pi}, \qquad \Gamma = \{z : |z| = 1\}.$$

Definition 3.6 *We say that a measurable on Γ function $\omega(t) = \omega(e^{i\gamma})$ belongs to the function space $SW_2^1(\Gamma)$, if $\omega(t)$ is the trace of a function $\widetilde{\omega}(z) \in W_2^1(\Omega)$.*

Later on we will use the same notation $\omega(z) = \widetilde{\omega}(z)$. Let us introduce the one-dimensional singular operators

$$Kg = \frac{1}{2\pi} \int_0^{2\pi} g(s) \operatorname{ctg} \frac{s - \gamma}{2} ds \quad \text{and} \quad K_{\omega} g = e^{\omega(\gamma)} K(e^{-\omega} g), \tag{3.13}$$

with $\gamma \in [0, 2\pi]$. It is well known [8], [11], [15], that the operator K is bounded in the space $L_p(\Gamma)$, $1 < p < \infty$ and

$$\left\| Kg \right\|_{p,\,\Gamma} \le C(p) \left\| g \right\|_{p,\,\Gamma},\ 1 < p < \infty. \tag{3.14}$$

Theorem 3.7 *Let $Re\, \omega(t) \in SW_2^1(\Gamma)$, then the operator K_{ω} is bounded in the space $L_p(\Gamma)$, $1 < p < \infty$, and the following estimate holds*

$$\left\| K_{\omega} g \right\|_{p,\,\Gamma} \le C(p) \left\| g \right\|_{p,\,\Gamma},\ 1 < p < \infty.$$

Proof. Evidently, it suffices to consider the case of real functions $\omega(t) = Re\, \omega(t)$. Let us consider the function

$$W(z) = \frac{e^{\omega(z)}}{2\pi} \left[\int_0^{2\pi} e^{-\omega(\gamma)} g(\gamma) \frac{e^{i\gamma} + z}{e^{i\gamma} - z} d\gamma + iv_0 \right] \tag{3.15}$$

where the real constant v_0 is defined by

$$v_0 = \frac{1}{2\pi}\left(\int_0^{2\pi} e^{\omega} ds\right)^{-1} \int_0^{2\pi} e^{\omega(s)} \int_0^{2\pi} e^{-\omega(\gamma)} g(\gamma) \mathrm{ctg}\, \frac{\gamma - s}{2} d\gamma ds.$$

In view of the construction, the function $W(z)$ solves the following problem

$$W_{\overline{z}} = AW,\ z \in \Omega, \qquad \text{with} \qquad A \equiv \omega_{\overline{z}} \in L_2(\Omega), \tag{3.16}$$

$$Re\, W(e^{i\gamma}) = g(\gamma), \qquad \int_0^{2\pi} Im\, W(e^{i\gamma}) d\gamma = 0. \tag{3.17}$$

Any solution of problem (3.16)-(3.17) can be represented by the formula

$$W(z) = i(T_0\varphi)(z) + \Phi(z) \equiv i(T_0\varphi)(z) + \frac{1}{2\pi}\int_0^{2\pi} g(\gamma)\frac{e^{i\gamma} + z}{e^{i\gamma} - z} d\gamma, \tag{3.18}$$

where $T_0\varphi$ is defined in (2.10) and $Im(T_0\varphi)(e^{i\gamma}) = 0$, $\Phi(z)$ is an analytic function such that

$$\|\Phi\|_{p,\,\Gamma} \leq C\,\|g\|_{p,\,\Gamma}\ , 1 < p < \infty \tag{3.19}$$

and φ satisfies the equation

$$\varphi = A(T_0\varphi)(z) - iA\Phi(z).$$

According to Theorem 3.1 and Lemma 3.5 we have

$$\|\varphi\|_{\frac{2p}{p+1},\,\Omega} \leq C\,\|A\Phi\|_{\frac{2p}{p+1},\,\Omega} \leq C\,\|A\|_{2,\,\Omega}\,\|\Phi\|_{2p,\,\Omega} \leq \widetilde{C}\,\|\Phi\|_{p,\,\Gamma}\,.$$

Using estimate (3.3) and representation (3.18) we obtain

$$\|W\|_{p,\,\Gamma} \leq C\,\|\varphi\|_{\frac{2p}{p+1},\,\Omega} + \|\Phi\|_{p,\,\Gamma} \leq \widetilde{C}\,\|\Phi\|_{p,\,\Gamma}\,. \tag{3.20}$$

On the other hand, a limit representation for $W(e^{i\gamma})$ according to (3.15) gives

$$W(e^{i\gamma}) = g(\gamma) + iK_\omega(g|\gamma) + ie^{\omega(\gamma)}v_0 \tag{3.21}$$

or

$$K_\omega(g|\gamma) = -iW(e^{i\gamma}) + ig(\gamma) - e^{\omega(\gamma)}v_0, \tag{3.22}$$

where

$$\left\|e^{\omega(\gamma)}v_0\right\|_{p,\,\Gamma} \leq C\,\|g\|_{p,\,\Gamma}\,,\ 1 < p < \infty. \tag{3.23}$$

Gathering (3.19), (3.20)-(3.23), we finally arrive at the desired estimate

$$\|K_\omega(g|\gamma)\|_{p,\,\Gamma} = \left\|-iW(e^{i\gamma}) + ig(\gamma) - e^{\omega(\gamma)}v_0\right\|_{p,\,\Gamma} \leq C\,\|g\|_{p,\,\Gamma}$$

which proves the theorem. □

It should be pointed out that the weight operators of type (3.13) arise in the consideration of classical conjugate boundary value problems. Such operators with different types of weights were investigated in [7], [9], [17], [18]. Let us consider one classical problem of this type, namely the Riemann problem for analytic functions:

$$\Phi^+(t) = G(t)\Phi^-(t) + g(t),\ t \in \Gamma = \{z : |z| = 1\},\ (\Phi^-(\infty) = 0). \tag{3.24}$$

Here $\Phi^\pm(z)$ are analytic functions in the domains $\Omega^\pm = \{z : |z| \lessgtr 1\}$, respectively.

The next lemma is an immediate consequence of Theorem 3.7 related to problem (3.24).

Lemma 3.8 *Let $g(t) \in L_p(\Gamma)$, $1 < p < \infty$, $\ln G(t) \in SW_2^1(\Gamma)$. Then problem (3.24) has a unique solution $\Phi^\pm(z) \in H_q(\Omega^\pm)$, $1 < q \le p$, which is represented in the form*

$$\Phi^\pm(z) = \frac{X^\pm(z)}{2i\pi} \int_\Gamma \frac{g(t)dt}{X^+(t)(t-z)},\ \ z \in \Omega^\pm$$

where,

$$\ln\ X^\pm(z) = \frac{1}{2i\pi} \int_\Gamma \frac{\ln G(t)dt}{(t-z)},\ \ z \in \Omega^\pm$$

and

$$\left\|\Phi^+\right\|_{p,\,\Gamma} \le C\left\|g\right\|_{p,\,\Gamma}, \qquad \left\|G\Phi^-\right\|_{p,\,\Gamma} \le C\left\|g\right\|_{p,\,\Gamma}.$$

Remark 3.9 The case where

$$G(t) = G_0(t)G_1(t),\ \ln G_0(t) \in SW_2^1(\Gamma), \ln\left(t^k G_1(t)\right) \in C^\alpha, 0 < \alpha < 1,$$

may be reduced to that considered in the usual way in [8]. It should be pointed out that various cases of problem (3.24) which were considered earlier, contain, in general, the following assumption on the function $G(t)$: $0 < m \le |G(t)| \le M < \infty$, or $G(t)$ has isolated zeros or singularities.

4. Boundary value problems for generalized analytic functions

Let us consider the Hilbert problem for a generalized analytic function w :

$$w_{\overline{z}} = A(z)w + B(z)\overline{w},\ z \in \Omega = \{z : |z| < 1\}, \tag{4.1}$$
$$Re\left[G(t)w(t)\right] = g(t),\ t \in \Gamma = \{z : |z| = 1\}. \tag{4.2}$$

In the case $G(t) \in C^{\alpha}(\Gamma)$, $0 < \alpha < 1$, $0 < m \leq |G(t)| \leq M < \infty$, we can assume without loss of generality [8] that $G(t) = t^{-\varkappa}$, ($\varkappa$ is an integer).

Theorem 4.1 *Let $g(t) \in L_p(\Gamma)$, $1 < p < \infty$, $G(t) = t^{-\varkappa}$, and $A(z)$, $B(z) \in L_2^{\alpha}(\Omega)$, $0 \leq \alpha < 1$. Then problem (4.1)-(4.2) with $\varkappa \geq 0$ has $2\varkappa + 1$ linearly independent solutions $w(z) \in W^1_{\gamma-\varepsilon}(\Omega) \cap L_{p-\varepsilon}(\Gamma)$, $1 < \gamma < \frac{2p}{p+1} < 2$, where $\varepsilon = \varepsilon(\alpha) > 0$ is a sufficiently small number, such that $\varepsilon(0) = 0$. Problem (4.1)-(4.2) with $\varkappa < 0$ is solved in the above-mentioned space if the function $g(t)$ satisfies some $|\varkappa| - 1$ integral conditions.*

Proof. In the case $\varkappa \geq 0$ we are looking for a solution in the usual form

$$W(z) = (T_{\varkappa}\varphi)(z) + \frac{z^{2\varkappa}}{2\pi} \int_0^{2\pi} g(\gamma) \frac{e^{i\gamma} + z}{e^{i\gamma} - z} d\gamma + P_{\varkappa}(z),$$

see [19], where the operator $T_{\varkappa}$ is defined by (2.10) and

$$P_{\varkappa}(z) = i \sum_{k=0}^{\varkappa} \left(C_k z^{k+\varkappa} + \overline{C_k} z^{-k+\varkappa} \right),$$

C_k being arbitrary constants. Hence, the function φ must satisfy the integral equation considered above. This implies the assertions of the theorem for $\varkappa \geq 0$. In the case $\varkappa < 0$ we use the operator $T_{-\varkappa}$ defined by (2.11) in a similar way.□

5. Smoothness of quasi-conformal mappings

Let $w(z) \in W^1_p(\Omega)$, $p > 2$ be a homeomorphism of the domain Ω onto itself such that

$$\partial_{\bar{z}} w + \mu_1(z) \partial_z w + \mu_2(z) \overline{\partial_z w} = 0, \ z \in \Omega; \quad w(0) = 0, \ w(1) = 1, \tag{5.1}$$

where

$$|\mu_1(z)| + |\mu_2(z)| \leq \mu_0 < 1. \tag{5.2}$$

It is well known [19] that such a homeomorphism exists and

$$z(w) \in W^1_p(\Omega), \ p(\mu_0) > 2,$$

Moreover,

$$w(z) \in W^2_q(\Omega),$$

with some $q > 2$, if

$$\mu_i(z) \in W^1_q(\Omega), \ q > 2.$$

Let us consider properties of the second derivatives $w_{zz}, w_{z\bar{z}}, w_{\bar{z}\bar{z}}$ supposing that

$$\partial_{\bar{z}}\mu_i, \partial_z\mu_i \in L_2(\Omega),\ i = 1, 2. \tag{5.3}$$

Theorem 5.1 *Assume that* (5.2) *and* (5.3) *hold. Then*

$$w(z) \in W^2_{2-\varepsilon}(\Omega) \cap W^1_{\frac{2(2-\varepsilon)}{\varepsilon}}(\Omega) \cap W^1_{\frac{2-\varepsilon}{\varepsilon}}(\Gamma_r),$$

$$z(w) \in W^2_{2-\varepsilon}(\Omega) \cap W^1_{\frac{2(2-\varepsilon)}{\varepsilon}}(\Omega) \cap W^1_{\frac{2-\varepsilon}{\varepsilon}}(\Gamma_r),$$

where $\varepsilon > 0$ *is an arbitrary small number.*

Proof. It is well known [19], [15] that the derivatives $u_1 = \partial_{\bar{z}}w$ and $u_2 = \partial_z w$ satisfy the following equations

$$\partial_{\bar{z}}u_i + \mu(z)\partial_z u_i = A_i(z)u_i,\ i = 1, 2,\ z \in \Omega,$$

where

$$|\mu(z)| \le \mu_1(\mu_0) < 1,\ |A_i| \le C(|\partial_{\bar{z}}\mu_i| + |\partial_z\mu_i|) \in L_2(\Omega).$$

Applying the known representations w, see [15], and the results of Theorem 3.4 we arrive at the desired assertion. □

References

[1] Antontsev, S. N., Generalized Cauchy-Riemann system with almost regular coefficients. In: *Metric questions of the theory of functions and mappings*, **7**, 3–17. Naukova Dumka, Kiev, 1975 (in Russian).

[2] Antontsev, S. N. and Monakhov, V. N., The Riemann-Hilbert boundary-value problem with discontinuous boundary conditions for systems of quasilinear elliptic equations, *Soviet Math. Dokl.*, **8** (1967), 868–870.

[3] Antontsev, S. N. and Monakhov, V. N., Boundary value problems of conjugation with a shift in multiply connected domains for quasilinear elliptic systems of first order equations. In: *Proceedings of the Symposium on Continuum Mechanics and Related Problems of Analysis* (Tbilisi, 1971), Vol. 2, 20–35. Mecniereba, Tbilisi, 1974 (in Russian).

[4] Besov, O. V., Il'in, V. P., and Nikolskii, S. M., *Integral Representations of Functions and Embedding Theorems.* John Wiley, New York, 1978 (Vol. 1), 1979 (Vol. 2).

[5] Bliev, N., *Generalized Analytic Functions in Fractional Spaces*, Pitman Monographs and Surveys in Pure and Applied Mathematics, **86**. Longman, Harlow, 1997.

[6] Bojarskii, B. V., Some boundary value problems for systems of $2n$ elliptic equations in a plane, *Dokl. Akad. Nauk SSSR*, **124** (1959), 15–18 (in Russian).

[7] Daniluk, I. I., The boundedness of a singular integral operator in L_p spaces with weight, *Trudy Tbilis. Mat. Inst. Razmadze*, **33** (1967), 32–44 (in Russian).

[8] Gakhov, F. D., *Boundary Value Problems.* Pergamon Press, Oxford, New York, Paris, 1966.

[9] Gegelia, T. G., On the boundedness of singular operators, *Soobshch. Akad. Nauk Gruz. SSR*, **20** (1958), 517–523.

[10] Hoffman, K., *Banach Spaces of Analytic Functions*, Prentice-Hall Series in Modern Analysis. Prentice-Hall, Inc., Englewood Cliffs, N. J., 1962. Russian translation: IL, Moscow, 1963.

[11] Krasnosel'skii, M. A., Zabreiko, P. P., Pustyl'nik, E. I., and Sobolevskii, P. E., *Integral Operators in Spaces of Summable Functions*. Nauka, Moscow, 1966 (in Russian).

[12] Krasnosel'skii, M. A. and Rutickii, Ja. B., *Convex Functions and Orlicz Spaces*. Nauka, IFML, Moscow, 1958 (in Russian). English translation: Noordhooff, Groningen, 1961.

[13] Markushevich, A. I., *Theory of Functions of a Complex Variable*, Vol II. Prentice-Hall, Inc., Englewood Cliffs, N.J. 1965.

[14] Mikhailov, L. G., *New Class of Singular Integral Equations and its Application to Differential Equations with Singular Coefficients*. Akademie-Verlag, Berlin, 1970.

[15] Monakhov, V. N., *Boundary Value Problems with Free Boundaries for Elliptic Systems of Equations*, Translations of Mathematical Monographs, **57**. AMS, Providence, R.I., 1983.

[16] Pokhozhaev, S. I., On the Sobolev imbedding theorem for $pl = n$. In: *Doclady Conference, Section Math.*, Moscow Power Institute, 158–170, 1965.

[17] Simonenko, I. B., The Riemann boundary-value problem for n pairs of functions with measurable coefficients and its application to the study of singular integrals in L_p spaces with weights, *Izv. Akad. Nauk SSSR Ser. Mat.*, **28** (1964), 277–306 (in Russian).

[18] Simonenko, I. B., Interpolation and extrapolation of linear operators in Orlicz spaces, *Mat. Sb. (N.S.)*, **63 (105)** (1964), 536–553 (in Russian).

[19] Vekua, I. N., *Generalized Analytic Functions*. Pergamon Press, London, 1962.

TOEPLITZ MATRICES WITH SLOWLY GROWING PSEUDOSPECTRA

Albrecht Böttcher
Faculty of Mathematics, TU Chemnitz, Germany
aboettch@mathematik.tu-chemnitz.de

Sergei Grudsky*
Faculty of Mechanics and Mathematics, Rostov on Don State University, Russia
and
Departamento de Matemáticas, CINVESTAV del I.P.N., México
grudsky@aaanet.ru

To Georgii Semenovitch Litvinchuk on His 70th Birthday

Abstract Let $T(a)$ be the infinite Toeplitz matrix with the symbol a and let $T_n(a)$ denote the $n \times n$ principal submatrix of $T(a)$. The pseudospectra of $T_n(a)$ are known to converge to the pseudospectrum of $T(a)$ as $n \to \infty$ provided a is piecewise continuous. Only recently, Mark Embree, Nick Trefethen, and one of the authors observed that this convergence may be spectacularly slow in case a has a jump. The main result of this paper says that such a slow convergence of pseudospectra is generic even within the class of continuous symbols.

Keywords: Toeplitz matrix, Toeplitz determinant, pseudospectrum.

1. Introduction and Main Results

Let a be a complex-valued function in L^∞ on the complex unit circle $\mathbf{T}$ and let $\{a_k\}_{k\in\mathbf{Z}}$ be the sequence of the Fourier coefficients of a,

$$a_k = \frac{1}{2\pi}\int_0^{2\pi} a(e^{i\theta})e^{-ik\theta}d\theta.$$

*This author was partially supported by DAAD grant A/01/06411 and by CONACYT grant, Cátedra Patrimonial, Nivel II, No. 010286.

The matrices $T(a) = (a_{j-k})_{j,k=0}^{\infty}$ and $T_n(a) = (a_{j-k})_{j,k=0}^{n-1}$ are referred to as the infinite and the $n \times n$ Toeplitz matrices with the symbol a. The matrix $T(a)$ is regarded as a (bounded) operator on $\ell^2(\mathbf{Z}_+)$, while $T_n(a)$ is thought of as an operator on $\mathbf{C}^n$ with the ℓ^2 norm.

It is well known that the eigenvalues of $T_n(a)$ do not in general mimic the spectrum of $T(a)$ as $n \to \infty$ (see, e.g., [7, 13, 16]). In physical language this means that when considering spectra of Toeplitz matrices, the passage from the finite volume case to the infinite volume case is discontinuous. Things change drastically when replacing spectra by pseudospectra. For $\varepsilon > 0$, the ε-pseudospectrum $\mathrm{sp}_\varepsilon A$ of a bounded linear operator A is defined as

$$\mathrm{sp}_\varepsilon A = \mathrm{sp}\, A \cup \{\lambda \notin \mathrm{sp}\, A : \|(A - \lambda I)^{-1}\| \geq 1/\varepsilon\}$$

(see [8, 14, 15]). For piecewise continuous symbols a, it was shown in [1, 11, 12] that $\mathrm{sp}_\varepsilon T_n(a)$ converges in the Hausdorff metric to $\mathrm{sp}_\varepsilon T(a)$ as $n \to \infty$. Thus, in this context the passage from the finite volume case to the infinite volume case is continuous.

This paper addresses the question of whether the convergence of the sets $\mathrm{sp}_\varepsilon T_n(a)$ to their limit $\mathrm{sp}_\varepsilon T(a)$ is convincingly fast. In case a is a rational function, $\mathrm{sp}_\varepsilon T_n(a)$ approaches $\mathrm{sp}_\varepsilon T(a)$ very rapidly. This is caused by the circumstance that the resolvent norm $\|(T_n(a) - \lambda I)^{-1}\|$ goes to infinity with exponential speed if $\lambda \in \mathrm{sp}\, T(a) \setminus a(\mathbf{T})$ (see [4, 12]). In [3], it was observed that for piecewise continuous symbols a the sets $\mathrm{sp}_\varepsilon T_n(a)$ may converge to $\mathrm{sp}_\varepsilon T(a)$ very slowly, which results from the fact that $\|(T_n(a) - \lambda I)^{-1}\|$ may grow to infinity only polynomially. We here prove that the same may happen to Toeplitz matrices with *continuous* symbols. Two precise results are as follows.

Let C denote the set of all continuous functions on $\mathbf{T}$. If $a \in C$, then $\mathrm{sp}\, T(a)$ is the union of the range $a(\mathbf{T})$ of a and all points in the plane encircled by $a(\mathbf{T})$ with nonzero winding number. Also notice that $T_n(a) - \lambda I = T_n(a - \lambda)$, so that $(T_n(a) - \lambda I)^{-1} = T_n^{-1}(a - \lambda)$.

Theorem 1.1 *Given any number $\beta > 1$, there exists a function $a \in C$ such that*

$$\|T_n^{-1}(a - \lambda)\| = O(n^\beta)$$

for some point $\lambda \in \mathrm{sp}\, T(a) \setminus a(\mathbf{T})$.

We denote by C^m the set of m times continuously differentiable functions on $\mathbf{T}$. Furthermore, given a function a with a Fourier coeffcients $\{a_n\}_{n=-\infty}^{\infty}$, we define Pa and Qa by

$$(Pa)(t) = \sum_{n=0}^{\infty} a_n t^n, \quad (Qa)(t) = \sum_{n=-\infty}^{-1} a_n t^n \quad (t \in \mathbf{T}).$$

Notice that if a is in C^2, then $a_n = O(1/n^2)$ as $|n| \to \infty$, which implies that Pa and Qa are well-defined continuous functions.

Theorem 1.2 *Let $a \in C^2$ and let $\lambda \in \mathbf{C}$ be a point whose winding number with respect to $a(\mathbf{T})$ is -1 (resp. 1). Then $\|T_n^{-1}(a-\lambda)\|$ increases faster than every polynomial,*

$$\lim_{n\to\infty} \|T_n^{-1}(a-\lambda)\|\, n^{-\beta} = \infty \quad \textit{for each} \quad \beta > 0, \tag{1.1}$$

if and only if Pa (resp. Qa) is in C^∞.

Theorems 1.1 and 1.2 will be proved in Section 3 on the basis of some results on Toeplitz determinants which will be established in Section 2.

2. Toeplitz determinants

Let W be the Wiener algebra of all functions on $\mathbf{T}$ with absolutely convergent Fourier series and let $F\ell^2_\alpha$ denote the set of all functions f on $\mathbf{T}$ for which

$$\sum_{n=-\infty}^{\infty} (|n|+1)^{2\alpha}|f_n|^2 < \infty.$$

For $\alpha \geq 0$, the set $W \cap F\ell^2_\alpha$ is an algebra under pointwise multiplication, and if $f \in W \cap F\ell^2_\alpha$ has no zeros on $\mathbf{T}$ and winding number zero about the origin, then f has a canonical Wiener-Hopf factorization in $W \cap F\ell^2_\alpha$:

$$f = f_- f_+ \quad \text{with} \quad f_-^{\pm 1} \in W \cap F\ell^2_\alpha \cap \overline{H^\infty}, \quad f_+^{\pm 1} \in W \cap F\ell^2_\alpha \cap H^\infty$$

(see [6, pp. 272–273]). Here $\overline{H^\infty}$ and H^∞ are the usual Hardy spaces,

$$\overline{H^\infty} = \{g \in L^\infty : g_n = 0 \ \text{for} \ n > 0\}$$
$$H^\infty = \{g \in L^\infty : g_n = 0 \ \text{for} \ n < 0\}.$$

The classical Szegö limit theorem (in the version by I. I. Hirschman) states that if $f \in W \cap F\ell^2_{1/2}$ has no zeros on $\mathbf{T}$ and winding number zero, then the determinants $D_n(f) := \det T_n(f)$ have the asymptotic behavior

$$D_n(f) = G(f)^n E(f)(1 + o(1)) \quad \text{as} \quad n \to \infty,$$

where the constants $G(f)$ and $E(f)$ are given by

$$G(f) = \exp(\log f)_0, \quad E(f) = \exp \sum_{k=1}^{\infty} k(\log f)_k (\log f)_{-k},$$

$(\log f)_j$ being the jth Fourier coefficient of $\log f$; see [6] for details.

We now turn to an extension of the Szegö limit theorem to symbols whose winding number about the origin is -1 or 1. For $1 \le p \le \infty$, let ℓ^p_β be the Banach space of all sequences $x = \{x_n\}_{n=0}^\infty$ for which

$$\|x\|^p := \sum_{n=0}^{\infty} |x_n|^p (n+1)^{p\beta} < \infty \quad (1 \le p < \infty),$$
$$\|x\| := \sup_{n \ge 0} |x_n|(n+1)^\beta < \infty \quad (p = \infty).$$

Lemma 2.1 *If $h_- \in W \cap \overline{H^\infty}$, then the Toeplitz operator $T(h_-)$ is bounded on ℓ^p_β for every $p \in [1, \infty]$ and every $\beta \ge 0$.*

Proof. This result (the "failure of the Brown-Halmos theorem for spaces with weights") is well known; see [6, Remark 6.3]. The proof follows from the observation that we can write

$$T(h_-) = \sum_{n=0}^{\infty} h_{-n} [V^{(-1)}]^n, \quad \sum_{n=0}^{\infty} |h_{-n}| < \infty,$$

where $V^{(-1)}$ sends $\{x_0, x_1, x_2, \dots\}$ to $\{x_1, x_2, x_3, \dots\}$, and that $V^{(-1)}$ has norm 1. □

The following result is not terribly new, similar results can be found in [5, 6, 9, 16], for example, but we need the result in exactly the form it is stated and for just this form we have no explicit reference.

Theorem 2.2 *Let $f \in W \cap F\ell^2_{1/2}$ and $Pf \in F\ell^2_\alpha$ for some $\alpha > 1/2$. Suppose f has no zeros on $\mathbf{T}$ and winding number zero about the origin. Let $f = f_- f_+$ be a Wiener-Hopf factorization in $W \cap F\ell^2_{1/2}$ and set $b = f_- f_+^{-1}$. Put $a(t) = t^{-1} f(t)$. Then $Pb \in F\ell^2_\alpha$, and, as $n \to \infty$,*

$$D_n(a) = (-1)^n G(f)^n E(f) G(b)^{-1} \left(b_n + o\left(\frac{1}{n^{2\alpha - 1/2}} \right) \right) (1 + o(1)). \quad (2.1)$$

Proof. We have $f_+ = P(f_-^{-1} \cdot Pf)$, and using Lemma 2.1 with $h_- = f_-^{-1}$, we deduce that $f_+ \in F\ell^2_\alpha$. It follows that $f_+^{-1} \in F\ell^2_\alpha$, and hence Lemma 2.1 and the equality $Pb = P(f_- \cdot f_+^{-1})$ imply that $Pb \in F\ell^2_\alpha$.

Let P_n be the orthogonal projection of $\ell^2(\mathbf{Z}_+)$ onto $\ell^2(\{0, \dots, n-1\})$ and put $Q_n = I - P_n$. The Hankel operators $H(g)$ and $H(\tilde{g})$ generated by a function $g \in L^\infty$ are given by the matrices

$$H(g) = (g_{j+k+1})_{j,k=0}^\infty, \quad H(\tilde{g}) = (g_{-j-k-1})_{j,k=0}^\infty.$$

We denote $b^{-1} = f_+ f_-^{-1}$ by c. In [5] (also see [2] or [6, pp. 442–443]), it is shown that, for all sufficiently large n,

$$T_n^{-1}(f) = P_n T(f_+^{-1}) T(f_-^{-1}) P_n - P_n T(f_+^{-1}) P_n T(c) \sum_{j=0}^{\infty} (Q_n K Q_n)^j Q_n T(b) P_n T(f_-^{-1}) P_n, \quad (2.2)$$

where $K = H(b)H(\tilde{c})$. The square of the Hilbert-Schmidt norm of $Q_n H(b)$ is

$$\|Q_n H(b)\|_2^2 = \sum_{k=1}^{\infty} k |b_{n+k}|^2 = \sum_{k=1}^{\infty} \frac{k n^{2\alpha-1}}{n^{2\alpha-1}} |b_{n+k}|^2,$$

and since $Pb \in F\ell_\alpha^2$, we have

$$\|Q_n H(b)\|_2^2 \leq \sum_{k=1}^{\infty} \frac{(n+k)^{2\alpha}}{n^{2\alpha-1}} |b_{n+k}|^2 = o\left(\frac{1}{n^{2\alpha-1}}\right). \quad (2.3)$$

This reveals that $\|Q_n K Q_n\| \leq \|Q_n H(b)\| \, \|H(\tilde{c})\| < 1$ and hence that the series on the right of (2.2) converges for all sufficiently large n. The $(n,1)$ entry of both sides of (2.2) is obtained by multiplying the identity by $P_n - P_{n-1}$ from the left and by P_1 from the right. This gives

$$-(P_n - P_{n-1}) P_n T(f_+^{-1}) P_n T(c) Q_n T(b) P_n T(f_-^{-1}) P_1 =: X_0$$

for the $j = 0$ term of the series. As $P_n T(f_+^{-1}) P_n = P_n T(f_+^{-1})$, we get

$$X_0 = -(P_n - P_{n-1}) T(f_+^{-1}) T(c) Q_n T(b) P_n T(f_-^{-1}) P_1.$$

Since $T(f_+^{-1})T(c) = T(f_+^{-1}c) - H(f_+^{-1})H(\tilde{c})$ and $f_+^{-1}c = f_-^{-1}$, it follows that

$$\begin{aligned} X_0 &= -(P_n - P_{n-1}) T(f_-^{-1}) Q_n T(b) P_n T(f_-^{-1}) P_1 \\ &\quad + (P_n - P_{n-1}) H(f_+^{-1}) H(\tilde{c}) Q_n T(b) P_n T(f_-^{-1}) P_1. \end{aligned}$$

Using that $Q_n = I - P_n$, we obtain

$$\begin{aligned} X_0 &= -(P_n - P_{n-1}) T(f_-^{-1}) T(b) P_n T(f_-^{-1}) P_1 \\ &\quad + (P_n - P_{n-1}) T(f_-^{-1}) P_n T(b) P_n T(f_-^{-1}) P_1 \\ &\quad + (P_n - P_{n-1}) H(f_+^{-1}) H(\tilde{c}) Q_n T(b) P_n T(f_-^{-1}) P_1. \end{aligned}$$

The equality $T(f_-^{-1})T(b) = T(f_-^{-1}b) = T(f_+^{-1})$ gives

$$\begin{aligned} X_0 &= -(P_n - P_{n-1}) T(f_+^{-1}) P_n T(f_-^{-1}) P_1 \\ &\quad + (P_n - P_{n-1}) T(f_-^{-1}) P_n T(b) P_n T(f_-^{-1}) P_1 \\ &\quad + (P_n - P_{n-1}) H(f_+^{-1}) H(\tilde{c}) Q_n T(b) P_n T(f_-^{-1}) P_1. \end{aligned}$$

Finally, since $P_nT(f_-^{-1})P_1 = T(f_-^{-1})P_1 = P_1T(f_-^{-1})P_1$, we see that

$$\begin{aligned} X_0 &= -(P_n - P_{n-1})T(f_+^{-1})T(f_-^{-1})P_1 \\ &\quad + (P_n - P_{n-1})T(f_-^{-1})P_nT(b)P_1T(f_-^{-1})P_1 \\ &\quad + (P_n - P_{n-1})H(f_+^{-1})H(\tilde{c})Q_nT(b)P_1T(f_-^{-1})P_1. \end{aligned} \tag{2.4}$$

Inserting (2.4) in (2.2) and making use of the identity

$$(P_n - P_{n-1})T(f_-^{-1})P_n = (P_n - P_{n-1})T(f_-^{-1})(P_n - P_{n-1}),$$

we arrive at the formula

$$\begin{aligned} &(P_n - P_{n-1})T_n^{-1}(f)P_1 \\ &= (P_n - P_{n-1})T(f_+^{-1})T(f_-^{-1})P_1 \\ &\quad -(P_n - P_{n-1})T(f_+^{-1})T(f_-^{-1})P_1 \\ &\quad + (P_n - P_{n-1})T(f_-^{-1})(P_n - P_{n-1})T(b)P_1T(f_-^{-1})P_1 \\ &\quad + (P_n - P_{n-1})H(f_+^{-1})H(\tilde{c})Q_nT(b)P_1T(f_-^{-1})P_1 \\ &\quad - (P_n - P_{n-1})T(f_+^{-1})P_nT(c)\sum_{j=1}^{\infty}(Q_nKQ_n)^jQ_nT(b)P_nT(f_-^{-1})P_1 \\ &= (P_n - P_{n-1})T(f_-^{-1})(P_n - P_{n-1})T(b)P_1T(f_-^{-1})P_1 \\ &\quad + (P_n - P_{n-1})H(f_+^{-1})H(\tilde{c})Q_nT(b)P_1T(f_-^{-1})P_1 \\ &\quad - (P_n - P_{n-1})T(f_+^{-1})P_nT(c)\sum_{j=1}^{\infty}(Q_nKQ_n)^jQ_nT(b)P_1T(f_-^{-1})P_1 \\ &=: Y_0 + Y_1 - Y_2. \end{aligned}$$

Considering Hilbert-Schmidt norms and taking into account that f_+^{-1} and Pb are in $F\ell^2_\alpha$, it can be readily verified that

$$\|(P_n - P_{n-1})H(f_+^{-1})\| = O(1/n^\alpha), \quad \|Q_nT(b)P_1\| = O(1/n^\alpha). \tag{2.5}$$

This shows that $Y_1 = O(1/n^{2\alpha})$. From (2.3) we infer that

$$\begin{aligned} \|Q_nKQ_n\| &\le \|Q_nH(b)\|\,\|H(\tilde{c})\| \\ &\le \|Q_nH(b)\|_2\,\|H(\tilde{c})\| = o\left(\frac{1}{n^{\alpha-1/2}}\right). \end{aligned} \tag{2.6}$$

In particular, $\|Q_nKQ_n\| \le 1/2$ for all sufficiently large n. For these n we have

$$\begin{aligned} &\sum_{j=1}^{\infty}\|Q_nKQ_n\|^j\,\|Q_nT(b)P_1\| = \frac{\|Q_nKQ_n\|}{1 - \|Q_nKQ_n\|}\,\|Q_nT(b)P_1\| \\ &\le 2\,\|Q_nKQ_n\|\,\|Q_nT(b)P_1\| = o\left(\frac{1}{n^{\alpha-1/2}}\right)O\left(\frac{1}{n^\alpha}\right) = o\left(\frac{1}{n^{2\alpha-1/2}}\right), \end{aligned}$$

the last estimate resulting from (2.6) and the second estimate of (2.5). This implies that $Y_2 = o(1/n^{2\alpha-1/2})$. As

$$Y_0 = (f_-^{-1})_0\, b_{n-1}(f_-^{-1})_0 = G(f_-)^{-2}b_{n-1},$$

we eventually obtain that

$$[T_n^{-1}(f)]_{n,1} = G(f_-)^{-2}b_{n-1} + o\left(\frac{1}{n^{2\alpha-1/2}}\right).$$

On the other hand, Cramer's rule implies that

$$[T_n^{-1}(f)]_{n,1} = (-1)^{n+1}\frac{D_{n-1}(a)}{D_n(f)}. \tag{2.7}$$

Szegö's theorem says that

$$D_n(f) = G(f)^n E(f)(1+o(1)).$$

Consequently, after replacing n by $n+1$,

$$(-1)^{n+2}\frac{D_n(a)}{G(f)^{n+1}E(f)}(1+o(1)) = G(f_-)^{-2}b_n + o\left(\frac{1}{n^{2\alpha-1/2}}\right),$$

and because

$$G(f)G(f_-)^{-2} = G(ff_-^{-2}) = G(f_+f_-^{-1}) = G(b)^{-1},$$

we arrive at the asserted formula (2.1) for $D_n(a)$. □

The following theorem provides us with estimates for the norms of the inverses of finite Toeplitz matrices whose symbol has winding number -1 or 1 about the origin.

Theorem 2.3 *Let $f \in C$ and suppose $T(f)$ is invertible. Put $a(t) = t^{-1}f(t)$. Then there exist an n_0 and a constant $M < \infty$ such that*

$$\|T_n^{-1}(f)\| \le M \quad \textit{for all} \quad n \ge n_0 \tag{2.8}$$

and

$$\left|\frac{D_{n-1}(f)}{D_n(a)}\right| \le \|T_n^{-1}(a)\| \le (1+M\|a\|_2)^2\left|\frac{D_{n-1}(f)}{D_n(a)}\right| \tag{2.9}$$

for all $n > n_0$, where $\|a\|_2$ is the ℓ^2 norm of the sequence $\{a_k\}_{k=-\infty}^{\infty}$.

Proof. Estimate (2.8) is well known; see [10] or [7]. To prove (2.9), fix $n \ge n_0+1$. If $D_n(a) = 0$, then (2.9) holds. So assume that $D_n(a) \ne 0$. By Cramer's rule, we have

$$[T_n^{-1}(a)]_{1,n} = (-1)^n\frac{D_{n-1}(f)}{D_n(a)},$$

and since $\|T_n^{-1}(a)\| \ge |[T_n^{-1}(a)]_{1,n}|$, we get the lower estimate in (2.9).

It remains to prove the upper estimate in (2.9). By vitue of (2.8), the equation

$$T_{n-1}(f)\begin{pmatrix} c_1 \\ \vdots \\ c_{n-1} \end{pmatrix} = -\begin{pmatrix} a_0 \\ \vdots \\ a_{n-2} \end{pmatrix} \tag{2.10}$$

has a unique solution $(c_1 \dots c_{n-1})^\top$. Put

$$C_n = \begin{pmatrix} 1 & 0 & \dots & 0 \\ c_1 & 1 & \dots & 0 \\ \vdots & \vdots & \ddots & \vdots \\ c_{n-1} & 0 & \dots & 1 \end{pmatrix}.$$

It is easily seen that

$$T_n(a)C_n = \left(\begin{array}{c|ccc} 0 & & & \\ \vdots & & T_{n-1}(f) & \\ 0 & & & \\ \hline e_n & a_{n-2} & \dots & a_0 \end{array}\right)$$

with some number e_n. Again by virtue of (2.8), there is a unique solution $(d_1 \dots d_{n-1})$ of the equation

$$(d_1 \dots d_{n-1})T_{n-1}(f) = -(a_{n-2} \dots a_0), \tag{2.11}$$

and on setting

$$D_n = \begin{pmatrix} 1 & \dots & 0 & 0 \\ \vdots & \ddots & \vdots & \vdots \\ 0 & \dots & 1 & 0 \\ d_1 & \dots & d_{n-1} & 1 \end{pmatrix},$$

we obtain

$$D_nT_n(a)C_n = \begin{pmatrix} 0 & T_{n-1}(f) \\ e_n & 0 \end{pmatrix}.$$

Finally, let $J_n = \begin{pmatrix} 0 & 1 \\ I_{n-1} & 0 \end{pmatrix}$. Then

$$D_nT_n(a)C_nJ_n = \begin{pmatrix} T_{n-1}(f) & 0 \\ 0 & e_n \end{pmatrix}. \tag{2.12}$$

Since $\det D_n = \det C_n = 1$ and $\det J_n = (-1)^{n+1}$, identity (2.12) implies that

$$e_n = (-1)^{n+1} \frac{D_n(a)}{D_{n-1}(f)}. \tag{2.13}$$

Furthermore, from (2.12) we get

$$T_n^{-1}(a) = C_n J_n \begin{pmatrix} T_{n-1}^{-1}(f) & 0 \\ 0 & e_n^{-1} \end{pmatrix} D_n$$

and thus

$$\|T_n^{-1}(a)\| \le \|C_n\| \, \|D_n\| \, \max\left(\|T_{n-1}^{-1}(f)\|, |e_n^{-1}| \right). \tag{2.14}$$

The matrix C_n is the identity matrix plus a matrix whose Frobenius norm is

$$(|c_1|^2 + \ldots + |c_{n-1}|^2)^{1/2}.$$

This in conjunction with (2.8) and (2.10) shows that

$$\begin{aligned} \|C_n\| &\le 1 + (|c_1|^2 + \ldots + |c_{n-1}|^2)^{1/2} \\ &\le 1 + M(|a_0|^2 + \ldots + |a_{n-2}|^2)^{1/2} \le 1 + M\|a\|_2. \end{aligned}$$

Analogously one can derive from (2.8) and (2.11) that $\|D_n\| \le 1 + M\|a\|_2$. From (2.8) and (2.14) we therefore obtain

$$\|T_n^{-1}(a)\| \le (1 + M\|a\|_2)^2 \max(M, |e_n^{-1}|). \tag{2.15}$$

If the sequence $\{|e_n^{-1}|\}_{n=n_0+1}^{\infty}$ contains a bounded subsequence, then (2.15) implies that $\liminf \|T_n^{-1}(a)\| < \infty$, which in turn gives the invertibility of $T(a)$ (see [7] or [10]). As $T(a)$ is not invertible, it follows that $|e_n^{-1}| \to \infty$. Consequently,

$$\|T_n^{-1}(a)\| \le (1 + M\|a\|_2)^2 |e_n^{-1}| \tag{2.16}$$

for all sufficiently large n, say $n \ge n_1$. Inserting (2.13) in (2.16) and replacing n_0 by $\max(n_0, n_1)$, we arrive at the upper estimate in (2.9). □

3. Slow growth of the resolvent norm

In this section we prove Theorems 1.1 and 1.2. Theorem 1.1 results from the following more general result by letting $k = 0$ and $R_n = n^\beta$ ($\beta > 1$).

Theorem 3.1 *Let k be a nonnegative integer and let $\{R_n\}_{n=1}^{\infty}$ be any sequence of positive numbers such that*

$$\sum_{n=1}^{\infty} \frac{n^k}{R_n} < \infty.$$

Then there exists a function $a \in C^k$ without zeros on $\mathbf{T}$ and with winding number -1 about the origin such that

$$\delta_1 R_n \le \|T_n^{-1}(a)\| \le \delta_2 R_n$$

for all $n \ge 1$, where $\delta_1, \delta_2 \in (0, \infty)$ are constants independent of n.

Proof. Put

$$b_+(t) = b_0 + \sum_{n=1}^{\infty} \frac{t^n}{R_n} \quad (t \in \mathbf{T}).$$

Clearly, $b_+ \in C^k \cap W$. By choosing $b_0 > 0$ large enough, we can guarantee that b_+ has no zeros on $\mathbf{T}$ and winding number zero about the origin. Let $f(t) = 1/b_+(t)$ and $a(t) = t^{-1} f(t)$. Obviously, $D_n(f) = (b_+^{-1})_0^n = 1/b_0^n$ for all $n \ge 1$. Formula (2.7) implies that

$$\begin{aligned} |D_n(a)| &= |D_{n+1}(f)|\,|[T_{n+1}^{-1}(f)]_{n+1,1}| \\ &= b_0^{-n-1} |[T_{n+1}(b_+)]_{n+1,1}| = b_0^{-n-1} R_n^{-1} \end{aligned}$$

for all $n \ge 1$. Thus, with $D_0(f) := 1$,

$$\left| \frac{D_{n-1}(f)}{D_n(a)} \right| = b_0^{-n+1} b_0^{n+1} R_n = b_0^2 R_n$$

for all $n \ge 1$. The assertion is now immediate from Theorem 2.3. □

We now turn to the proof of Theorem 1.2.

Theorem 3.2 *Let $a \in W \cap F\ell^2_{1/2}$. Suppose that $Pa \in C^\infty$ and that a has the winding number -1 about some point $\lambda \in \mathbf{C}$. Then $\|T_n^{-1}(a - \lambda)\|$ increases faster than any power of n, that is, (1.1) is valid.*

Proof. We can write $a(t) - \lambda = t^{-1} f(t)$ where $f \in W \cap F\ell^2_{1/2}$ has no zeros on $\mathbf{T}$ and winding number zero about the origin. Clearly, $Pf \in W \cap F\ell^2_\alpha$ for all $\alpha > 1/2$. From Theorems 2.2 and 2.3 we infer that $Pb \in F\ell^2_\alpha$ and

$$\|T_n^{-1}(a - \lambda)\| \ge \left| \frac{D_{n-1}(f)}{D_n(a)} \right| = \left| \frac{G(b)}{G(f)} \right| \frac{1 + o(1)}{|b_n + o(1/n^{2\alpha - 1/2})|}$$

for all $\alpha > 1/2$. In particular, $Pb \in C^\infty$, and hence, for all $\alpha > 1/2$, $b_n = o(1/n^{2\alpha-1/2})$ and thus $\|T_n^{-1}(a - \lambda)\| \ge \delta n^{2\alpha - 1/2}$ with some constant $\delta > 0$. Choosing $2\alpha - 1/2 > \beta$, we arrive at the assertion. □

Theorem 3.3 *Let $a \in W \cap F\ell^2_{1/2}$ and $Pa \in F\ell^2_\alpha$ for some $\alpha > 1$. Suppose that a has the winding number -1 about $\lambda \in \mathbf{C}$. If $\|T_n^{-1}(a - \lambda)\|$ increases faster than every power of n, that is, if (1.1) holds, then Pa is in C^∞.*

Proof. By assumption, there is a constant $\delta_1 > 0$ such that

$$\|T_n^{-1}(a-\lambda)\| \geq \delta_1 n^{2\alpha-1/2}$$

for all $n \geq 1$. On writing $a(t) - \lambda = t^{-1}f(t)$ and employing Theorems 2.2 and 2.3 as in the proof of Theorem 3.2, we obtain that

$$\frac{1}{|b_n + o(1/n^{2\alpha-1/2})|} \geq \delta_2 n^{2\alpha-1/2}$$

with some constant $\delta_2 > 0$ for all sufficiently large n. Thus, $b_n = O(1/n^{2\alpha-1/2})$ as $n \to +\infty$, which implies that $Pb \in F\ell^2_\beta$ for all $\beta < 2\alpha - 1$. Since $f_+^{-1} = P(f_-^{-1} \cdot Pb)$, we deduce from Lemma 2.1 that $f_+^{-1} \in F\ell^2_\beta$, which implies that $f_+ \in F\ell^2_\beta$, and as $Pf = P(f_- \cdot f_+)$, we can again invoke Lemma 2.1 to conclude that Pf and hence also Pa is in $F\ell^2_\beta$ for all $\beta < 2\alpha - 1$.

In summary, we have proved that if $a \in W \cap F\ell^2_{1/2}$ and $Pa \in F\ell^2_\alpha$ for some $\alpha > 1$, then $Pa \in F\ell^2_\beta$ for all $\beta < 2\alpha - 1$. As the sequence $\alpha_0, \alpha_1, \dots$ given by $\alpha_0 = \alpha > 1$ and

$$\alpha_{n+1} = 2\alpha_n - 1/2$$

increases monotonically to ∞, we arrive at the conclusion that Pa is in C^∞. □

For the winding number -1, Theorem 1.2 follows from Theorems 3.2 and 3.3. Passing to transpose operators, we obtain Theorem 1.2 in the case where the winding number is 1.

Acknowledgements

We thank Mark Embree and Alexei Karlovich for valuable comments on an earlier version of this paper.

References

[1] Böttcher, A., Pseudospectra and singular values of large convolution operators, *J. Integral Equations Appl.*, **6** (1994), 267–301.

[2] Böttcher, A., One more proof of the Borodin-Okounkov formula for Toeplitz determinants, *Integr. Equat. Oper. Theor.*, **41** (2001), 123–125.

[3] Böttcher, A., Embree, M., and Trefethen, L. N., Piecewise continuous Toeplitz matrices and operators: slow approach to infinity, *SIAM J. Matrix Anal. Appl.*, **24** (2002), 484–489.

[4] Böttcher, A. and Grudsky, S., Toeplitz band matrices with exponentially growing condition numbers, *Electron. J. Linear Algebra*, **5** (1999), 104–125.

[5] Böttcher, A. and Silbermann, B., Notes on the asymptotic behavior of block Toeplitz matrices and determinants, *Math. Nachr.*, **98** (1980), 183–210.

[6] Böttcher, A. and Silbermann, B., *Analysis of Toeplitz Operators*. Springer-Verlag, Berlin, 1990.

[7] Böttcher, A. and Silbermann, B., *Introduction to Large Truncated Toeplitz Matrices*. Springer-Verlag, New York, 1999.

[8] Embree, M. and Trefethen, L. N., *Pseudospectra Gateway*. Web site: http://www.comlab.ox.ac.uk/pseudospectra.

[9] Fisher, M. E. and Hartwig, R. E., Asymptotic behavior of Toeplitz matrices and determinants, *Arch. Rat. Mech. Anal.*, **32** (1969), 190–225.

[10] Gohberg, I. and Feldman, I. A., *Convolution Equations and Projection Methods for Their Solution*, Translations of Mathematical Monographs, **41**. AMS, Providence, R.I., 1974.

[11] Landau, H., On Szegö's eigenvalue distribution theorem and non-Hermitian kernels, *J. Analyse Math.*, **28** (1975), 335–357.

[12] Reichel, L. and Trefethen, L. N., Eigenvalues and pseudo-eigenvalues of Toeplitz matrices, *Linear Algebra Appl.*, **162/164** (1992), 153–185.

[13] Schmidt, P. and Spitzer, F., The Toeplitz matrices of an arbitrary Laurent polynomial, *Math. Scand.*, **8** (1960), 15–38.

[14] Trefethen, L. N., Pseudospectra of matrices. In: *Numerical analysis 1991* (Dundee, 1991), Pitman Res. Notes Math. Ser., **260**, 234–266. Longman Scientific & Technical, Harlow, 1992.

[15] Trefethen, L. N., Pseudospectra of linear operators, *SIAM Review*, **39** (1997), 383–406.

[16] Widom, H., Eigenvalue distribution of nonselfadjoint Toeplitz matrices and the asymptotics of Toeplitz determinants in the case of nonvanishing index. In: Operator Theory: Advances and Applications, **48**, 387–421. Birkhäuser Verlag, 1990.

A NUMERICAL PROCEDURE FOR THE INVERSE STURM-LIOUVILLE OPERATOR

B. Malcolm Brown
Department of Computer Science, University of Wales, Cardiff, UK
malcolm@cs.cf.ac.uk

Valentin S. Samko
Department of Computer Science, University of Wales, Cardiff, UK
val@digiways.com

Ian W. Knowles
Department of Mathematics, The University of Alabama at Birmingham, USA
iwk@vorteb.math.uab.edu

Marco Marletta
Department of Mathematics, The University of Leicester, UK
Marco.Marletta@cs.cardiff.ac.uk

Abstract A method is introduced for the reconstruction of the potential of a Sturm-Liouville operator L generated by $l(y) = -y''(x) + q(x)y(x)$ and separated boundary conditions in $L^1[0,b]$, $0 < b < \infty$ from a spectrum and a set of normalizing constants. These constants can be calculated from the second spectrum with the same boundary conditions at zero.

Keywords: Sturm-Liouville, Inverse spectral problems, ill posed problems.

1. Introduction

Recently (cf. [23], [9], [21], [3]) there has been much interest in the inverse spectral problem for the Sturm-Liouville equation

$$-y'' + Qy = \lambda y, \quad x \in [0,b] \ ,0 < b < \infty, \quad \lambda \in \mathbb{C}\backslash\mathbb{R}, \quad Q \in L^1[0,b]. \qquad (1.1)$$

Uniqueness results for the potential Q have been proved, in terms of the associated Titchmarsh-Weyl function, some of which provide a local version of the celebrated Borg-Marchenko uniqueness theorem, first mentioned by Borg in 1946 [4]. As Levitan says in the preface to his book [14, p. 1] Borg was the first to undertake a systematic investigation of this problem. In particular he showed that a knowledge of one spectra is in general insufficient to determine the potential Q.

In the paper of Gel'fand and Levitan [7] it is shown that when b is finite, Q is determined by an integral equation and this apparently provides a means to determine Q numerically from two sets of spectral data. Algorithms based on this result are known [22] as is a method based on finite elements, (an up to date account of known methods may be found in [17]). However the recovery of a potential from spectral data is a difficult and ill-conditioned numerical problem.

We present a new approach to the numerical recovery of Q from spectral data. It is based on a method of Knowles [10] developed for the recovery of coefficients in a PDE from boundary data. The data used to recover the potential are a subset of the eigenvalues together with the multiplier constants. The method consists of first defining a Banach space functional whose unique zero is found at Q. Starting with an arbitrary guess Q_0 for Q, a gradient descent is performed on the functional until the minimum is achieved. An extended version of this article has been submitted for publication [5].

2. Formulation of the method

2.1 Background and algorithm setup

Let $\{\lambda_i\}_{i=0}^{\infty}$ be the eigenvalues of the Sturm-Liouville equation (1.1) with boundary conditions

$$\begin{aligned} y(0) + Ay'(0) &= 0 \,, \\ y(1) + By'(1) &= 0 \,. \end{aligned} \tag{2.1}$$

Also, let $u_q(x,\lambda)$ and $v_q(x,\lambda)$ be the solutions of (1.1) with q replacing Q with boundary conditions

$$\begin{aligned} u_q(0,\lambda) &= A\,, \quad & v_q(1,\lambda) &= B\,, \\ u_q'(0,\lambda) &= -1\,, \quad & v_q'(1,\lambda) &= -1\,. \end{aligned} \tag{2.2}$$

Then there exist multiplier constants C_n, $n \geq 1$ such that for $0 \leq x \leq 1$

$$u_Q(x,\lambda_n) = C_n v_Q(x,\lambda_n) \,. \tag{2.3}$$

Levinson [12] gives a formula to calculate these constants knowing two spectra with separated boundary conditions.

If we define a functional $G(q)$ such that $G(Q) = 0$ and $G(q) \geq 0$ for all q in $L^1[0,1]$, $q(x) > 0$, then we can use the gradient descent algorithm to find the minimum of $G(q)$.

Let the spectrum $\{\lambda_i\}_{i=0}^{\infty}$ and the multiplier constants $\{C_i\}_{i=0}^{\infty}$ be given. Denote $u_{q,n}(x) = u_q(x, \lambda_n)$ and $v_{q,n}(x) = v_q(x, \lambda_n)$. Then we can define $G_n(q)$ as

$$G_n(q) = \int_0^1 (u'_{q,n} - C_n v'_{q,n})^2 + q(u_{q,n} - C_n v_{q,n})^2 \tag{2.4}$$

and the functional $\tilde{G}(q)$ as a formal sum

$$\tilde{G}(q) = \sum_{n=0}^{\infty} G_n(q) \; . \tag{2.5}$$

It can be shown that for functions q of bounded variation on $[0, 1]$

$$G_n(q) = \frac{A^2}{8} \left(\int_0^1 (q - Q)(\omega)\, d\omega \right)^2 + O\left(\frac{1}{n^2}\right) \, , \tag{2.6}$$

so $\tilde{G}(q)$ diverges unless $A = 0$. Thus we introduce $G(q) = \sum_{n=0}^{N} \omega_n G_n(q)$ for some $N > 0$, both working with a finite number of terms in (2.5) and a suitable weight $\omega_n > 0$.

By construction $G(Q) = 0$, $G(q) > 0$ for $q \neq Q$. In order to proceed with the gradient descent algorithm, we next calculate the Gâteaux derivative of $G(q)$.

By definition of the Gâteaux derivative

$$G'_n(q)[h] = \lim_{\varepsilon \to 0} \frac{G_n(q + \varepsilon h) - G_n(q)}{\varepsilon} \; .$$

A calculation based on integrating by parts, noting the difference of two squares, using the ODE to eliminate u'' and v'' and then using properties of the Green's function we get

$$G'_n(q)[h] = \int_0^1 h \times H_n \tag{2.7}$$

where

$$\begin{aligned} H_n(x) \;\; = \;\; & \Bigg[2 \left(u_q \left(v_q - B\, \frac{u_q + v_q\, u'_q(1)}{B\, u'_q(1) + u_q(1)} \right) (u'_q(1) + C) \right. \\ & + \left. v_q \left(u_q - A\, \frac{u_q\, v'_q(0) + v_q}{A\, v'_q(0) + v_q(0)} \right) (C v'_q(0) + 1) C \right) \\ & + (u_q - C v_q)^2 + 2\lambda(-u_q z + C v_q t) \Bigg] \end{aligned}$$

where the dependence of u_q, v_q and C on λ_n is suppressed and z and t are solutions of

$$\begin{aligned} -z'' + (q - \lambda) z \;\; &= \;\; u_q - C v_q \, , \quad z(1) = 0 \, , \quad z'(1) = 0 \, , \\ -t'' + (q - \lambda) t \;\; &= \;\; u_q - C v_q \, , \quad t(0) = 0 \, , \quad t'(0) = 0 \, . \end{aligned}$$

2.2 Minimization algorithm

Knowing $G'(q)[h]$ we can use the gradient descent algorithm in order to find the minimum of $G(q)$. Since we have

$$G'(q)[h] = \int_0^1 h(x)H(x)dx \ , \quad H(x) = \sum \omega_n H_n \ , \tag{2.8}$$

then, by the Riesz representation theorem the L^2 gradient is $H(x)$ and

$$G(q+h) = G(q) + G'(q)[h] + O(h^2) \ .$$

So, there exists an α such that $G(q - \alpha H) < G(q)$. Thus we can set up a recovery algorithm for Q, producing a sequence $G(Q_{m-1} - \alpha_m H_m) < G(Q_{m-2} - \alpha_{m-1} H_{m-1}) < \ldots < G(Q_{initial} - \alpha_1 H_1)$.

Also, since the higher eigenvalues are likely to contribute less to the recovery procedure than the lower ones, we introduce a set of weights $\omega_n > 0$, decreasing with n, which reduce the effect of the higher eigenvalues on the procedure. These also have the effect of replacing the divergent series (2.5) by a convergent one.

Our method requires the knowledge of the following data:

- the boundary condition parameters A, B (see (2.2));
- number of given eigenvalues and multiplier constants $N+1$;
- the spectrum $\{\lambda_n\}_{n=0}^N$ of the problem together with corresponding multiplier constants $\{C_n\}_{n=0}^N$.

The method also allows computations having several sets of input data, i.e., several spectra and sets of multiplier constants corresponding to different boundary conditions. This has numerical advantages especially in cases when the given eigenvalues and multiplier constants contain noise. In these cases there are several schemes for the part of the algorithm responsible for calculating the gradient and minimization of $F(\alpha) = G(q + \alpha H)$.

Algorithm

1. Set initial Q_0, usually, some positive constant.
2. Form H_n, $n = 0, \ldots, N$ as defined by (2.7) and (2.8).
3. Form weighted sum $H = \sum_{n=0}^N \omega_n H_n$.
4. *(optional)* Calculate Sobolev type gradient.
5. Minimize $F(\alpha) = G(q - \alpha H)$ to get α_{min}.
6. Set $Q_{j+1} = Q_j + \alpha_{min} h$ where j is the number of the iteration and enforce $Q_{j+1} > 0$ by cut-off .
7. If **stopping criterion is not satisfied** go to 2.

We use the relative stopping criterion $G(Q_n) - G(Q_{n+1}) < G(Q_n) * \varepsilon$, so the iterative process only stops when the difference of the value of the functional between two iterations is small enough relative to the value of the functional. When using the absolute stopping criterion, minimization sometimes stops when $G(Q_n)$ is small but still decreasing at a reasonable rate.

2.3 Regularization criterion

Our algorithm produces a sequence of functions $Q_n(x)$, $n = 0, ..., M$ where M is the number of iterations completed for the stopping criterion to be satisfied. These functions are approximations to the true potential Q and by construction, $G(Q_0) > G(Q_1) > ... > G(Q_M)$. Although we know that $G(Q) = 0$, in practice we may either get trapped in a local minimum, or slow down on a plateau enough to satisfy the stopping criterion; thus $G(Q_N)$ will be greater than zero. If $G(Q_i) < G(Q_j)$ it does not yet mean that Q_i is a "better" approximation to Q than Q_j. This depends on how we decide which approximation is the best. We define a method to determine which potential from the sequence of recovered potentials suits the initial data best of all. The straightforward way to do this is to recover the spectra and multiplier constants corresponding to all the potentials from the sequence, then to decide the best based on some measure of the "distance" of the recovered spectrum from the true spectrum. We have the following procedure.

Let $\{\lambda_n\}_{n=0}^{N}$ and $\{C_n\}_{n=0}^{N}$ be the initial data.

For each $0 \leq k \leq M$ calculate N eigenvalues $\{\lambda_{k,n}\}_{n=0}^{N}$ and N multiplier constants $\{C_{k,n}\}_{n=0}^{N}$ corresponding to the potential $Q_k(x)$. Now we can define:

a) the best recovered potential in term of the multiplier constants is Q_m where m is an integer such that for any $0 \leq l \leq M$

$$\sup_{n=0}^{N} |C_{l,n} - C_n| \geq \sup_{n=0}^{N} |C_{m,n} - C_n| \ ;$$

b) the best recovered potential in term of eigenvalues is Q_m where m is an integer such that for any $0 \leq l \leq M$

$$\sup_{n=0}^{N} |\lambda_{l,n} - \lambda_n| \geq \sup_{n=0}^{N} |\lambda_{m,n} - \lambda_n| \ .$$

3. Some examples

We shall discuss the recovery of the following 3 problems (see Fig. 1):

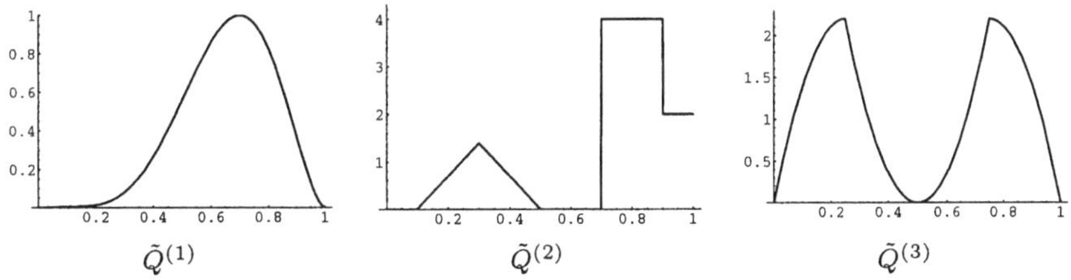

$\tilde{Q}^{(1)}$ $\tilde{Q}^{(2)}$ $\tilde{Q}^{(3)}$

Figure 1. True potentials

Potentials recovered using the L^2 gradient can be seen in Fig. 2. We shall discuss alternative gradients in section 4.

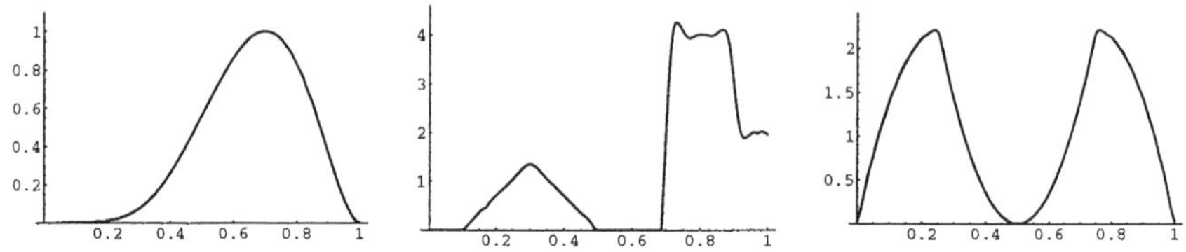

$\tilde{Q}^{(1)}$, 30 eigenvalues $\tilde{Q}^{(2)}$, 30 eigenvalues $\tilde{Q}^{(3)}$, 30 eigenvalues

Figure 2. Recovered potentials

3.1 Comparison with other methods

In [7] Gel'fand and Levitan show that the potential Q can be uniquely recovered from the knowledge of two spectra of (1.1) by an integral equation. There is a known algorithm by Rundell and Sacks [22] based on results of Gel'fand and Levitan, based on finite elements. Although the method requires knowledge of the spectral data and $\int_0^1 q(t)dt$, this mean value can be obtained from the spectrum from the asymptotic formula for eigenvalues. In [14] p. 67, and [7], p. 299, (1.14) asymptotic expansions of eigenvalues of the Sturm-Liouville problem are proved. Using only the first term of the series and the boundary conditions (2.1) we get

$$\lambda_n = \pi^2 n^2 + 2\left(\frac{1}{B} - \frac{1}{A}\right) + \int_0^1 q(t)dt + O\left(\frac{1}{n^2}\right) . \tag{3.1}$$

In practice we only know a finite number of eigenvalues, thus we can only calculate $\int_0^1 q(t)dt$ up to some error. Knowing the exact mean value Rundell and Sacks recover a smooth potential using only a few eigenvalues.

4. Using different gradients

We use two types of gradients in the recovery procedure, the L^2 gradient and a Sobolev type gradient. The experience of the authors in numerical experiments is that the L^2 gradient produces the best results when recovering both continuous and discontinuous potentials, although the Sobolev type gradient sometimes produces better results when the initial data contains noise.

4.1 L^2 gradient

The results in this section (and those in section 3) were obtained using the L^2 gradient. Some additional data on some potentials recovered with the L^2 gradient can be seen in Fig. 3.

We measure error in the recovered potential as a supremum norm $\sup_{i=0}^{M} |\lambda_i - \tilde{\lambda}_i|$ between the true eigenvalues and eigenvalues generated by the recovered potential. This error is shown beneath each picture.

Problem 1, 1515 iterations | Problem 2, 365 iterations | Problem 3, 393 iterations

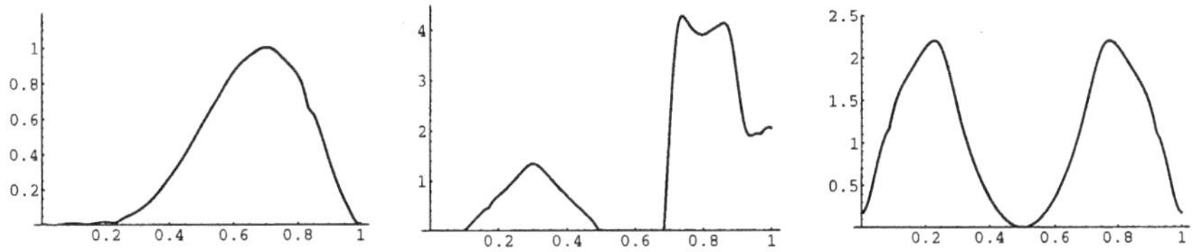

$G(\tilde{Q}^{(1)}) = 10^{-12}$, 10 eigenvalues used, error: 0.00000002

$G(\tilde{Q}^{(2)}) = 0.120359$, 20 eigenvalues used, error: 0.00136696

$G(\tilde{Q}^{(3)}) = 0.007742$, 15 eigenvalues used, error: 0.00215579

Figure 3. Potential recovered using the L^2 gradient.

4.2 Sobolev type gradient

The Sobolev type gradient (see [18]) is very smooth and because of that is sometimes used to avoid noise in the recovered potential and to recover smooth potentials in general.

Let

$$\mathcal{H}^1 = \{g : g, g' \in L^2[0,1]\} . \tag{4.1}$$

where g' is the distributional derivative of g. One can develop a formula for computing the Sobolev type gradient from the known L^2 gradient. $H^\star$, the solution of $-H^{\star\prime\prime} + H^\star = H, \quad H^{\star\prime}(0) = H^{\star\prime}(1) = 0$, is the Sobolev type gradient. Both the weakness and the strength of the Sobolev type gradient is its smoothness. Thus it is usually hard to recover discontinuous or non-smooth potentials using the Sobolev gradient. Our experiments show that only in the case of very smooth potentials (for example $\sin(\frac{x}{10})$) or when noise is present in the initial data does the Sobolev type gradient sometimes recover the potential better than the L^2 gradient. Fig. 4 shows what can be recovered in the three considered problems using the Sobolev type gradient.

4.3 Noise in initial data

When using a spectrum and multiplier constants with noise added, the algorithm tends to recover a potential whose spectrum is closer to the asymptotic formulae for eigenvalues of the Sturm-Liouville equation. We have the following theorem that shows how much the spectrum may differ when we alter the potential and also, how much noise we can introduce into the eigenvalues.

Theorem 4.1 *For all $Q(x), \tilde{Q}(x)$ on $[0,1]$, let $\{\lambda_n\}_{n=0}^{\infty}$ and $\{\tilde{\lambda}_n\}_{n=0}^{\infty}$ be the eigenvalues of the problem* (1.1).

Then:

$|\lambda_n - \tilde{\lambda}_n| = \left|\int_0^1 (Q(s) - \tilde{Q}(s))\, ds\right| + O\left(\frac{1}{n}\right) , \ n \to \infty.$

Problem 1	Problem 2	Problem 3
3569 iterations	588 iterations	698 iterations

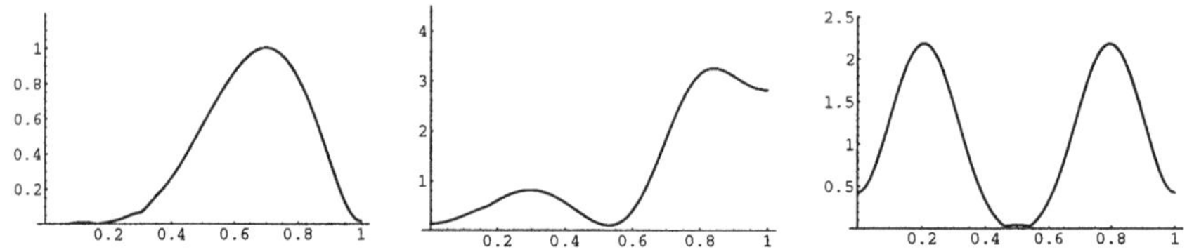

$G(\tilde{Q}^{(1)}) = 0.003961$	$G(\tilde{Q}^{(2)}) = 1000$	$G(\tilde{Q}^{(3)}) = 2.538096$
10 eigenvalues used	20 eigenvalues used	15 eigenvalues used
error: 0.000297145	error: 0.070912396	error: 0.006496933

Figure 4. Potential recovered using the Sobolev type gradient.

When generating initial data with noise, one might add absolute noise, but that technique does not work well since it perturbs the first few eigenvalues too much. Since the first few eigenvalues represent much information about the potential, this method produces the initial data corresponding to a totally different potential. Also if we add relative noise, then the resulting spectrum will not satisfy theorem 4.1. Fig. 5 shows the potentials recovered from the initial data with a relative noise 0.01% or with an absolute noise 0.01, whichever is smaller. All results in this figure were produced using the L^2 gradient.

Problem 1	Problem 2	Problem 3

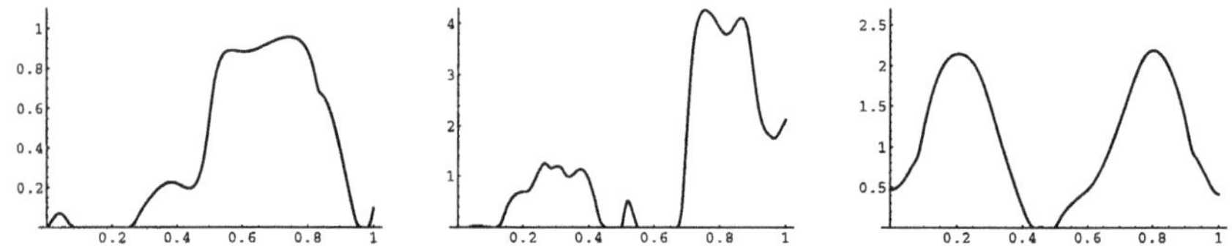

$G(\tilde{Q}^{(1)}) = 5.15356$	$G(\tilde{Q}^{(2)}) = 4.86506$	$G(\tilde{Q}^{(3)}) = 14.6753$
558 iterations	163 iterations	6 iterations
error: 0.12733663	error: 0.48855866	error: 0.22138664

Figure 5. Potential recovered from the perturbed data.

Noise in initial data usually results in oscillatory perturbations in the recovered potential. In such cases the Sobolev type gradient sometimes produces a better result than the L^2 gradient due to its smoothness. In Fig. 6 we use 5 different spectra (10 eigenvalues in each) and corresponding multiplier constants to demonstrate how we can recover the potential from multiple perturbed spectra. In this example we use the same noise as in the previous one, relative noise 0.01% or with an absolute noise 0.01, whichever is smaller.

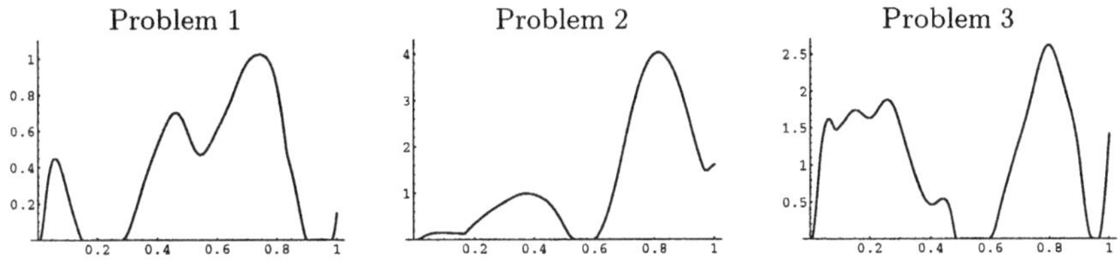

recovered using the Sobolev type gradient	recovered using the L^2 gradient	recovered using the L^2 gradient
$G(\tilde{Q}^{(1)}) = 146.94604$	$G(\tilde{Q}^{(2)}) = 515.38239$	$G(\tilde{Q}^{(3)}) = 441.57086$
35 iterations	28 iterations	15 iterations
error: 0.30575277	error: 1.19901678	error: 0.89176274

Figure 6. Potential recovered from multiple perturbed spectra.

5. Conclusion

• Practical experiments have demonstrated that when having exact initial data, $G(Q_n)$ tends to zero as n tends to ∞ and

$$\sup_{n=0}^{N} |\lambda_n - \tilde{\lambda}_{n,k}| < \sup_{n=0}^{N} |\lambda_n - \tilde{\lambda}_{n,l}|, \quad k < l ,$$

where $\lambda_{n,i}$ is the n'th eigenvalue of spectrum corresponding to Q_i. As we introduce noise in the given spectrum and multiplier constants, the best recovered potential in terms of eigenvalues and in terms of multiplier constants (see section 2.3) may be somewhere in the middle of the iterative process.

• Perturbed initial data results in oscillatory noise in the recovered potential. Sometimes the Sobolev type gradient may be used to avoid such noise, but it only appears to work well with smooth potentials, since in practice we cannot recover non-smooth or discontinuous potentials with the Sobolev type gradient. Alternatively we can use our regularization criterion as described in section 2.3.

• In the general case if we do not know the nature of the potential we want to recover, we would rather use the L^2 gradient since in numerous experiments it gave the best results.

• Even with a slight noise in the spectrum and the multiplier constants, the recovered potential differs a lot from the true one, but we still can recover the shape of the true potential.

• In future work we hope to explore potentials with internal singularities.

• It is possible that other minimization schemas could reduce the iteration count, however we have not explored this.

References

[1] Bennewitz, C., On the local Borg-Marchenko theorem, *Comm. Math. Phys.*, to appear.

[2] Bennewitz, C., A uniquenes theorem in inverse spectral theory, Preprint 2000:15, **112**.

[3] Bennewitz, C., A proof of the local Borg-Marchenko theorem, *Comm. Math. Phys.*, **218** (2001), 131–132.

[4] Borg, G., Eine Umkehrung der Sturm-Liouvilleschen Eigenwertaufgabe. Bestimmung der Differentialgleichung durch die Eigenwerte, *Acta Math.*, **78** (1946), 1–96.

[5] Brown, B. M., Samko, V. S., Knowles, I. W., and Marletta, M., Inverse spectral problem for the Sturm-Liouville equation, in preparation.

[6] Chadan, K., Colton, D., Päivärinta, L., and Rundell, W., *An Introduction to Inverse Scattering and Inverse Spectral Problems*, SIAM, Philadelphia, PA, 1997.

[7] Gel'fand, I. M. and Levitan, B. M., On the determination of a differential equation from its spectral function, *Amer. Math. Soc. Transl. (2)*, **1** (1955), 253–304.

[8] Gesztesy, F. and Simon, B., A new approach to inverse spectral theory. II. General real potentials and the connection to the spectral measure, *Ann. of Math. (2)*, **152** (2000), no. 2, 593–643.

[9] Gesztesy, F. and Simon, B., On local Borg-Marchenko uniqueness results, *Comm. Math. Phys.*, **211** (2000), no. 2, 273–287.

[10] Knowles, I., A variational algorithm for electrical impedance tomography, *Inverse Problems*, **14** (1998), no. 6, 1513–1525.

[11] Knowles, I. W., Descent methods for inverse problems, *Nonlinear Analysis*, **47** (2001), 3235–3245.

[12] Levinson, N., The inverse Sturm-Liouville problem, *Mat. Tidsskr. B.*, **1949** (1949), 25–30.

[13] Levitan, B. M., Determination of a Sturm-Liouville differential equation in terms of two spectra, *Izv. Akad. Nauk SSSR Ser. Mat.*, **28** (1964), 63–78 (in Russian).

[14] Levitan, B. M., *Inverse Sturm-Liouville Problems.* VSP, Zeist, 1987.

[15] Marchenko, V., Some questions of the theory of one-dimensional linear differential operators of the second order, *Trudy Moscov. Mat. Obsch.*, **1** (1952), 327–420 (in Russian). English translation: *Amer. Math. Soc. Transl. (2)*, **101** (1973), 1–104.

[16] Marchenko, V., Certain problems in the theory of second-order differential operators, *Dokl. Akad. Nauk SSSR*, **72** (1950), 457–460 (in Russian).

[17] McLaughlin, J. R., Solving inverse problems with spectral data. In: *Surveys on solution methods for inverse problems*, 169–194. Springer, Vienna, 2000.

[18] Neuberger, J. W., *Sobolev Gradients and Differential Equations.* Springer-Verlag, Berlin, 1997.

[19] Pöschel, J. and Trubowitz, E., *Inverse Spectral Theory.* Academic Press Inc., Boston, MA, 1987.

[20] Pryce, J. D., *Numerical Solution of Sturm-Liouville Problems.* The Clarendon Press, Oxford University Press, New York, 1993.

[21] Ramm, A. and Simon, B., A new approach to inverse spectral theory. III. Short-range potentials, *J. Anal. Math.*, **80** (2000), 319–334.

[22] Rundell, W. and Sacks, P. E., Reconstruction techniques for classical inverse Sturm-Liouville problems, *Math. Comp.*, **58**(197) (1992), 161–183.

[23] Simon, B., A new approach to inverse spectral theory. I. Fundamental formalism, *Ann. of Math. (2)*, **150** (1999), no. 3, 1029–1057.

[24] Titchmarsh, E. C., *The Theory of Functions*, 2nd ed. Oxford University Press, London, 1975.

[25] Yurko, V. A., Inverse problems for differential equations with singularities lying inside the interval, *J. Inverse Ill-Posed Probl.*, **8** (2000), no. 1, 89–103.

A GEOMETRICAL PROOF OF A THEOREM OF CRUM*

Ramón Bruzual
Escuela de Matemática, Fac. Ciencias, Universidad Central de Venezuela, Venezuela
rbruzual@euler.ciens.ucv.ve, rbruzual@reacciun.ve

Marisela Domínguez
Escuela de Matemática, Fac. Ciencias, Universidad Central de Venezuela, Venezuela
mdomin@euler.ciens.ucv.ve

Abstract Using Hilbert space geometrical arguments and basic results of measure theory we obtain a proof of a result of M. Crum, which says that a complex valued measurable positive definite function on the real line, can be decomposed as the sum of a continuous positive definite function and a positive definite function null at almost every point.

Keywords: measurable, positive definite.

1. Introduction

A function $f : \mathbb{R} \longrightarrow \mathbb{C}$ is said to be *positive definite* if for any positive integer n and any $x_1, \ldots, x_n$ in $\mathbb{R}$, $\lambda_1, \ldots, \lambda_n$ in $\mathbb{C}$ we have

$$\sum_{i,j=1}^{n} f(x_i - x_j)\lambda_i\overline{\lambda_j} \geq 0.$$

The Bochner theorem says that if f is a continuous positive definite function on $\mathbb{R}$ then f is the Fourier transform of a positive finite Borel measure on $\mathbb{R}$.

F. Riesz ([4]) extended Bochner's theorem, by proving that every measurable positive definite function on $\mathbb{R}$ is equal almost everywhere to the Fourier transform of a positive finite Borel measure on $\mathbb{R}$.

*Both authors were supported in part by the CDCH of the Univ. Central de Venezuela and by CONICIT grant G-97000668. Both authors were visitors at IVIC during the realization of this paper.

That is, if $f : \mathbb{R} \to \mathbb{C}$ is a measurable positive definite function then

$$f = f^c + f^o,$$

where f^c is continuous and positive definite and $f^o = 0$ almost everywhere.

M. Crum ([2]) proved that also the function f^o is positive definite.

Our purpose is to give a proof of the result of Crum, the idea of our proof is the following: We are going to construct a Hilbert space $\mathcal{H}$ and a strongly continuous group of unitary operators on $\mathcal{H}$; from this unitary group and the Lebesgue differentiation theorem we will obtain the function f^c, and then with geometrical arguments related to another Hilbert space, we will show that the function f^o is positive definite.

2. The strongly continuous unitary group associated to f

Let $f : \mathbb{R} \to \mathbb{C}$ be a measurable positive definite function.

It is well known that

$$f(x) = \overline{f(-x)} \quad \text{and} \quad |f(x)| \leq f(0)$$

for all $x \in \mathbb{R}$.

Lemma 2.1 *Let $\mathcal{N}$ be the set of the complex valued continuous functions with domain $\mathbb{R}$ and compact support.*

If for $h_1, h_2 \in \mathcal{N}$ we define

$$\langle h_1, h_2 \rangle = \int_{-\infty}^{+\infty} \int_{-\infty}^{+\infty} h_1(x)\overline{h_2(y)} f(x-y)dxdy,$$

then $\langle \ , \ \rangle$ is a non-negative sesquilinear form.

Remark 2.2 This assertion is not immediate since f is not supposed to be continuous and the usual argument based on Riemann sums cannot be used here.

For the proof of this Lemma we will need some additional results.

Proposition 2.3 *There exits a Hilbert space $(\mathcal{E}, \langle \ , \ \rangle_{\mathcal{E}})$ such that:*

(a) *The elements of $\mathcal{E}$ are complex valued measurable bounded functions defined on the real line. If for $z \in \mathbb{R}$ we define $f_z : \mathbb{R} \to \mathbb{C}$ by $f_z(x) = f(x-z)$, then the function f_z belongs to $\mathcal{E}$.*

(b) *If $\mathcal{M}$ is the linear span of the functions of the form $f_z(\cdot)$ with $z \in \mathbb{R}$, then $\mathcal{M}$ is a dense subspace of $\mathcal{E}$.*

(c) *If* $p, q \in \mathcal{M}$ *and*

$$p(x) = \sum_{i=1}^{n} p_i f(x - x_i) \qquad q(x) = \sum_{j=1}^{m} q_j f(x - y_j), \tag{2.1}$$

where $m, n \in \mathbb{N}$, $p_1, ..., p_n, q_1, ..., q_m \in \mathbb{C}$, $x_1, ..., x_n, y_1, ..., y_m \in \mathbb{R}$, *then*

$$\langle p, q \rangle_{\mathcal{E}} = \sum_{j=1}^{m} \sum_{i=1}^{n} p_i \overline{q_j} f(y_j - x_i).$$

(d) *If* $\varphi \in \mathcal{E}$ *then*

$$\varphi(x) = \langle \varphi, f_x \rangle_{\mathcal{E}} \quad \text{for all } x \in \mathbb{R}.$$

Proof. If $p, q \in \mathcal{M}$ are as in (2.1), we define

$$B(p, q) = \sum_{j=1}^{m} \sum_{i=1}^{n} p_i \overline{q_j} f(y_j - x_i)$$

The elements of $\mathcal{M}$ are bounded measurable functions. From the positive definiteness of f, it follows that B is a non-negative sesquilinear form on $\mathcal{M}$, so $\mathcal{M}$ is a pre-Hilbert space. Let $(\mathcal{E}, \langle\ ,\ \rangle_{\mathcal{E}})$ be the Hilbert space obtained by completing $\mathcal{M}$ and let $\|\ \|_{\mathcal{E}}$ denote the associated norm in $\mathcal{E}$.

Let $p \in \mathcal{M}$, from the definition of B it follows that

$$p(x) = \langle p, f_x \rangle_{\mathcal{E}} \quad \text{for all } x \in \mathbb{R}, \tag{2.2}$$

so we have

$$|p(x)| = |\langle p, f_x \rangle_{\mathcal{E}}| \leq \|p\|_{\mathcal{E}} \|f_x\|_{\mathcal{E}} = \|p\|_{\mathcal{E}} f(0), \quad \text{for all } x \in \mathbb{R}. \tag{2.3}$$

From this inequality if follows that convergence in $\mathcal{M}$ implies uniform convergence in the whole line.

Therefore $\mathcal{E}$ elements are measurable bounded functions and (2.2) and (2.3) are valid for $\mathcal{E}$ elements, so the Proposition is proved. □

Remark 2.4 Note that $F(x, y) = f(x - y)$ defines a reproducing kernel for $\mathcal{E}$ in the sense of Aronszajn ([1]).

Proof of Lemma 2.1 If $\varphi \in \mathcal{E}$, the function $u \mapsto \overline{\varphi(-u)}$ is measurable and it is bounded by $\|\varphi\|_{\mathcal{E}} f(0)$.

Therefore, for $h \in \mathcal{N}$, the antilinear functional from $\mathcal{E}$ to $\mathbb{C}$ given by

$$\varphi \longmapsto \int_{-\infty}^{+\infty} h(u) \overline{\varphi(-u)}\, du$$

is continuous.

By the Riesz representation theorem there exists an element Th in $\mathcal{E}$ such that

$$\langle Th, \varphi\rangle_{\mathcal{E}} = \int_{-\infty}^{+\infty} h(u)\overline{\varphi(-u)}\, du$$

for all $\varphi \in \mathcal{E}$. Note that this relation defines a linear operator T from $\mathcal{N}$ to $\mathcal{E}$.

If $x \in \mathbb{R}$,

$$\begin{aligned} (Th)(x) = \langle Th, f_x\rangle_{\mathcal{E}} &= \int_{-\infty}^{+\infty} h(u)\overline{f_x(-u)}\, du \\ &= \int_{-\infty}^{+\infty} h(u)\overline{f(-u-x)}\, du \\ &= \int_{-\infty}^{+\infty} h(u) f(x+u)\, du. \end{aligned} \tag{2.4}$$

Finally, if $h \in \mathcal{N}$, then

$$\begin{aligned} 0 \le \langle Th, Th\rangle_{\mathcal{E}} &= \int_{-\infty}^{+\infty} h(u)\,\overline{(Th)(-u)}\, du \\ &= \int_{-\infty}^{+\infty} h(u)\left(\int_{-\infty}^{+\infty} \overline{h(v) f(v-u)}\, dv\right) du \\ &= \int_{-\infty}^{+\infty}\int_{-\infty}^{+\infty} h(u)\overline{h(v)} f(u-v)\, du\, dv. \end{aligned} \tag{2.5}$$

□

Let $(\mathcal{H}, \langle\ ,\ \rangle_{\mathcal{H}})$ denote the Hilbert space obtained by completing $\mathcal{N}$, after the natural quotient.

$\|\ \|_{\mathcal{H}}$ will denote the norm associated with the product $\langle\ ,\ \rangle_{\mathcal{H}}$.

Proposition 2.5 *There exists a linear isometric operator A from $\mathcal{H}$ to $\mathcal{E}$ such that*

$$(Ah)(x) = \int_{-\infty}^{+\infty} h(u) f(x+u)\, du$$

for all $h \in \mathcal{N}$.

Proof. From the proof of Lemma 2.1 it follows that the function

$$h \mapsto Th$$

from $\mathcal{N}$ to $\mathcal{E}$ is linear and isometric. Therefore it can be extended to an isometric operator A from $\mathcal{H}$ to $\mathcal{E}$. □

Now we construct a group of unitary operators $(U_r)_{r\in(-\infty,+\infty)}$ on $\mathcal{H}$ in the following way:

For $h \in \mathcal{N}$ and $r \in \mathbb{R}$ we define U_r by

$$(U_r h)(x) = h(x - r).$$

It is clear that $U_r : \mathcal{N} \longrightarrow \mathcal{N}$ is a unitary operator, therefore it can be extended to a unitary operator on $\mathcal{H}$, this extension will be denoted by U_r also.

Proposition 2.6 *The family $(U_r)_{r\in(-\infty,+\infty)}$ is a strongly continuous group of unitary operators on the Hilbert space $\mathcal{H}$.*

Proof. Clearly $(U_r)_{r\in(-\infty,+\infty)}$ is a group of unitary operators. Since $\mathcal{N}$ is dense in $\mathcal{H}$, in order to prove the strong continuity, it is enough to show that

$$\lim_{t\to 0}\langle U_t h, h\rangle_{\mathcal{H}} = \langle h, h\rangle_{\mathcal{H}} \quad \text{for all} \quad h \in \mathcal{N}.$$

Take $h \in \mathcal{N}$, since f is bounded and h has bounded support we have that

$$\begin{aligned}
\lim_{t\to 0}\langle U_t h, h\rangle_{\mathcal{H}} &= \lim_{t\to 0}\int_{-\infty}^{+\infty}\int_{-\infty}^{+\infty} h(x-t)\overline{h(y)}f(x-y)dx \\
&= \int_{-\infty}^{+\infty}\int_{-\infty}^{+\infty} h(x)\overline{h(y)}f(x-y)dx \\
&= \langle h, h\rangle_{\mathcal{H}}.
\end{aligned}$$

□

3. Construction of the function f^c

For $n \in \mathbb{N}$ let φ_n be the function defined by

$$\varphi_n(x) = \begin{cases} n & \text{if } -\frac{1}{2n} < x < \frac{1}{2n}; \\ 0 & \text{in another case.} \end{cases}$$

It is easy to check that $\varphi_n \in \mathcal{H}$ for all n.

For $t \in \mathbb{R}$ let $\varphi_{n,t}$ be the function defined by

$$\varphi_{n,t}(x) = \varphi_n(x - t).$$

Making use of a classical result of measure theory (see, for example, Dunford and Schwartz, page 216, Corollary 7, 1957), if $F : \mathbb{R} \longrightarrow \mathbb{C}$ is a bounded and measurable function then

$$\lim_{n\to\infty}\int_{-\infty}^{+\infty} \varphi_{n,t}(x)F(x)dx = F(t) \tag{3.1}$$

at almost every point $t \in \mathbb{R}$. If F is also continuous then the equality holds at every point of $\mathbb{R}$.

Proposition 3.1 *For $t \in \mathbb{R}$ the sequence $\{\varphi_{n,t}\}_{n=1}^{+\infty}$ is weakly convergent in $\mathcal{H}$.*

Proof. Since $\mathcal{N}$ is dense in $\mathcal{H}$ it is enough to show that the sequence $\{\| \varphi_{n,t} \|_{\mathcal{H}}\}_{n=1}^{+\infty}$ is bounded and that $\lim_{n\to\infty}\langle \varphi_{n,t}, h\rangle_{\mathcal{H}}$ exists for all $h \in \mathcal{N}$.

Since $|f(x)| \leq f(0)$ we have that

$$\| \varphi_{n,t} \|_{\mathcal{H}}^2 = \int_{-\infty}^{+\infty}\int_{-\infty}^{+\infty} \varphi_{n,t}(x)\overline{\varphi_{n,t}(y)} f(x-y)dxdy \leq f(0).$$

If $h \in \mathcal{N}$ then

$$\begin{aligned}\lim_{n\to\infty}\langle \varphi_{n,t}, h\rangle_{\mathcal{H}} &= \lim_{n\to\infty}\int_{-\infty}^{+\infty}\int_{-\infty}^{+\infty} \varphi_{n,t}(x)\overline{h(y)} f(x-y)dxdy \\ &= \int_{-\infty}^{+\infty} f(t-y)\,\overline{h(y)}\,dy \qquad (3.2)\end{aligned}$$

□

Let δ_t denote the weak limit of the sequence $\varphi_{n,t}$.

Proposition 3.2

(a) *If $h \in \mathcal{N}$ and $t \in \mathbb{R}$ then*

$$\langle \delta_t, h\rangle_{\mathcal{H}} = \int_{-\infty}^{+\infty} f(t-y)\,\overline{h(y)}\,dy.$$

(b) $f(t) = \langle \delta_t, \delta_0\rangle_{\mathcal{H}}$ *at almost every point $t \in \mathbb{R}$.*

Proof. (a) follows immediately from (3.2).

In order to prove (b) note that

$$\begin{aligned}\langle \delta_t, \delta_0\rangle_{\mathcal{H}} &= \lim_{n\to\infty}\langle \delta_t, \varphi_{n,0}\rangle_{\mathcal{H}} \\ &= \lim_{n\to\infty}\int_{-\infty}^{+\infty} f(t-y)\,\overline{\varphi_{n,0}(y)}\,dy \\ &= f(t)\end{aligned}$$

at almost every point $t \in \mathbb{R}$. □

It is clear that for every $t \in \mathbb{R}$

$$U_t\delta_0 = \delta_t.$$

Therefore

$$f(t) = \langle U_t\delta_0, \delta_0\rangle_{\mathcal{H}}$$

at almost every point $t \in \mathbb{R}$.

It is easy to check that the function f^c defined by

$$f^c(t) = \langle U_t\delta_0, \delta_0\rangle_{\mathcal{H}}$$

is positive definite and, from the strong continuity of (U_t), it follows that f^c is continuous.

So, we have proved

$$f = f^c + f^o$$

where f^c is positive definite and continuous, and f^o is null almost everywhere.

It remains only to prove that f^o is positive definite.

4. Positive definiteness of f^o

Since $f = f^c$ at almost every point we have that

$$(Ah)(x) = \int_{-\infty}^{+\infty} h(u)f^c(x+u)du$$

for all $h \in \mathcal{N}$ and $x \in \mathbb{R}$.

Therefore if $t, x \in \mathbb{R}$ then

$$(A\delta_{-t})(x) = f^c(x-t).$$

Define f_t^c and f_t^o by

$$f_t^c(x) = f^c(x-t), \qquad f_t^o(x) = f^o(x-t).$$

Since $A\delta_{-t} = f_t^c$ we have that $f_t^c \in \mathcal{E}$ and from $f = f^c + f^o$ it follows that $f_t^o \in \mathcal{E}$.

Let $x, y \in \mathbb{R}$ then

$$\begin{aligned}
\langle f_x^c, f_y^c\rangle_{\mathcal{E}} &= \langle A\delta_{-x}, A\delta_{-y}\rangle_{\mathcal{E}} \\
&= \langle \delta_{-x}, \delta_{-y}\rangle_{\mathcal{H}} \\
&= \langle U_{-x}\delta_0, U_{-y}\delta_0\rangle_{\mathcal{H}} \\
&= \langle U_{y-x}\delta_0, \delta_0\rangle_{\mathcal{H}} \\
&= f^c(y-x).
\end{aligned}$$

From (d) of Proposition 2.3 it follows that

$$\langle f_x^c, f_y\rangle_{\mathcal{E}} = f_x^c(y) = f^c(y-x).$$

Since $f_y = f_y^c + f_y^o$ it must be

$$\langle f_x^c, f_y^o\rangle_{\mathcal{E}} = 0,$$

and therefore

$$\langle f_x^o, f_y^o\rangle_{\mathcal{E}} = \langle f_x - f_x^c, f_y^o\rangle_{\mathcal{E}} = \overline{f_y^o(x)} = f^o(y-x).$$

Finally let $n \in \mathbb{N}$, $x_1, \ldots, x_n \in \mathbb{R}$, $\lambda_1, \ldots, \lambda_n \in \mathbb{C}$

$$\sum_{i,j=1}^{n} f^o(x_i - x_j)\lambda_i\overline{\lambda_j} = \left\langle \sum_{j=1}^{n} \overline{\lambda_j}\, f^o_{x_j}, \sum_{i=1}^{n} \overline{\lambda_i}\, f^o_{x_i} \right\rangle_{\mathcal{E}} \geq 0.$$

References

[1] Aronszajn, N., Theory of reproducing kernels, *Trans. Amer. Math. Soc.*, **68** (1950), 337–404.

[2] Crum, M., On positive definite functions, *Proc. London Math. Soc.*, **6** (1956), 584–560.

[3] Dunford, N. and Schwartz, J., *Linear Operators. I. General Theory*, Pure and Applied Mathematics, **7**. Interscience Publishers, Inc., New York, 1958.

[4] Riesz, F., Über Sätze von Stone und Bochner, *Acta Sci. Math.*, **6** (1932–1934), 184–198.

LOCALIZATION AND MINIMAL NORMALIZATION OF SOME BASIC MIXED BOUNDARY VALUE PROBLEMS

Luís P. Castro
University of Aveiro, Portugal
lcastro@mat.ua.pt

Roland Duduchava
A. Razmadze Mathematical Institute, Tbilisi, Georgia
duduch@rmi.acnet.ge

Frank–Olme Speck
Instituto Superior Técnico, Technical University of Lisbon, Portugal
fspeck@math.ist.utl.pt

To Professor Georgii S. Litvinchuk on the occasion of his 70th birthday

Abstract We consider a class of mixed boundary value problems in spaces of Bessel potentials. By localization, an operator L associated with the BVP is related through operator matrix identities to a family of pseudodifferential operators which leads to a Fredholm criterion for L. But particular attention is devoted to the non-Fredholm case where the image of L is not closed. Minimal normalization, which means a certain minimal change of the spaces under consideration such that either the continuous extension of L or the image restriction, respectively, is normally solvable, leads to modified spaces of Bessel potentials. These can be characterized in a physically relevant sense and seen to be closely related to operators with transmission property (domain normalization) or to problems with compatibility conditions for the data (image normalization), respectively.

Keywords: Normalization, boundary value problems, localization, pseudo-differential operators, Wiener-Hopf operators, Fredholm property, Bessel potential spaces.

1. Introduction to mixed boundary value problems and normalization

We confine our attention to the following model boundary value problem (BVP) based on considerations in [38, p. 186 ff.] and [37]. Let $\Omega \subset \mathbb{R}^2$ be a bounded domain with smooth boundary $\Gamma = \partial\Omega$ divided into two simply connected parts and their common boundary points, i.e. (see Figure 1.1)

$$\Gamma = \Gamma^1 \cup \Gamma^2 \cup \{x^1, x^2\}. \tag{1.1}$$

Let A be a linear differential operator with smooth coefficients in Ω of order $2m$ where $m \in \mathbb{N}_1$ and B^1, B^2 are vectors of linear boundary operators both with smooth coefficients on Γ (extendible to $\overline{\Omega}$) of order $m^1 = (m_1^1, \dots, m_m^1)$ and $m^2 = (m_1^2, \dots, m_m^2)$, respectively, such that $0 \le m_j^1, m_j^2 \le 2m - 1$. More precisely we have B^k with components

$$b_j^k = b_j^k(x, D) = \sum_{|s| \le m_j^k} b_{j,s}^k(x) T_0^k(D^s \varphi) = T_0^k \left(\sum_{|s| \le m_j^k} b_{j,s}^k(x) D^s \varphi \right), \quad k = 1, 2 \tag{1.2}$$

where T_0^k denotes the (usual) trace operator on Γ^k.

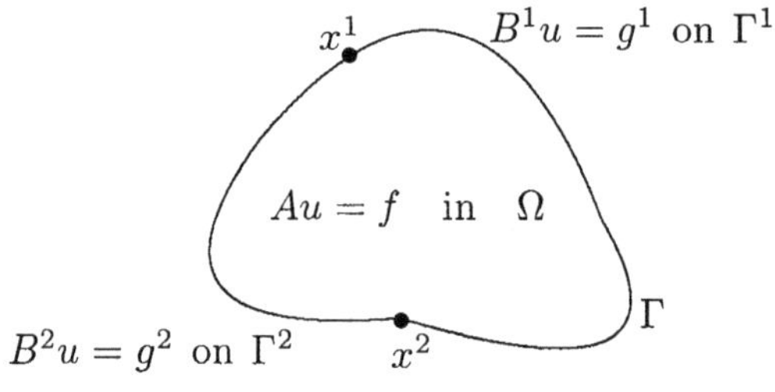

Figure 1.1: Mixed boundary value problem.

We look for all solutions $u \in H^{2m+l}(\Omega)$, $l \ge 0$, such that

$$\begin{aligned} Au(x) &= A(x, D)u(x) = f(x), && x \in \Omega \\ B^k u(x) &= \left(b_1^k(x, D)u(x), \dots, b_m^k(x, D)u(x)\right) && \\ &= \left(g_1^k(x), \dots, g_m^k(x)\right), && x \in \Gamma^k, \quad k = 1, 2, \end{aligned} \tag{1.3}$$

where $f \in H^l(\Omega)$, $g_j^k \in H^{2m+l-m_j^k-1/2}(\Gamma^k)$ are (arbitrarily) given, $j = 1, \dots, m$, and refer, for short, to the *mixed BVP* (1.3). It is called *piecewise elliptic*, if B^k ($k = 1, 2$) have extensions $\widetilde{B^k}$ to the whole Γ such that (1.3) with B^k replaced

by $\widetilde{B^k}$ and with Γ^k, $k = 1, 2$, replaced by Γ are both elliptic ($k = 1$ or 2), i.e., cf. [38, p. 187],
I: the principle part of the Fourier symbol of A does not degenerate:

$$A^{pr}(x, \xi) \neq 0, \qquad 0 \neq \xi \in \mathbb{R}^2, \quad x \in \overline{\Omega}; \tag{1.4}$$

II: the Shapiro-Lopatinsky condition is satisfied for A^{pr}, $\widetilde{B^1}^{pr}$ and for A^{pr}, $\widetilde{B^2}^{pr}$ as well [38, §11]. This means that the following two initial value problems have only the trivial solution in $H^{2m}(\mathbb{R} \times \mathbb{R}_+)$

$$\begin{aligned} A^{pr}\left(x_0, \xi, \tfrac{1}{i}\tfrac{d}{dt}\right) v(t) &= 0 \\ \widetilde{B^k}^{pr}\left(x_0, \xi, \tfrac{1}{i}\tfrac{d}{dt}\right) v(t) &= 0 \end{aligned} \tag{1.5}$$

for $k = 1, 2$, $x_0 \in \Gamma$, $\xi \in \mathbb{R}\backslash\{0\}$, where (in brief) the operators result from (1.3) by fixing the coefficients, taking the principal parts, applying locally a linear transformation such that x_0 becomes the origin, the x_1-axis is tangential to Γ and Ω lies locally left at zero, and replacing the differential operators $D^{(1,0)}$ by ξ and $D^{(0,1)}$ by $\frac{1}{i}\frac{d}{dt}$, respectively, see [38, §11] for details and for other equivalent formulations.

Remark 1.1 Smooth transformations do not destroy the Shapiro-Lopatinsky condition. This fact will be important for the localization principle presented in Section 3.

Later, in Section 8, we shall discuss modifications and generalizations of the model problem.

Examples 1.2 1. The very basic mixed Dirichlet/Neumann problem for the Laplace equation

$$\begin{aligned} \Delta u &= 0 && \text{in} \quad \Omega \\ u &= g^1 && \text{on} \quad \Gamma^1 \\ \frac{\partial u}{\partial n} &= g^2 && \text{on} \quad \Gamma^2 \end{aligned} \tag{1.6}$$

where $u \in H^1(\Omega)$ is unknown and $g^1 \in H^{1/2}(\Gamma^1)$, $g^2 \in H^{-1/2}(\Gamma^2)$ are given, is well-posed [37]. Normalization is not needed in this space setting.

2. BVPs (as well as transmission problems) for the Helmholtz equation in half-spaces in various cases of basic boundary conditions [23, 33], yield *compatibility conditions* as to be necessary for the solvability or for the Fredholm property in the initial spaces (H^1 formulation) [21, 25, 26, 27]. For instance

the mixed impedance BVP for the upper half-plane

$$\begin{aligned} (\Delta + k_0^2)\, u &= 0 && \text{in} && \Omega = \{x = (x_1, x_2) \in \mathbb{R}^2 : x_2 > 0\} \\ \frac{\partial u}{\partial x_2} + iz^1 u &= g^1 && \text{on} && \Gamma^1 = \mathbb{R}_+ \times \{0\} \\ \frac{\partial u}{\partial x_2} + iz^2 u &= g^2 && \text{on} && \Gamma^2 = \mathbb{R}_- \times \{0\} \end{aligned} \tag{1.7}$$

where $k_0, z^1, z^2 \in \mathbb{C}$, $\Im m\, k_0 > 0$, $u \in H^1(\Omega)$, $g^k \in H^{-1/2}(\Gamma^k)$ implies that

$$g^1 - Rg^2 \in r_{\Gamma_1} H_{\overline{\Gamma_1}}^{-1/2}(\Gamma) \subsetneq H^{-1/2}(\Gamma_1) \tag{1.8}$$

where R is the reflection operator with respect to x_1 and $H_{\overline{\Gamma_1}}^{-1/2}(\Gamma)$ denotes the $H^{-1/2}$ distributions defined on $\Gamma = \partial\Omega = \mathbb{R} \times \{0\}$ and supported on $\overline{\Gamma_1}$, which represents, as restricted on Γ_1, a proper dense subspace of $H^{-1/2}(\Gamma_1) = r_{\Gamma_1} H^{-1/2}(\Gamma)$. In other words: the distributions of $r_{\Gamma_1} H_{\overline{\Gamma_1}}^{-1/2}(\Gamma)$ are extendible by zero to Γ within $H^{-1/2}(\Gamma)$ but the elements of $H^{-1/2}(\Gamma_1)$ are not, in general.

Under this additional compatibility condition, the problem is already well-posed, which can be understood in the following sense: the operator defined by

$$\begin{aligned} L_0 \,:\; X_0 = \{u \in H^1(\Omega) : (\Delta + k_0^2)u = 0\} &\to H^{-1/2}(\Gamma^1) \times H^{-1/2}(\Gamma^2) \\ u &\mapsto (g^1, g^2) \end{aligned} \tag{1.9}$$

is not normally solvable. However, restricting the image of L_0 by (1.8) and changing the norm correspondingly, e.g. by using the zero extension operator $\ell^0_{\Gamma_1 \to \Gamma} : r_{\Gamma_1} H_{\overline{\Gamma_1}}^{-1/2}(\Gamma) \to H^{-1/2}(\Gamma)$

$$\|(g_1, g_2)\| = \|g_1\|_{H^{-1/2}(\Gamma_1)} + \left\|\ell^0_{\Gamma_1 \to \Gamma}(g^1 - Rg^2)\right\|_{H^{-1/2}(\Gamma)} \tag{1.10}$$

the (image) restricted operator becomes normally solvable and, moreover, bounded invertible (up to some parameters of z^k in a set of $\mathbb{C}$–measure zero [22]). We call this the *image normalization* of L_0, cf. [27] and Section 6 for the general concept.

Instead of restricting the image of L_0, we can extend the domain. Let us explain this in the special case of $z^1 = z^2 = z$ (which numbers do not effect the compatibility condition (1.8)). The solutions of the Helmholtz equation in $H^1(\Omega)$ can be written as [23, 33]

$$u(x) = F^{-1}_{\xi \mapsto x_1} \{\exp[-x_2 t(\xi)] \cdot \widehat{u_0}(\xi)\}, \qquad x_1 \in \mathbb{R}, \quad x_2 > 0 \tag{1.11}$$

where (for the trace u_0 of u) $Fu_0(\xi) = \widehat{u_0}(\xi)$ denotes the one-dimensional Fourier transformation and $t(\xi) = (\xi^2 - k_0^2)^{1/2}$ with vertical branch cut over infinity. Thinking of smooth data in a dense subspace $\mathcal{S}(\mathbb{R}_+)$ instead of $H^{-1/2}(\Gamma^k)$ we can write the boundary conditions as

$$F^{-1}(iz-t)\cdot Fu_0 \quad = \quad g = \begin{cases} g^1 & \text{on} \quad \Gamma^1 \\ g^2 & \text{on} \quad \Gamma^2 \end{cases} \tag{1.12}$$

i.e. we understand the boundary value due to the boundary conditions of (1.7) in the sense of

$$H^{-1/2}(\mathbb{R}) \subset H^{-1/2}(\Gamma^1) \oplus H^{-1/2}(\Gamma^2) \tag{1.13}$$

which embedding is dense, proper and continuous. Evidently X_0 is [19] isomorphic to $H^{-1/2}(\mathbb{R})$ by the formulas (1.11)–(1.13) (provided $iz-t$ has no zeros on $\mathbb{R}$). If we equip X_0 with the norm induced by the direct sum and take its completion $\overline{X_0}$, we evidently get a well-posed problem associated with the *domain normalization* of L_0, cf. [27].

Remark 1.3 The above examples are treated in the sense of weak solutions ($m=1$, $l=-1$, $2m+l=1$) in contrast to the assumptions before ($l \geq 0$). We plan to come back to these questions later.

2. Associated operators

The following operators shall be considered. They are fundamental for the Fredholm property and normalization, respectively, as well.

$$L = \begin{pmatrix} A \\ B^1 \\ B^2 \end{pmatrix} : H^{2m+l}(\Omega) \to \begin{cases} H^l(\Omega) \\ \times \prod_{j=1}^m H^{2m+l-m_j^1-1/2}(\Gamma^1) \\ \times \prod_{j=1}^m H^{2m+l-m_j^2-1/2}(\Gamma^2) \end{cases} \tag{2.1}$$

$$L_{x_0}^{pr} = \begin{pmatrix} A_{x_0}^{pr} \\ B_{x_0}^{1\,pr} \\ B_{x_0}^{2\,pr} \end{pmatrix} \quad : \quad H^{2m+l}(\Omega) \to \begin{cases} H^l(\Omega) \\ \times \prod_{j=1}^m H^{2m+l-m_j^1-1/2}(\Gamma^1) \\ \times \prod_{j=1}^m H^{2m+l-m_j^2-1/2}(\Gamma^2) \end{cases}$$

where $A_{x_0}^{pr}$ denotes the principal part of A with frozen coefficients in $x_0 \in \overline{\Omega}$ and $B_{x_0}^{k\,pr}$ is defined in the same way if $x_0 \in \Gamma^k \cup \{x^1, x^2\}$; otherwise we put $B_{x_0}^{k\,pr} = 0$, i.e. for all $x_0 \in \Omega \cup \Gamma^{3-k}$. This means that $L_{x_0}^{pr}$ is defined for each $x_0 \in \overline{\Omega}$ as a mapping in the same space setting, but has a different form, namely

$$L_{x_0}^{pr} = \begin{pmatrix} A_{x_0}^{pr} \\ 0 \\ 0 \end{pmatrix}, \begin{pmatrix} A_{x_0}^{pr} \\ B_{x_0}^{1\,pr} \\ 0 \end{pmatrix}, \begin{pmatrix} A_{x_0}^{pr} \\ 0 \\ B_{x_0}^{2\,pr} \end{pmatrix}, \begin{pmatrix} A_{x_0}^{pr} \\ B_{x_0}^{1\,pr} \\ B_{x_0}^{2\,pr} \end{pmatrix} \tag{2.2}$$

for $x_0 \in \Omega$, $x_0 \in \Gamma^1$, $x_0 \in \Gamma^2$ or $x_0 \in \{x^1, x^2\}$, respectively.

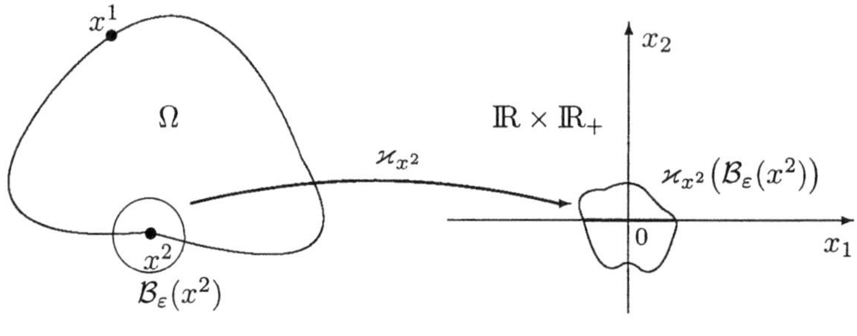

Figure 2.1: Local transformation.

Further, denoting by $\mathcal{B}_\varepsilon(x_0)$ the ε-ball centered at $x_0 \in \overline{\Omega}$, we consider smooth transformations $\varkappa_{x_0} : \mathcal{B}_\varepsilon(x_0) \to \mathbb{R}^2$ of neighborhoods of x_0 into $\mathbb{R}^2$ such that (see Figure 2.1 for the case $x_0 = x^2$)

$$\begin{aligned} &\varkappa_{x_0}(x_0) = 0, && \text{for} \quad x_0 \in \overline{\Omega} \\ &\varkappa_{x_0}(\Omega \cap \mathcal{B}_\varepsilon(x_0)) \subset \{x \in \mathbb{R}^2 : x_2 > 0\} && \\ &\varkappa_{x_0}(\Gamma \cap \mathcal{B}_\varepsilon(x_0)) \subset \{x \in \mathbb{R}^2 : x_2 = 0\} && \text{for} \quad x_0 \in \Gamma \end{aligned} \tag{2.3}$$

and the transformed operator

$$L^{\langle pr\rangle}_{x_0,0} = \begin{pmatrix} A^{\langle pr\rangle}_{x_0,0} \\ B^{k\,pr}_{x_0,0} \\ B^{(3-k)\,pr}_{x_0,0} \end{pmatrix} : H^{2m+l}(\mathbb{R} \times \mathbb{R}_+) \to \begin{cases} H^l(\mathbb{R} \times \mathbb{R}_+) \\ \times \prod_{j=1}^m H^{2m+l-m_j^k-1/2}(\mathbb{R}_-) \\ \times \prod_{j=1}^m H^{2m+l-m_j^{3-k}-1/2}(\mathbb{R}_+) \end{cases} \tag{2.4}$$

where $A^{pr}_{x_0,0}$ (etc.) are the principal parts of the transformed operators and $A^{\langle pr\rangle}_{x_0,0} = F^{-1}\Phi(\langle\cdot\rangle) \cdot F$ corresponds with the PDO with Fourier symbol of $A^{pr}_{x_0,0}$ replacing the variable $\xi \in \mathbb{R}^2$ by $\left(\frac{1+\xi^2}{\xi^2}\right)^{1/2}\xi$ which makes $A^{\langle pr\rangle}_{x_0,0}$ a bijection provided (1.4) is satisfied, cf. [15, Lemma 7.1].

Remark 2.1 A similar idea, namely the replacement of ξ by $\frac{1+|\xi|}{|\xi|}\xi$, was used in [15, §7], which is modified here for computational reasons, see Remark 5.5 and Corollary 5.6. Again $B^{k\,pr}_{x_0,0} = 0$ for $x_0 \in \Omega \cup \Gamma^{3-k}$.

3. Localization

We like to apply a modification of the local principle [9, 16, 32] in a convenient form. Following those well-known concepts, the above-mentioned operators L (etc.) are called *operators of local type*, if the composition $\varphi \cdot L\psi\cdot$ is compact for any $\varphi, \psi \in C^\infty(\dot{\mathbb{R}}^2)$ with $\operatorname{supp}\varphi \cap \operatorname{supp}\psi = \emptyset$. An equivalent condition can be formulated as follows: the commutator $[L, g\cdot] = Lg \cdot - g \cdot L$ is a compact operator for all $g \in C_0^\infty(\dot{\mathbb{R}}^2)$. Here $\dot{\mathbb{R}}^2$ is the one-point compactification of $\mathbb{R}^2$ and $\varphi\cdot$ denotes the multiplication operator generated by the function φ in the corresponding spaces, i.e. restricted to Ω and Γ^k, respectively. This is obviously the case under the assumptions of Section 1. Note that the multiplication operator $g\cdot$ acts in a scalar way on the right side of L and as a diagonal 3×3 matrix on the left side of L in the sense of the image space of (2.1). In other words

$$g \cdot L = \begin{pmatrix} g_{|\Omega}\cdot & 0 & 0 \\ 0 & g_{|\Gamma^1}\cdot & 0 \\ 0 & 0 & g_{|\Gamma^2}\cdot \end{pmatrix} \circ L \tag{3.1}$$

and we shall work with the short notation on the left side of this identity for convenience (also for smooth functions defined at least on $\overline{\Omega}$).

To define local equivalence (see below) we need localizing classes, which must be covering, to provide proper localization, see [16, Chapter 5].

Let $g \in C^\infty(\Omega)$, $0 \le g(x) \le 1$, $g(x) = 1$ for $x \in \mathcal{B}_1(0)$ and $g(x) = 0$ for $x \notin \mathcal{B}_2(0)$. Further we define the family of functions

$$M_{x_0} = \left\{ g_{x_0,\delta} \in C^\infty(\overline{\Omega}) : g_{x_0,\delta}(x) = g\left(\frac{x - x_0}{\delta}\right), 0 < \delta < d(x_0) \right\} \text{ for } x_0 \in \overline{\Omega}, \tag{3.2}$$

where

$$d(x_0) = \frac{1}{2} \begin{cases} \operatorname{dist}(x_0, \Gamma), & x_0 \in \Omega \\ \operatorname{dist}(x_0, \Gamma^{3-k}), & x_0 \in \Gamma^k, \quad k = 1, 2 \\ \operatorname{dist}(x^1, x^2), & x_0 \in \{x^1, x^2\} \end{cases} \tag{3.3}$$

Note, that for $x_0 \in \Omega$ the class M_{x_0} consists of functions which are supported only inside Ω, i.e. $\operatorname{supp} g_{x_0,\delta} \cap \Gamma = \emptyset$. For $x_0 \in \Gamma$ we have $g_{x_0,\delta}(x^k) = 0$ if $x_0 \neq x^k$ and $g_{x_0,\delta}(x^k) \neq 0$ if $x_0 = x^k$.

We denote by $\mathfrak{M}_{x_0} = \mathfrak{M}_{x_0}(X)$ the class of multiplication operators of the form $G_{x_0,\delta} = g_{x_0,\delta}\cdot = g_{x_0,\delta} I_X$ for all spaces appearing in Section 2 (scalar and vectorial as well, cf. (3.1)). It is obvious that $M = \cup_{x_0 \in \overline{\Omega}} M_{x_0}$ has the characteristics of (a family of functions that generates) a *covering system of localizing classes for mixed BVPs* (1.3), i.e.

(j) For all $x_0 \in \overline{\Omega}$ and $g \in M_{x_0}$, a multiplication operator $G = gI_X \in \mathfrak{M}_{x_0}(X)$ is bounded but not compact for any admissible space X;

(jj) If $g_1, g_2 \in M_{x_0}$, there exists a function $g \in M_{x_0}$ such that $g_1 g = g_2 g = g$;

(jjj) If $S \subset M$ such that, for every $x_0 \in \overline{\Omega}$ there exists a function $g \in S$ with $g(x_0) \neq 0$, then there exist numbers $N, N_1 \in \mathbb{N}$ such that $g = g_1 + \cdots + g_N$ is invertible in $C^\infty(\overline{\Omega})$ where $g_1 \in M_{x^1}$, $g_2 \in M_{x^2}$, $g_3 + \cdots + g_{N_1} \in M_{\Gamma^1 \cup \Gamma^2} = \cup_{x_0 \in \Gamma^1 \cup \Gamma^2} M_{x_0}$, and $g_{N_1+1} + \cdots + g_N \in M_\Omega = \cup_{x_0 \in \Omega} M_{x_0}$, i.e. $g_3(x) = \cdots = g_N(x) = 0$ for $x \in \{x^1, x^2\}$, $g_{N_1+1}(x) = \cdots = g_N(x) = 0$ for $x \in \Gamma$.

Remark 3.1 All conditions can be reformulated in terms of multiplication operators on the admissible spaces, e.g.

(jj′) If $G_1, G_2 \in \mathfrak{M}_{x_0}(X)$, there exists another $G \in \mathfrak{M}_{x_0}(X)$ such that $G_1 G = G G_1 = G_2 G = G G_2 = G$.

and (jjj′) can be written by analogy.

If we had dropped the (redundant) zeros in (2.2), the different sizes of multiplication operators in different x_0's do not allow this reformulation of (jjj). In that case, relations resulting from the partition of unity had a more complicated form.

Since we shall have a finite number of local transformations

$$\varkappa_j : U_j \to V_j, \qquad U_j \subset \Omega, \qquad V_j \subset \mathbb{R}^2, \qquad \mathrm{j} = 1, 2, \ldots, \mathrm{K}$$

(see Fig. 2.1 and (2.3)), we will use, for convenience, $\varkappa(x)$ which coincides with $\varkappa_j(x)$ for considerations in U_j. Similarly we use $\varkappa^{-1}(z)$, dropping the index j, for the inverse transformation.

The operators L and $L^{pr}_{x_0}$ are said to be *locally equivalent in* $x_0 \in \overline{\Omega}$, if for all $\varepsilon > 0$ there exists a $\delta > 0$ such that

$$|||\left(L - L^{pr}_{x_0}\right) G_{x_0,\delta}||| < \varepsilon, \quad |||G_{x_0,\delta}\left(L - L^{pr}_{x_0}\right)||| < \varepsilon \tag{3.4}$$

where $|||\cdot|||$ denotes the norm in the quotient space of bounded linear modulo compact operators. The relation (3.4) is obviously true for the operators introduced before, cf. [38]. Moreover, we have the *local quasi-equivalence* of $L^{pr}_{x_0}$ in $x_0 \in \Gamma$ and $L^{\langle pr \rangle}_{x_0,0}$ as well as $L^{pr}_{x_0,0}$ in 0 (we will apply both concepts later):

$$\begin{aligned} |||G_{x_0,\delta}\left(L^{pr}_{x_0} - T_\varkappa L^{\langle pr \rangle}_{x_0,0} T_{\varkappa^{-1}}\right)||| &= |||\left(L^{pr}_{x_0} - T_\varkappa L^{\langle pr \rangle}_{x_0,0} T_{\varkappa^{-1}}\right) G_{x_0,\delta}||| < \varepsilon, \\ |||G_{x_0,\delta}\left(L^{pr}_{x_0} - T_\varkappa L^{pr}_{x_0,0} T_{\varkappa^{-1}}\right)||| &= |||\left(L^{pr}_{x_0} - T_\varkappa L^{pr}_{x_0,0} T_{\varkappa^{-1}}\right) G_{x_0,\delta}||| < \varepsilon, \end{aligned} \tag{3.5}$$

where we define the operators $T_\varkappa$ in the following way.

If $x_0 \in U_j$, $\varkappa = \varkappa_j$ and $\varphi \in X = H^{2m+l}(\Omega)$, let

$$\mathcal{T}_\varkappa \varphi(x) = \begin{cases} \varphi(\varkappa_j(x)), & x \in \Omega \cap U_j \\ 0, & x \in \Omega \setminus U_j \end{cases} \tag{3.6}$$

If X is the image space of L, we modify $\mathcal{T}_\varkappa$ in the sense of (3.1) and $\mathcal{T}_\varkappa^{-1}$ is defined similarly in the corresponding space settings.

Further we call the above operators *locally Fredholm* in x_0 if there exist *local left* and *right regularizers* $R^l_{x_0}$, $R^r_{x_0}$ such that

$$\begin{aligned} &|||\left(R^l_{x_0} L - I\right) G_{x_0,\delta}||| = 0 \\ &|||G_{x_0,\delta}\left(L R^r_{x_0} - I\right)||| = 0 \end{aligned} \tag{3.7}$$

for suitable $G_{x_0,\delta} \in \mathfrak{M}_{x_0}$.

On the plane $\mathbb{R}^2$ we consider the "pull-back" localizing classes

$$\mathfrak{M}'_{z_0} = \{G'_{z_0,\delta}(z) = G_{x_0,\delta}(\varkappa^{-1}(z)) \ : \ G_{x_0,\delta} \in \mathfrak{M}_{x_0}\}, \quad z_0 = \varkappa(x_0) \in \mathbb{R}^2 \tag{3.8}$$

and define locally equivalent and locally Fredholm operators by analogy.

Theorem 3.2 *Let L be the operator* (2.1) *associated to a mixed BVP* (1.3). *The following assertions are equivalent:*

(i) L is Fredholm;

(ii) L is locally Fredholm in x_0, for all $x_0 \in \overline{\Omega}$;

(iii) $L^{\langle pr \rangle}_{x_0,0}$ is locally Fredholm in $\varkappa(x_0)$, for all $x_0 \in \overline{\Omega}$.

Proof. In brief: The equivalence of the conditions (i) and (ii) follows from a general localization principle described in [9] – modified for the present settings. Namely, if (i) L is Fredholm, there exist (global) regularizers R^l and R^r acting in spaces inversely to (2.1) such that $R^l L - I$ and $L R^r - I$ are compact in the corresponding spaces. This implies (3.7) for any $g \in C^\infty(\dot{\mathbb{R}}^2)$, i.e. condition (ii). Conversely, if (ii) is satisfied, since $\{\mathfrak{M}_{x_0}\}_{x_0 \in \overline{\Omega}}$ is a covering system of localizing classes in all Bessel potential spaces, from the collection $\{G_{x_0,\delta}\}_{x_0 \in \overline{\Omega}, \delta > 0}$ of operators $G_{x_0,\delta} \in \mathfrak{M}_{x_0}$, which provide localization in (3.7), we can select a finite collection $\{G_{x_j,\delta_j}\}_{j=1,\dots,N}$, such that $\sum_{j=1,\dots,N} G_{x_j,\delta_j}$ is invertible. Therefore,

$$R^l = \sum_{j=1}^N R^l_{x_j} G_{x_j,\delta_j}$$

is a left regularizer of L because

$$R^l L - I = \sum_{j=1}^N R^l_{x_j} G_{x_j,\delta_j} L - I = \sum_{j=1}^N \left(R^l_{x_j} L G_{x_j,\delta_j} - G_{x_j,\delta_j}\right) + \text{compact}$$

is compact. Here we used that $G_{x_j,\delta_j}L = LG_{x_j,\delta_j}$ up to lower order terms which are compact according to the well-known embedding theorems in spaces of Bessel potentials. The same argument holds for the right global and local regularizers.

The equivalence of (ii) and (iii) is an obvious consequence of the fact that operators which are of local type and locally quasi-equivalent in $x_0 \in \overline{\Omega}$ and $z_0 = \varkappa(x_0) \in \mathbb{R}^2$, like L and $L_{x_0}^{\langle pr \rangle}$, are simultaneously locally Fredholm in x_0 and in $z_0 = \varkappa(x_0)$ or not, respectively. □

Remark 3.3 The same reasoning implies the equivalence of the local Fredholm property of the operators L_{x_0}, $L_{x_0}^{pr}$, $L_{x_0}^{\langle pr \rangle}$ and $L_{x_0,0}^{pr}$, respectively, in each $x_0 \in \overline{\Omega}$. But these conditions are less important for our purposes.

Theorem 3.4 *If the BVP* (1.3) *is piecewise elliptic, the associated operator* L *is Fredholm if and only if* $L_{x_0,0}^{\langle pr \rangle}$ *is Fredholm for* $x_0 \in \{x^1, x^2\}$.

Proof. First, for the operator $L_{x_0,0}^{\langle pr \rangle}$, the Fredholm property and the local Fredholm property in 0 amount to the same. This is a consequence of the fact that $L_{x_0,0}^{\langle pr \rangle}$ is locally Fredholm anyway in all other points of $\mathbb{R} \times \mathbb{R}_+$ and at infinity due to the assumption of piecewise ellipticity [38].

Considering Theorem 3.2 we have to prove the "if" part, i.e., assuming that the BVP (1.3) is piecewise elliptic and that $L_{x_0,0}^{\langle pr \rangle}$ is Fredholm for $x_0 \in \{x^1, x^2\}$, we like to construct a two-sided regularizer of L.

First we think of local regularizers in $x_0 \in \Omega$ and $x_0 \in \Gamma^1$. The BVP

$$\begin{aligned} Au = f \quad &\in \quad H^l(\Omega) = Y_0 \\ \widetilde{B^1}u = g \quad &\in \quad \textstyle\prod_{j=1}^m H^{2m+l-m_j^1-1/2}(\Gamma) = Y_1 \end{aligned} \tag{3.9}$$

is elliptic for $u \in H^{2m+l}(\Omega) = X$. Thus

$$\widetilde{L^1} = \begin{pmatrix} A \\ \widetilde{B^1} \end{pmatrix} : X \to Y_0 \times Y_1 \tag{3.10}$$

has a regularizer $\widetilde{R^1} = \left(\widetilde{R_{11}}, \widetilde{R_{12}}\right) : Y_0 \times Y_1 \to X$. Putting

$$R^1 = \left(\widetilde{R_{11}}, \widetilde{R_{12}}E_{\Gamma_1 \to \Gamma}, 0\right) : Y_0 \times Y_{1|\Gamma^1} \times Y_{2|\Gamma^2} \to X \tag{3.11}$$

where $E_{\Gamma_1 \to \Gamma} : Y_{1|\Gamma^1} \to Y_1$ is a continuous extension operator (of local type), with obvious space identifications, we obtain a local regularizer $R_{x_0} = R^1$ of L for all $x_0 \in \Omega \cup \Gamma^1$.

By analogy we have local regularizers for $x_0 \in \Gamma^2$. Together with the second assumption and Theorem 3.2, we get a family of local regularizers for all $x_0 \in \overline{\Omega}$ and can construct a global regularizer as in the proof of Theorem 3.2. □

4. Reduction to semi-homogeneous problems

In the present section we study the localized operator $L^{\langle pr \rangle}_{x_0,0}$ for $x_0 \in \{x^1, x^2\}$ in Theorem 3.4 and (2.4) in detail.

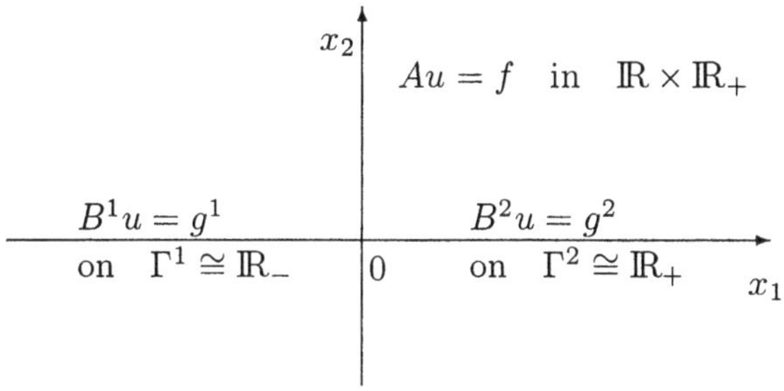

Figure 4.1: Canonical mixed BVP.

It is the operator associated with the *canonical mixed BVP* for the upper half-plane where now, abbreviating the previous notation, the operators A and B^k have constant coefficients. Figure 4.1 corresponds with the quasi-localization in $x_0 = x^1$ in Figure 1.1 and the roles of B^1 and B^2 are exchanged, if we consider $x_0 = x^2$.

We also consider the two BVPs where the mixed boundary condition is replaced either by $B^1_{\mathbb{R}} u = g$ on the full line $\Gamma = \{(x_1, x_2) : x_2 = 0\} = \Gamma^1 \cup \{(0,0)\} \cup \Gamma^2 \cong \mathbb{R}$ or by $B^2_{\mathbb{R}} u = g$ in the corresponding spaces, i.e. associated with the operators (in short notation)

$$L^k_{\mathbb{R}} = \begin{pmatrix} A \\ B^k_{\mathbb{R}} \end{pmatrix} : H^{2m+l}(\mathbb{R} \times \mathbb{R}_+) \to \left\{ \begin{array}{c} H^l(\mathbb{R} \times \mathbb{R}_+) \\ \times \prod_{j=1}^m H^{2m+l-m^k_j-1/2}(\mathbb{R}). \end{array} \right. \tag{4.1}$$

Both are elliptic (see Section 1) and admit therefore the following results.

Proposition 4.1 *The elements of* $\ker A$ *are represented by*

$$\begin{aligned} u(x_1, x_2) = F^{-1}_{\xi \mapsto x_1} \Big\{ & \Big(\widehat{u_{11}}(\xi) + x_2 \widehat{u_{12}}(\xi) + \cdots + x_2^{\mu_1 - 1} \widehat{u_{1\mu_1}}(\xi) \Big) \exp[-t_1(\xi) x_2] \\ & + \cdots + \Big(\widehat{u_{\kappa 1}}(\xi) + x_2 \widehat{u_{\kappa 2}}(\xi) + \cdots + x_2^{\mu_\kappa - 1} \widehat{u_{\kappa \mu_\kappa}}(\xi) \Big) \exp[-t_\kappa(\xi) x_2] \Big\} \end{aligned} \tag{4.2}$$

where κ is the number of square roots $t_1, \dots, t_\kappa$ with positive real part of the characteristic polynomial $A(i\xi, z)$ with respect to z, the numbers $\mu_1, \dots, \mu_\kappa$ are the corresponding multiplicities, $\sum_{\lambda=1}^{\kappa} \mu_\lambda = m$, and

$$u_{\lambda,\nu} \in H^{2m+l-\nu+1/2}, \qquad \nu = 1, \dots, \mu_\lambda, \quad \lambda = 1, \dots, \kappa. \tag{4.3}$$

Proof. Making use of the characterization of elliptic PDOs [38, p. 150], the representation formula (4.2) is a consequence of partial Fourier transformation with respect to the first variable and solution of a homogeneous ODE by an exponential ansatz. (4.3) results from estimates in spaces of Bessel potentials, see [33] for the case of second order PDEs. □

Proposition 4.2 *The operators*

$$B^k_{\mathbb{R}} : \ker A \mapsto Y^k_1 = \prod_{j=1}^{m} H^{2m+l-m^k_j-1/2}(\mathbb{R}) \tag{4.4}$$

and $L^k_{\mathbb{R}}$ *in* (4.1) *are homeomorphisms.*

Proof. $\ker A$ as a closed subspace of H^{2m+l} is toplinear isomorphic to the topological product space of the data (4.3), which are in one-to-one (homeomorphic) correspondence with the data space Y^k_1 by means of linear algebra and elementary estimates in Bessel potential spaces. Together with the Fredholm property of L^k [38] and elementary space decomposition, this implies the statement. □

These results allow the following interpretations.

Corollary 4.3 *In short, the operators* $L^k_{\mathbb{R}}$ *can be written as*

$$L^k_{\mathbb{R}} = \begin{pmatrix} A \\ B^k_{\mathbb{R}} \end{pmatrix} : X \to \left\{ \begin{matrix} Y_0 \\ \times\ Y^k_1 \end{matrix} \right. = Y^k \tag{4.5}$$

where $B^k_{\mathbb{R}}$ *is a retraction* [30], *i.e. right invertible by an extension (or coretraction) operator* E^k.

The data (product) space Y^0_1 *given by* (4.3) *is the image of a particular boundary operator* $B^0_{\mathbb{R}}$ *of an elliptic BVP due to* $L^0 = (A, B^0_{\mathbb{R}})^T$.

The spaces Y^k_1, $k = 0, 1, 2$, *are isomorphically connected by translation invariant operators*

$$\begin{matrix} Y^1_1 & \underset{B_-}{\longleftarrow} & Y^0_1 & \underset{B_+}{\longrightarrow} & Y^2_1 \end{matrix} \tag{4.6}$$

where the Fourier symbols $\Psi_\pm$ *of* $B_\pm = F^{-1}\Psi_\pm \cdot F$ *are rational* $m \times m$ *matrix functions of the variables* ξ, $t_1(\xi), \ldots, t_\kappa(\xi)$.

Proof. This is a consequence of the well-posedness of the two elliptic BVPs in combination with Proposition 4.1 and Proposition 4.2. □

Coming to the question of reducing BVPs to semi-homogeneous BVPs, where $Au = 0$ instead of $Au = f$, we study bounded linear operators in Banach spaces of the form

$$L = \begin{pmatrix} A \\ B \end{pmatrix} : X \to Y = \begin{pmatrix} Y_0 \\ Y_1 \end{pmatrix} \tag{4.7}$$

and like to relate L to

$$L^0 = B_{|X_0} : X_0 \to Y_1 \tag{4.8}$$

where $X_0 = \ker A$. The question is, if the two operators are *toplinear equivalent after extension* (in brief: equivalent after extension) i.e. [2, 4, 5] if there exist Banach spaces Z_1, Z_2 and homeomorphisms E_1, E_2 such that

$$\begin{pmatrix} L^0 & 0 \\ 0 & I_{Z_1} \end{pmatrix} = E_1 \begin{pmatrix} L & 0 \\ 0 & I_{Z_2} \end{pmatrix} E_2 \tag{4.9}$$

which yields that the two operators have similar properties (invertibility etc.) and various explicit results (representation of generalized inverses and normalization, e.g.) can be obtained from corresponding results for the other (simpler) one. Here we have:

Lemma 4.4 *Assuming* (4.7) *where A is right invertible, i.e. surjective and $X_0 = \ker A$ is complemented, then the operators* (4.7) *and* (4.8) *are toplinear equivalent after extension.*

Proof. It is sufficient to observe that

$$\begin{aligned} \begin{pmatrix} L^0 & 0 \\ 0 & I_{Y_0} \end{pmatrix} &= \begin{pmatrix} L^0 & 0 \\ 0 & A \end{pmatrix} \begin{pmatrix} I & 0 \\ 0 & \widetilde{R} \end{pmatrix} \\ &: \begin{pmatrix} Y_1 \\ Y_0 \end{pmatrix} \leftarrow \begin{pmatrix} X_0 \\ X_1 \end{pmatrix} \leftarrow \begin{pmatrix} X_0 \\ Y_0 \end{pmatrix} \end{aligned} \tag{4.10}$$

provided $A\widetilde{R} = I_{Y_0}$, $X_1 = \operatorname{im} R = RAX$, $\widetilde{R} = \operatorname{Rst} R : Y_0 \to X_1$ (i.e. image restricted) and the first factor on the right is equivalent to L (coincides up to homeomorphisms). □

Remark 4.5 For this step, only properties of A (not of B) are important. Here a right inverse is obtained by continuous extension of $H^l(\mathbb{R}^2_+)$ to $H^l(\mathbb{R}^2)$, inversion of the PDO A (with constant coefficients) on $\mathbb{R}^2$, and restriction to the half-space.

Evidently the same arguments hold when B stands for a "mixed boundary operator".

Proposition 4.6 *Consider the operator in Theorem* 3.4 *defined by* (2.4) *and now abbreviated by*

$$L_M = \begin{pmatrix} A \\ B^1 \\ B^2 \end{pmatrix} : X \to \begin{pmatrix} Y_0 \\ r_-Y_1^1 \\ r_+Y_1^2 \end{pmatrix}. \tag{4.11}$$

This operator is toplinear equivalent after extension to each of the following:

$$L_M^0 = \begin{pmatrix} B^1 \\ B^2 \end{pmatrix}_{|X_0} : X_0 = \ker A \to \begin{pmatrix} r_- Y_1^1 \\ r_+ Y_1^2 \end{pmatrix} \tag{4.12}$$

$$L_0^0 = B^2{}_{|X_0^0} : X_0^0 = X_0 \cap \ker B^1 \to r_+ Y_1^2. \tag{4.13}$$

Proof. The first equivalence relation is clear from the foregoing consideration. A right inverse of the differential operator A with constant coefficients, see (2.4) and (4.1), is obtained by

$$R = rF^{-1}\Phi_A^{-1} \cdot F\ell$$

where $\ell : H^l(\mathbb{R} \times \mathbb{R}_+) \to H^l(\mathbb{R}^2)$ denotes a continuous extension operator, Φ_A is the Fourier symbol of A, and $r : H^{2m+l}(\mathbb{R}^2) \to H^{2m+l}(\mathbb{R} \times \mathbb{R}_+)$ stands for the restriction to the half-plane.

The second relation follows similarly by Lemma 4.4. B^1 is right invertible, since a right inverse is obtained by continuous extension from $r_- Y_1^1$ to Y_1^1, inversion of the operator $L^{1,0}$ associated to the (non-mixed, semi-homogeneous) BVP due to (A, B^1) in $\mathbb{R}_+^2$ by Corollary 4.3, and restriction to the half-line $\mathbb{R}_+$. □

5. The Fredholm property

Now we focus on the two operators $L_{x_0,0}^{\langle pr \rangle}$, $x_0 \in \{x^1, x^2\}$, in (2.4) which are both directly associated to the mixed canonical BVP described in Section 4.

Let W_{x^k} denote the two Wiener-Hopf operators of the form

$$W_{x^k} = r_+ B_+ B_-^{-1} : \prod_{j=1}^{m} H_+^{2m+l-m_j^k-1/2} \to H^{2m+l-m_j^{3-k}-1/2}(\mathbb{R}_+) \tag{5.1}$$

where $k = 1, 2$ and $B_\pm$ are the operators of (4.6) resulting from the operator $L_{x^k,0}^{\langle pr \rangle}$ of (2.4), $k = 1, 2$ (or the corresponding mixed canonical BVP described in Figure 3.1 for $k = 1$).

Proposition 5.1 *The operator L_{0,x^k}^0 defined by* (4.13) *in the case of $L_M = L_{x^k,0}^{\langle pr \rangle}$, is equivalent to W_{x^k} (in the sense of toplinear* [19] *equivalent or isomorphic* [9] *or just equivalent* [5] *operators in Banach spaces).*

Proof. Let us consider $k = 1$, the situation corresponding to Figure 4.1.

The elements of $X_0 = \ker A$ can be represented in the form (4.2) due to Proposition 4.1, in particular the elements of $X_0^0 = X_0 \cap \ker B^1$ can be written in this form. Proposition 4.2 implies that the restriction of the operator $L_{\mathbb{R}}^1$ given by (4.1) on the space X_0^0, acting into its image, i.e.

$$\operatorname{Rst} L_{\mathbb{R}}^1 = \begin{pmatrix} A \\ B_{\mathbb{R}}^1 \end{pmatrix} : X_0^0 \to \begin{cases} \{0\} \\ \times \prod_{j=1}^m H_+^{2m+l-m_j^1-1/2} \end{cases} \tag{5.2}$$

is a homeomorphism according to Proposition 4.2 and to the fact that X_0^0 is a complemented subspace of $H^{2m+l}(\mathbb{R}_+^2)$. With the corresponding restrictions of the boundary operators

$$\begin{aligned} \operatorname{Rst} B_{\mathbb{R}}^1 &: X_0^0 \to H_+^r, & r &= (2m+l-m_j^1-1/2)_{j=1,\dots,m} \\ \operatorname{Rst} B^2 &: X_0^0 \to H^s(\mathbb{R}_+), & s &= (2m+l-m_j^2-1/2)_{j=1,\dots,m} \end{aligned} \tag{5.3}$$

we have the composition of toplinear mappings

$$W_{x^1} = \operatorname{Rst} B^2 \circ \mathcal{P}^{-1} \circ \mathcal{P} \circ \left(\operatorname{Rst} B_{\mathbb{R}}^1\right)^{-1} : H_+^r \to X_0^0 \to Y_1^0 \to X_0^0 \to H^s(\mathbb{R}_+) \tag{5.4}$$

where $\mathcal{P}$ is the Poisson operator defined by (4.2) and we put $B_+ = \operatorname{Rst} B^2 \circ \mathcal{P}^{-1}$, $B_- = \operatorname{Rst} B_{\mathbb{R}}^1 \circ \mathcal{P}^{-1}$ in accordance with (4.6). □

Theorem 5.2 *Let L be the operator* (2.1) *associated with the mixed BVP* (1.3). *Then L is a Fredholm operator if and only if the two operators W_{x^k} of* (5.1) *are Fredholm.*

Proof. This result connects Theorem 3.2 with Proposition 5.1 via Proposition 4.6. □

Corollary 5.3 *The Fourier symbols Φ^k of*

$$W_{x^k} = r_+ F^{-1} \Phi^k \cdot F : H_+^{r^k} \to H^{s^k}(\mathbb{R}_+) \tag{5.5}$$

are regular, i.e. $F^{-1}\Phi^k \cdot F : H^{r^k} \to H^{s^k}$ are bijective, and they are $m \times m$ matrices of functions from $\mathcal{R}\,(\xi, t_1(\xi), \dots, t_\kappa(\xi))$, i.e. they are rational in these variables.

Proof. Let us assume firstly that $x^1 = (0,0)$ and $\partial\Omega$ is locally like Figure 4.1 near the origin. The principle parts of the Fourier symbols $\Psi_{\mathbb{R}}^k$ of $\operatorname{Rst} B_{\mathbb{R}}^1 \circ \mathcal{P}^{-1}$ and $\operatorname{Rst} B^2 \circ \mathcal{P}^{-1}$ are computed from (1.2) and (4.2) as

$$\left(\Psi_{\mathbb{R}}^k\right)_{(\lambda,\nu),j} = \sum_{|\sigma|=m_j^k} b_{j,\sigma}^k(0)\,(i\xi)^{\sigma_1}\, D_{x_2}^{\sigma_2}(x_2^\nu \exp(-t_\lambda x_2))_{|x_2=0} \tag{5.6}$$

showing the dependence of the Fourier transformed principal parts of the data in terms of the ansatz functions $\widehat{u_{\lambda,\nu}}$ by a multiplication operator.

Thus the orders of the mappings result from the increase orders of the symbols, bijectivity from the one-to-one correspondence in appropriate Banach spaces, see Proposition 4.2 and Corollary 4.3, and the algebraic form is obvious from (5.4) and (5.6). The localization does not effect the above properties and the case $k = 2$ holds by analogy. □

Corollary 5.4 *The two operators W_{x^k} are (toplinear) equivalent to the* lifted operators

$$W_{x^k,0} = r_+ F^{-1} \Phi_0^k \cdot F : \left(L_+^2\right)^m \to L^2(\mathbb{R}_+)^m \tag{5.7}$$

where

$$\begin{aligned} \Phi_0^k &= \lambda_-^{s^k} \Phi^k \lambda_+^{-r^k} \\ \lambda_\pm^s(\xi) &= \operatorname{diag}\left((\xi \pm i)^{s_1}, \dots, (\xi \pm i)^{s_m}\right) \end{aligned} \tag{5.8}$$

for $s = (s_1, \dots, s_m) \in \mathbb{R}^m$.

Proof. See e.g. [9, 15, 36]. □

Remark 5.5 Instead of $\lambda_\pm(\xi) = \xi \pm i$ one can apply $\xi \pm k_0$ where $\Im m k_0 > 0$ or different functions of this type in different places of the matrices (5.8) in order to reduce the "algebraic complexity" of Φ_0^k [14]. For instance, if $t_1 = t_2 = \cdots = t_\kappa$ in (4.2) (A^{pr} being a power of the Laplacian, e.g.) and $l \in \mathbb{N}_0$ (see the original BVP in Section 1), then

$$\Phi_0^k(\xi) \in \mathcal{R}\left(\xi, t_1(\xi)\right)^{m \times m}, \qquad t_1(\xi) = \left(\xi^2 + 1\right)^{1/2} \tag{5.9}$$

taking the usual branch due to a vertical cut between $\pm i$ over infinity. But, in any case, we have the following result:

$$\left(\frac{\xi - i}{\xi + i}\right)^{-l} \Phi_0^k(\xi) \in \mathcal{R}\left(\xi, t_1(\xi), \dots, t_\kappa(\xi)\right)^{m \times m}, \qquad k = 1, 2. \tag{5.10}$$

Corollary 5.6 *The Fourier symbols of the two lifted Wiener-Hopf operators $W_{x^k,0}$ due to the two locally quasi-equivalent operators $L_{x^k,0}^{\langle pr \rangle}$ of Theorem 3.2 are regular elements of the algebra of Hölder continuous $m \times m$ matrix functions defined on the two-point compactification of $\mathbb{R}$:*

$$\Phi_0^k \in \mathcal{G}C^\mu(\ddot{\mathbb{R}})^{m \times m}, \qquad \mu \in]0, 1[\tag{5.11}$$

with a possible jump at infinity.

Proof. This is an elementary consequence of the form of Φ_0^k discussed before, cf. [27]. □

Theorem 5.7 *Let the mixed BVP (1.3) be piecewise elliptic. Then it is Fredholm if and only if the two (for short)* assigned Wiener-Hopf operators $W_{x^k,0}$ *from (5.7) have symbols which satisfy*

$$\det\left(\lambda \Phi_0^k(-\infty) + (1 - \lambda)\Phi_0^k(+\infty)\right) \neq 0, \qquad \lambda \in [0, 1], \quad k = 1, 2. \tag{5.12}$$

If this condition is violated, the operator L associated with the mixed boundary value problem, is not normally solvable: $\operatorname{im} L$ *is not closed but the* defect numbers *are finite:*

$$\begin{array}{rcll}\alpha(L) & = & \dim\ker L & < \infty \\ \beta(L) & = & \operatorname{codim}\overline{\operatorname{im} L} & < \infty.\end{array} \tag{5.13}$$

Proof. The question of the BVP (1.3) to be Fredholm was already reduced before to the corresponding question for the operators (5.7) with symbols (5.11). But the Fredholm criterion is known for Wiener-Hopf (matrix) operators in L^2 spaces with piecewise continuous symbols [24]. Considering L in dependence of the smoothness parameter $l \geq 0$, say $L(l)$ in (2.1), we have evidently the monotony property

$$\left.\begin{array}{rcl}\alpha\left(L(l_1)\right) & \leq & \alpha\left(L(l_2)\right) \\ \beta\left(L(l_1)\right) & \geq & \beta\left(L(l_2)\right)\end{array}\right\} \quad \text{for} \quad l_1 > l_2 \tag{5.14}$$

and the set of non-Fredholm points of $L(l)$ on the l-semi-axis is locally finite due to (5.12), i.e.

$$\#S_{NF}\left([a,b]\right) = \#\left\{l \in [a,b] : L(l) \text{ is not Fredholm}\right\} < \infty \tag{5.15}$$

for any $0 \leq a < b < \infty$. This yields (5.13). □

6. Normalization – the basic idea

The normalization problem for a single bounded linear operator $T : X_0 \to Y_0$, which is not normally solvable in given Banach spaces X_0, Y_0 can be formulated as to find another pair of Banach spaces X_1, Y_1 such that T maps $X_0 \cap X_1$ into Y_1, $X_0 \cap X_1$ is dense in X_1 and the operator T restricted on $X_0 \cap X_1$, as an operator into Y_1, has a continuous extension on X_1, in brief

$$\overline{T} = \operatorname{Ext} T_{|X_0 \cap X_1} : X_1 \to Y_1 \tag{6.1}$$

and such that $\overline{T}$ is normally solvable. Then we say that the pair $(X_1, Y_1) \in \mathcal{N}(T)$ *solves the normalization problem for* T. See [13, 17, 18, 28] for details, modifications and generalizations of the concept.

For mixed BVPs (1.3) we get the following result immediately from Theorem 5.7.

Corollary 6.1 *Let $L(l)$ denote the operator associated to the mixed BVP (1.3) and $L(l_0)$ be not normally solvable for some $l_0 \geq 0$. Then there exists an $\epsilon > 0$ such that $L(l) : X(l) \to Y(l)$, abbreviating the spaces in (2.1), is Fredholm for $l \in I_\epsilon =]l_0 - \epsilon, l_0 + \epsilon[\cap [0, \infty[\setminus \{l_0\}$, i.e. $(X(l), Y(l)) \in \mathcal{N}(L(l_0))$. Further ϵ can be chosen such that both defect numbers $\alpha(L(l))$ and $\beta(L(l))$ have only one jump within I_ϵ, namely at l_0.*

As we saw in the second item of Examples 1.2, and for various other reasons [23, 25, 26], it may be interesting, not to change the topologies of both spaces, X_0 and Y_0, simultaneously. This leads to the question whether a normalization problem is solvable under the additional assumptions

$$X_1 = X_0, \quad Y_1 \subset Y_0 \qquad \text{or} \qquad X_1 \supset X_0, \quad Y_1 = Y_0 \tag{6.2}$$

respectively. We call these *minimal normalization problems* [25, 27] and denote the corresponding normalized operators by

$$\overset{<}{T} = \operatorname{Rst} T : X_0 \to Y_1 \qquad \text{and} \qquad \overset{>}{T} = \operatorname{Ext} T : X_1 \to Y_0 \tag{6.3}$$

provided $(X_0, Y_1) \in \mathcal{N}(T)$ or $(X_1, Y_0) \in \mathcal{N}(T)$ holds, respectively. The first one is referred to as *image normalization*, the second as *domain normalization.*

Proposition 6.2 *If a Banach space operator $T \in \mathcal{L}(X_0, Y_0)$ is not normally solvable, but* $\ker T$ *and* $\overline{\operatorname{im} T}$ *are complemented, then the two minimal normalization problems are uniquely solvable up to isomorphy of the spaces in question.*

Proof. For the image normalization decompose

$$\begin{aligned} X_0 &= \ker T \oplus \widetilde{X_0} \\ Y_0 &= \overline{\operatorname{im} T} \oplus \widetilde{Y_0}. \end{aligned} \tag{6.4}$$

Since $\operatorname{Rst} T : \widetilde{X_0} \to \operatorname{im} T$ is bijective (but not bounded invertible with respect to the given norms of X_0 and Y_0), we choose

$$Y_1 = \operatorname{im} T \oplus \widetilde{Y_0} \tag{6.5}$$

with the norm, for $y = z + \widetilde{y}$ due to this decomposition,

$$\|y\|_{Y_1} = \left\| (\operatorname{Rst} T)^{-1} z \right\|_{X_0} + \|\widetilde{y}\|_{Y_0}. \tag{6.6}$$

The desired properties of Y_1 are evident. Uniqueness follows from considering two analogous decompositions of normalized space pairs and the isomorphy of finite dimensional spaces with the same dimension and their complements, respectively. The case of domain normalization is left as an exercise. □

Remark 6.3 The assumptions of Proposition 6.2 are satisfied for Hilbert space operators (every closed subspace is complemented) or in the case where the defect numbers (5.13) of T are finite.

In order to solve the two concrete minimal normalization problems for mixed BVPs, we now start with normalizing the assigned Wiener-Hopf operators (5.7) in the scalar case $m = 1$ corresponding with a second order PDE in (1.3). So we study (dropping the dependence on x^k) operators of the form

$$W_0 = r_+ F^{-1} \Phi_0 \cdot F : L^2_+ \to L^2(\mathbb{R}_+) \tag{6.7}$$

where

$$\Phi_0 \in \mathcal{G}C^\mu(\ddot{\mathbb{R}}), \qquad \mu \in]0,1[. \tag{6.8}$$

It is convenient to study simultaneously the restrictions for $s > 0$ and continuous extensions for $s < 0$ of W_0, briefly denoted by

$$W_s = r_+ F^{-1} \Phi_0 \cdot F : H^s_+ \to H^s(\mathbb{R}_+). \tag{6.9}$$

Lemma 6.4 ([8, 27]) *Putting "the complex winding number"*

$$w = w(\Phi_0) = \frac{1}{2\pi i} \int_{\mathbb{R}} d \log \Phi_0 = \sigma + i\tau \tag{6.10}$$
$$\sigma = \frac{1}{2\pi} \int_{\mathbb{R}} d \arg \Phi_0, \qquad \tau = \frac{1}{2\pi} \log \left| \frac{\Phi_0(-\infty)}{\Phi_0(+\infty)} \right|$$

we can write any $\Phi_0 \in \mathcal{G}C^\mu(\ddot{\mathbb{R}})$ *in the form*

$$\Phi_0 = \left(\frac{\lambda_-}{\lambda_+} \right)^w \Psi \tag{6.11}$$

where

$$\begin{aligned} &\Psi \in \mathcal{G}C^\mu(\dot{\mathbb{R}}) \\ &\operatorname{ind} \Psi = \frac{1}{2\pi} \int_{\mathbb{R}} d \arg \Psi = 0 \\ &\Psi(+\infty) = \Psi(-\infty) = \Phi(+\infty). \end{aligned} \tag{6.12}$$

Let us recall that the spaces of Bessel potentials [36] were introduced by [1] and [3] as convolutions of L^p-integrable functions with modified Bessel functions of the third kind for $s > 0$. With the help of the Fourier transformation the spaces H^s can be written, more rigorously, as images of the so-called [12, 31, 34] *Bessel potential operators*

$$\Lambda^{-s} = F^{-1} \lambda^{-s} \cdot F : L^2 \to H^s, \qquad s \in \mathbb{R} \tag{6.13}$$

where $\lambda(\xi) = \left(\xi^2 + 1\right)^{1/2}$ and, moreover,

$$\Lambda^w_\pm = F^{-1} \lambda^w_\pm \cdot F : H^s \to H^{s - \Re e w} \tag{6.14}$$

where $\lambda_\pm(\xi) = \xi \pm i$, $s \in \mathbb{R}$, $w \in \mathbb{C}$ and

$$\begin{array}{lcccl} \operatorname{Rst}\Lambda_+^w & : & H^s & \to & H^{s-\Re e w} \\ r_+\Lambda_-^w \ell^{(s)} & : & H^s(\mathbb{R}_+) & \to & H^{s-\Re e w}(\mathbb{R}_+) \end{array} \tag{6.15}$$

are isomorphisms, $\ell^{(s)}\varphi$ denoting any extension of $\varphi \in H^s(\mathbb{R}_+)$ to $\ell^{(s)}\varphi \in H^s$, see [15]. So it is natural to consider the following modified spaces as images of combined Bessel potential operators which are not normally solvable in the above-mentioned spaces $H^s_\pm$ and $H^s(\mathbb{R}_\pm)$, cf. [27, Definition 2.1].

Definition 6.5 For every $w \in \mathbb{C}$, let

$$\begin{aligned} \overset{<}{H}{}^w(\mathbb{R}_+) &= r_+\Lambda_-^{-w-1/2} H_+^{-1/2} = r_+\Lambda_-^{-w-1/2}\Lambda_+^{1/2} L_+^2 \\ &\subset H^{\Re e w}(\mathbb{R}_+) \end{aligned} \tag{6.16}$$

equipped with the norm induced by $H_+^{-1/2}$, namely with

$$\|\psi\|_{\overset{<}{H}{}^w(\mathbb{R}_+)} = \|\ell^0\varphi\|_{H^{-1/2}} \tag{6.17}$$

where $\varphi = r_+\Lambda_-^{w+1/2}\ell^{(\Re e w)}\psi$ and $\ell^{(\Re e w)}\psi$ denotes any extension from $\mathbb{R}_+$ to $\mathbb{R}$ within the space $H^{\Re e w}$ (note that the extension of φ by zero within the space $H^{-1/2}$ is possible due to the given form of φ). Further, we define for $w \in \mathbb{C}$,

$$\overset{>}{H}{}_+^w = \operatorname{clos}\left\{\psi \in H_+^{\Re e w} : \|\psi\|_{\overset{>}{H}{}_+^w} = \|r_+\Lambda_+^{w-1/2}\psi\|_{H^{1/2}(\mathbb{R}_+)}\right\}. \tag{6.18}$$

It is known [27, Corollary 2.2, Corollary 2.4] **(i)** that the embeddings

$$\overset{<}{H}{}^w(\mathbb{R}_+) \subset H^{\Re e w}(\mathbb{R}_+), \qquad \overset{>}{H}{}_+^w \supset H_+^{\Re e w} \tag{6.19}$$

are proper, dense and continuous for any $w \in \mathbb{C}$, **(ii)** that

$$\overset{<}{H}{}^{-k-1/2}(\mathbb{R}_+) = r_+ H_+^{-k-1/2}, \qquad r_+\overset{>}{H}{}_+^{k+1/2} = H_0^{k+1/2}(\mathbb{R}_+) \tag{6.20}$$

for $k \in \mathbb{N}_0$ (where the last space is the closure of $\mathcal{D}(\mathbb{R}_+) = C_0^\infty(\mathbb{R}_+)$ in $H^{k+1/2}(\mathbb{R}_+)$), and **(iii)** that the following operators are homeomorphisms for any $w_1, w_2 \in \mathbb{C}$:

$$\begin{array}{lcccl} \operatorname{Rst} r_+\Lambda_-^{w_1-w_2}\ell^{(\Re e w_1)} & : & \overset{<}{H}{}^{w_1}(\mathbb{R}_+) & \to & \overset{<}{H}{}^{w_2}(\mathbb{R}_+) \\ \operatorname{Ext}\Lambda_+^{w_1-w_2}{}_{|H_+^{\Re e w_1}} & : & \overset{>}{H}{}_+^{w_1} & \to & \overset{>}{H}{}_+^{w_2}. \end{array} \tag{6.21}$$

Finally, we need the basic result about normalization from [27, Theorem 2.5]:

Theorem 6.6 *Let* $\Phi_0 \in \mathcal{G}C^\mu(\ddot{\mathbb{R}})$, $\mu \in]0,1[$ *and* $w = \sigma + i\tau$ *be defined by* (6.10). *Then* W_s *given by* (6.9) *is not normally solvable if and only if*

$$\kappa = s + \sigma + \frac{1}{2} \in \mathbb{Z}. \tag{6.22}$$

In this case

$$\left(H^s_+, \overset{<}{H}{}^{s-i\tau}(\mathbb{R}_+)\right), \left(\overset{>}{H}{}^{s-i\tau}(\mathbb{R}_+), H^s(\mathbb{R}_+)\right) \in \mathcal{N}(W_s). \tag{6.23}$$

Moreover, each of the normalized operators is one-sided invertible with index

$$\mathrm{Ind}\ \overset{<}{W}_s = -\kappa, \qquad \mathrm{Ind}\ \overset{>}{W}_s = -\kappa + 1 \tag{6.24}$$

respectively, and one-sided inverses are explicitly obtained by factorization of the Fourier symbols of $W_{s\pm\epsilon}$, $\epsilon \in]0,1[$, and extension or restriction of the corresponding one-sided inverses of $W_{s\pm\epsilon}$, respectively (see [27] for details).

Now, the basic idea is, to normalize the original mixed BVP "consequently" provided it is piecewise elliptic and not Fredholm.

Remark 6.7 In the system's case, there are corresponding solutions of the minimal normalization problem based upon a representation of the (multiplicative) jump of $\Phi_0 \in \mathcal{G}C^\mu(\ddot{\mathbb{R}})^{m\times m}$ at infinity in normal Jordan form

$$\Phi_0^{-1}(+\infty)\Phi_0(-\infty) = M^{-1} J_{\Phi_0} M \tag{6.25}$$

where $M \in \mathcal{G}\mathbb{C}^{m\times m}$ and the quasidiagonal matrix

$$J_{\Phi_0} = \mathrm{diag}\ (J_1, \ldots, J_{\widetilde{m}}) \tag{6.26}$$

has Jordan blocks of size $l_j \times l_j$ in the diagonal given by

$$J_j = \begin{pmatrix} \mu_j & 1 & 0 & \cdots & 0 \\ 0 & \mu_j & 1 & \cdots & 0 \\ \vdots & \vdots & \vdots & & \vdots \\ 0 & 0 & 0 & \cdots & 1 \\ 0 & 0 & 0 & \cdots & \mu_j \end{pmatrix} \tag{6.27}$$

which yields

$$\Phi_0(\xi) = \Phi_0(-\infty) M^{-1} \left(\mathrm{diag} \left(\left(\frac{\lambda_-(\xi)}{\lambda_+(\xi)} \right)^{\widetilde{\omega_j}} \right) J_{\Phi_0} + \Psi_0(\xi) \right) M \tag{6.28}$$

where the elements of $\Psi_0(\xi)$ are decreasing of order $|\xi|^{-\mu}$ at infinity, see [27, Section 6].

For the sake of shortness we shall not work out the system's case here in detail but like to point out that the corresponding results for the scalar (or diagonal) case can be obtained by analogy including explicit factorization, which is a different and subtle problem in general, see [7] and [20], and already in the case $m = 2$, see [14].

Remark 6.8 As shown in [11] there is a direct connection between an image normalization of an elliptic Wiener-Hopf equation in L_p space with a piecewise-continuous symbol (having a jump at infinity) which is not Fredholm (the Widom-Gohberg-Krupnik arc crosses the origin) and the corresponding image normalization of a related Wiener-Hopf equation with continuous symbol which has a zero of order 1 at a finite point. The normalized space is defined in both cases by integral operators of Cesaro type (for details see [11, §3.1–3.3]). These results can be modified for the present setting of Bessel potential spaces in order to obtain useful characterizations of the normalized spaces.

7. Minimal normalization in the scalar case

Here we consider the case $m = 1$, which is easily extended to the case $m > 1$ when we have diagonal (or triangular, at least) Wiener-Hopf operators such that we can normalize component-wise.

Let us focus on the image normalization of a mixed BVP (1.3) that is piecewise elliptic but not Fredholm, and work it out, step by step, starting with the assigned Wiener-Hopf operators $W_{x^k,0}$. At least one of them violates the Fredholm condition (5.12), say, only if $k = 1$, for simplicity.

As seen before, its symbol

$$\Phi_{x^1,0} \in \mathcal{GR}\left(\xi, \sqrt{\frac{\xi - i}{\xi + i}}\right), \tag{7.1}$$

because $\Phi_{x^1} \in \mathcal{GR}\left(\xi, \sqrt{\xi^2 + 1}\right)$ and W_{x^1} acts between spaces of order $r^1 = 2m + l - m^1 - 1/2$ and $s^1 = 2m + l - m^2 - 1/2$, i.e. of order $l \pm 1/2$ (since $m = 1$, $m^k \in \{0, 1\}$). Thus, $\Phi_{x^1,0} \in \mathcal{G}C^{\mu}(\ddot{\mathbb{R}})$ and we can apply Theorem 6.6 with $s = 0$ and $\sigma = 1/2 \bmod \mathbb{Z}$. Therefore $W_{x^1,0} : L^2_+ \to L^2(\mathbb{R}_+)$ is normalized by

$$\overset{<}{W}_{x^1,0} \quad : \quad L^2_+ \to \overset{<}{H}^{-i\tau}(\mathbb{R}_+) \tag{7.2}$$

where τ results from the modulus jump of $\Phi_{x^1,0}$ at infinity (the integer part of the winding number does not matter and the fractional real part is guilty for non-Fredholmness), see (6.10) and (5.12).

According to Corollary 5.4 and property (6.21) of the Bessel potential operators, the unlifted operator $W_{x^1} : H^{r^1}_+ \to H^{s^1}(\mathbb{R}_+)$ is normalized by

$$\overset{<}{W}_{x^1} = r_+\Lambda_-^{-s^1} \ell\, \overset{<}{W}_{x^1,0}\, \Lambda_+^{r^1} : H^{r^1}_+ \to r_+\Lambda_-^{-s^1+i\tau-1/2}\Lambda_+^{1/2} L^2_+ = \overset{<}{H}^{w}\,(\mathbb{R}_+) \tag{7.3}$$

where $w = s^1 - i\tau$.

Now we modify the boundary operator B^2 into

$$\overset{<}{B^2}\colon H^{2m+l}(\Omega) \to \overset{<}{H}{}^{s^1}(\Gamma^2, x^1) \tag{7.4}$$

imposing the local behavior of $\overset{<}{H}{}^{w}(\mathbb{R}_+)$ near zero by transformation in the image of B^2 near x^1; more precisely

$$\overset{<}{H}{}^{s^1}(\Gamma^2, x^1) = \left\{\varphi \in H^{s^1}\left(\Gamma^2\right) : \ell^0\left((\omega\varphi) \circ \varkappa_{x^1}^{-1}\right) \in \overset{<}{H}{}^{w}(\mathbb{R}_+)\right\} \tag{7.5}$$

where $\omega \in C^\infty\left(\Gamma^2\right)$, $0 \le \omega \le 1$, $\omega = 1$ in $\mathcal{B}_\epsilon(x^1) \cap \Gamma^2$, $\omega = 0$ in $\mathcal{B}_\epsilon(x^2) \cap \Gamma^2$, $\varkappa_{x^1}$ is the transformation from (2.3) and ℓ^0 denotes zero extension to $\mathbb{R}_+$ which represents a continuous operator according to the smooth cut-off by ω. The induced norm is given by

$$\|\varphi\|_{\overset{<}{H}{}^{s^1}(\Gamma^2,x^1)} = \left\|\ell^0\left((\omega\varphi) \circ \varkappa_{x^1}^{-1}\right)\right\|_{\overset{<}{H}{}^{w}(\mathbb{R}_+)} + \left\|\ell^0\left(((1-\omega)\varphi) \circ \varkappa_{x^1}^{-1}\right)\right\|_{H^{s^1}(\mathbb{R}_+)}. \tag{7.6}$$

Finally, the image of L has to be adapted, in order to end up with a normalized operator due to (2.1). Starting again with localization of

$$\overset{<}{L} = \begin{pmatrix} A \\ B^1 \\ \overset{<}{B^2} \end{pmatrix} : H^{2m+l}(\Omega) \to \begin{cases} H^l(\Omega) \\ \times\ H^{r^1}(\Gamma^1) \\ \times\ \overset{<}{H}{}^{s^1}(\Gamma^2, x^1) \end{cases} \tag{7.7}$$

we obtain similar results due to local behavior, continuous embedding and compactness criteria, ending up with the fact that $\overset{<}{L}$ is Fredholm in the described situation. The modification for the case where $W_{x^2,0}$ violates (5.12) is obvious. Domain normalization runs analogously. Various details follow directly from the fact that the normalized spaces are images of Bessel potential operators, e.g. the local type property [29, Prop. 1.3] of the normalized operators and compactness of lower order terms etc. So we proved the following result (up to analogous conclusions).

Theorem 7.1 *Consider the mixed BVP* (1.3) *that is piecewise elliptic, see* (1.4), (1.5), *suppose* $m = 1$ *(i.e. the PDE is of second order) and* $l \ge 0$. *Let* $W_{x^k,0}(l)$ *denote the two assigned (scalar) Wiener-Hopf operators from* (5.7) *and* $w^k(l) = \sigma^k(l) + i\tau^k(l)$ *the numbers defined in* (6.10) *due to the Fourier symbols* $\Phi_0^k(l)$ *of* $W_{x^k,0}(l)$. *Then the BVP* (1.3), *i.e. the associated operator* L *in* (2.1), *is Fredholm for all* $l \ge 0$ *up to a set*

$$S_{NF}(\mathbb{R}_+) = \left(l^1 + \mathbb{N}_1\right) \cup \left(l^2 + \mathbb{N}_1\right) \tag{7.8}$$

where $l^k \in [0,1[$. *More precisely*

$$\begin{aligned} w^k(l) &= w^k(0) + l \\ l^k &= -\left(\sigma^k(0) + \tfrac{1}{2}\right) \qquad \operatorname{mod} \mathbb{Z} \end{aligned} \tag{7.9}$$

for $k = 1, 2$. *For the critical orders* $l \in S_{NF}(\mathbb{R}_+)$ *(where the BVP is not Fredholm), the image of* L *is not closed, but the defect numbers are finite, and* L *can be normalized as follows: If* $l^1 \neq l^2$ *and* $l \in \left(l^1 + \mathbb{N}_1\right)$, *the image normalized operator* (7.7) *of* L *is Fredholm, and, for* $l \in l^2 + \mathbb{N}_1$, *an obvious normalization of* B^1 *helps. If* $l^1 = l^2$ *and* $l \in S_{NF}(\mathbb{R}_+)$ *the image normalization is given by simultaneous image normalization of* B^1 *and* B^2. *Completely analogous results hold for the domain normalization (based upon Theorem* 6.6*).*

Example 7.2 The full scale of normalized spaces $\overset{<}{H}{}^w(\mathbb{R}_+)$ (scalar case) appears already for the following class of mixed canonical problems given in Figure 7.1, where a, b and c are complex constants.

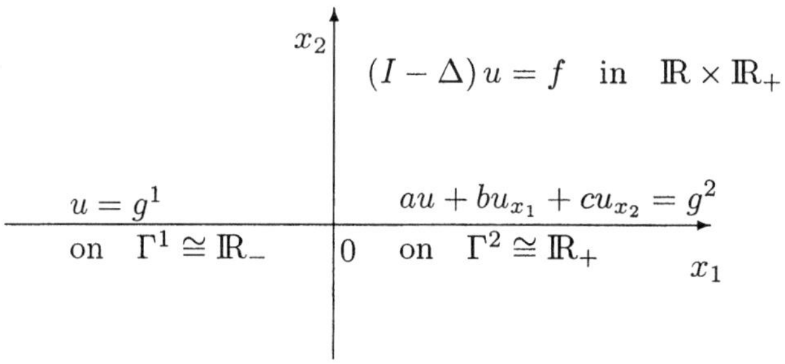

Figure 7.1: Mixed BVP with an oblique derivative boundary condition.

Consider only the semi-homogeneous problem due to $f = 0$, $g^1 = 0$ (cf. Section 4). The representation formula for $u \in \ker(I - \Delta)$ in the upper half-plane

$$u(x) = F^{-1}_{\xi \mapsto x_1} \widehat{u_0}(\xi) \exp[-x_2 t(\xi)] \tag{7.10}$$

gives us the Wiener-Hopf operator (see Section 5)

$$W = r_+ F^{-1} (a - ib\xi - ct) \cdot F : H_+^{2+l-1/2} \to H^{2+l-3/2}(\mathbb{R}_+) \tag{7.11}$$

and the lifted operator $W_0 : L^2_+ \to L^2(\mathbb{R}_+)$ with symbol

$$\Phi_0 = \left(\frac{\xi - i}{\xi + i}\right)^{l+1/2} \frac{a - ib\xi - ct}{\xi + i} \tag{7.12}$$

and the condition for (piecewise) ellipticity:

$$a - ib\xi - ct(\xi) \neq 0, \qquad \xi \in \mathbb{R}. \tag{7.13}$$

We compute the complex winding number

$$w = \frac{1}{2\pi i} \int_{\mathbb{R}} d \log \Phi_0 = \sigma + i\tau \tag{7.14}$$

and find a non-Fredholm operator for $\sigma = \frac{1}{2} \mod \mathbb{Z}$, i.e., thinking of $\Phi_0(+\infty) = -c - ib$, $\Phi_0(-\infty) = (c - ib) \exp[-2\pi i(l + \frac{1}{2})]$, for

$$l + \arg \frac{-c - ib}{c - ib} \in \mathbb{Z}. \tag{7.15}$$

This condition has plenty of realizations, e.g.:
Case 1. Let $l = 0$ and $b, c \in \mathbb{R}$, then $c = 0$ suffices, i.e. the principle part of the oblique derivative is tangential (and not really oblique);
Case 2. Let $l = 0$, $b, c \in \mathbb{C}$ and $(-c - ib)/(c - ib) = p > 0$; this is satisfied for $c = ib(p + 1)/(p - 1)$, i.e. an imaginary ratio of the coefficients b and c; and so on for other $l \notin \mathbb{N}_0$.

8. Concluding remarks

The case $m > 1$ can be treated with the idea of Remark 6.7 and yields, in general, rather complicated formulas. Then, the compatibility conditions for the data that result from image normalization, cf. (1.8), are not necessarily "local". For instance, in the canonical problem treated in Section 4, thinking of the Bi-Laplacian for A and higher order boundary operators, one can meet conditions which combine data after application of convolution type operators on the boundary.

Various generalizations are possible: for three - or n-dimensional configurations, systems of PDE (as resulting from Maxwell's and Lamé's equations), pseudodifferential equations, weak formulations ($l < 0$), less smooth boundaries and other spaces of Besov-Triebel-Lizorkin type [30, 36].

Beside of the interpretation of image normalization in terms of compatibility conditions, one can understand the domain normalization in some cases as imposing a transmission property [10]. A simple case is an operator in (5.1) of the form

$$W = r_+ B : H_+^{l+1/2} \to H^{l+1/2}(\mathbb{R}_+), \qquad l \in \mathbb{N}_0 \tag{8.1}$$

where $B = I + c \cdot \Lambda_+^{-\epsilon}$ with $\epsilon > 0$ which maps $H_+^{l+1/2}$ onto $r_+ H_+^{l+1/2} \subset H^{l+1/2}(\mathbb{R}_+)$ and is not Fredholm.

In the case $l = 0$ the embedding $r_+ H_+^{1/2} \subset H^{1/2}(\mathbb{R}_+)$ is proper and dense [27]. Thus $\overset{>}{W}$ can be identified with (is equivalent to)

$$\overset{>}{W} \ell^0 = \operatorname{Ext} W \ell^0 : H^{1/2}(\mathbb{R}_+) \to H^{1/2}(\mathbb{R}_+) \tag{8.2}$$

which obviously has the transmission property [10].

If $l = 1, 2, 3, \ldots$, the closure of $r_+ H_+^{l+1/2}$ in $H^{l+1/2}(\mathbb{R}_+)$ has codimension l [6], thus the operator

$$\widetilde{W} = r_+ B \ell^{l+1/2} : H^{l+1/2}(\mathbb{R}_+) \to H^{l+1/2}(\mathbb{R}_+) \tag{8.3}$$

with arbitrary extension $\ell^{l+1/2}$ into $H^{l+1/2}$, which has the transmission property, coincides with $\overset{>}{W}\ell$ (defined on that closure) up to an operator of characteristic l, i.e. is equivalent after extension by finite dimensional operators.

Further interesting studies may be based on the characterization of the normalizing conditions (compatibility or transmission property, respectively) in terms of integral conditions and resulting conclusions for the asymptotic behavior of solutions. For instance, due to (6.18), the elements $\psi \in \overset{>}{H}{}_+^{l+1/2}$, $l \in \mathbb{N}_0$, are characterized by

$$\left\| r_+ \Lambda_+^l \psi \right\|_{H^{1/2}(\mathbb{R}_+)} < \infty \tag{8.4}$$

which is equivalent to $\psi \in \ell^0 H^{1/2}(\mathbb{R}_+)$ with

$$\begin{aligned} &D^j \psi(0) = 0, \qquad\qquad j = 0, \ldots, l-1 \\ &\int_0^\infty \tfrac{1}{x} \left| D^l \psi(x) \right|^2 dx < \infty, \end{aligned} \tag{8.5}$$

see [35, p. 11].

Some of these questions will be treated in a forthcoming paper.

Acknowledgements

This article was started during the second author's visit to Instituto Superior Técnico, U.T.L., and Universidade de Aveiro, Portugal, in May–July 2001. The work was supported by "Fundação para a Ciência e a Tecnologia" through "Centro de Matemática Aplicada" and "UI&D Matemática e Aplicações", respectively.

References

[1] Aronszajn, N. and Smith, K. T., Theory of Bessel potentials I, *Ann. Inst. Fourier*, **11** (1961), 385–475.

[2] Bart, H. and Tsekanovskii, V. E., Matricial coupling and equivalence after extension. In: *Operator Theory and Complex Analysis*, Proc. Workshop in Sapporo, Japan, 1991, Operator Theory: Advances and Applications, **59**, 143–160. Birkhäuser, Basel, 1992.

[3] Calderón, A.-P., Lebesgue spaces of differentiable functions and distributions. In: *Proc. Sympos. Pure Math.*, Vol. IV, 33–49. AMS, Providence, R.I., 1961.

[4] Castro, L. P., *Relations Between Singular Operators and Applications*, Ph.D. Thesis in Mathematics, I.S.T., Technical University of Lisbon, 1998.

[5] Castro, L. P. and Speck, F.-O., Regularity properties and generalized inverses of delta-related operators, *Z. Anal. Anwendungen*, **17** (1998), 577–598.

[6] Castro, L. P. and Speck, F.-O., Relations between convolution type operators on intervals and on the half-line, *Integr. Equat. Oper. Theor.*, **37** (2000), 169–207.

[7] Clancey, K. and Gohberg, I., *Factorization of Matrix Functions and Singular Integral Operators*, Operator Theory: Advances and Applications, **3**. Birkhäuser Verlag, Basel, 1981.

[8] Duduchava, R., *Integral Equations with Fixed Singularities.* Teubner, Leipzig, 1979.

[9] Duduchava, R., On multidimensional singular integral operators. I: The half-space case; II: The case of compact manifolds, *J. Operator Theory*, **11** (1984), 41–76 (I); 199–214 (II).

[10] Duduchava, R., Wiener-Hopf equations with the transmission property, *Integr. Equat. Oper. Theor.*, **15** (1992), 412–426.

[11] Duduchava, R. and Silbermann, B., Boundary value problems in domains with peaks, *Mem. Differential Equations Math. Phys.*, **21** (2000), 1–122.

[12] Duduchava, R. and Speck, F.-O., Pseudodifferential operators on compact manifolds with Lipschitz boundary, *Math. Nachr.*, **160** (1993), 149–191.

[13] Dybin, V. B., Normalization of the Wiener-Hopf operator, *Dokl. Akad. Nauk SSSR*, **191** (1970), 759–762 (in Russian). English translation: *Soviet Math. Dokl.*, **11** (1970), 437–441.

[14] Ehrhardt, T. and Speck, F.-O., Transformation techiques towards the factorization of non-rational 2 × 2 matrix functions, *Linear Algebra Appl.*, **353** (2002), 53–90.

[15] Èskin, G. I., *Boundary Value Problems for Elliptic Pseudodifferential Operators*, Translations of Mathematical Monographs, **52**. AMS, Providence, R.I., 1981.

[16] Gohberg, I. and Krupnik, N., *One-Dimensional Linear Singular Integral Equations*, Vol. I, Operator Theory: Advances and Applications, **53**. Birkhäuser Verlag, Basel, 1992.

[17] Khaikin, M. I., On the regularization of operators with non-closed range, *Izv. Vyssh. Uchebn. Zaved. Mat.*, **8** (1970), 118–123 (in Russian).

[18] Kravchenko, V. G., On normalization of singular integral operators, *Dokl. Akad. Nauk SSSR*, **285** (1985), 1314–1317 (in Russian). English translation: *Sov. Math. Dokl.*, **32** (1985), 880–883.

[19] Lang, S., *Real and Functional Analysis*, Graduate Texts in Mathematics, **142**. Springer-Verlag, New York, 1993.

[20] Litvinchuk, G. S. and Spitkovskii, I. M., *Factorization of Measurable Matrix Functions*, Operator Theory: Advances and Applications, **25**. Birkhäuser Verlag, Basel, 1987.

[21] Lüneburg, E. and Hurd, R. A., On the diffraction problem of a half plane with different face impedances, *Canad. J. Phys.*, **62** (1984), 853–860.

[22] Meister, E. and Speck, F.-O., Diffraction problems with impedance conditions, *Appl. Anal.*, **22** (1986), 193–211.

[23] Meister, E. and Speck, F.-O., Modern Wiener-Hopf methods in diffraction theory. In: *Ordinary and partial differential equations*, Vol. II (Dundee, 1988), Pitman Res. Notes Math. Ser., **216**, 130–171. Longman Scientific & Technical, Harlow, 1989.

[24] Mikhlin, S. G. and Prössdorf, S., *Singular Integral Operators.* Springer-Verlag, Berlin, 1986.

[25] Moura Santos, A., *Minimal Normalization of Wiener-Hopf Operators and Applications to Sommerfeld Diffraction Problems*, Ph.D. Thesis in Mathematics, I.S.T., Technical University of Lisbon, 1999.

[26] Moura Santos, A., Speck, F.-O., and Teixeira, F. S., Compatibility conditions in some diffraction problems. In: *Direct and inverse electromagnetic scattering* (Gebze, 1995), Pitman Res. Notes Math. Ser., **361**, 25–38. Longman, Harlow, 1996.

[27] Moura Santos, A., Speck, F.-O., and Teixeira, F. S., Minimal normalization of Wiener-Hopf operators in spaces of Bessel potentials, *J. Math. Anal. Appl.*, **225** (1998), 501–531.

[28] Prössdorf, S., *Some Classes of Singular Equations*, North-Holland Mathematical Library, **17**. North-Holland Publishing Company, Amsterdam, 1978.

[29] Rabinovich, V. S., Pseudodifferential operators on a class of noncompact manifolds, *Math. USSR, Sbornik*, **18** (1972), 45–59.

[30] Runst, T. and Sickel, W., *Sobolev Spaces of Fractional Order, Nemytskij Operators and Nonlinear Partial Differential Equations*, de Gruyter Series in Nonlinear Analysis and Applications, **3**. De Gruyter, Berlin, 1996.

[31] Sadosky, C. and Cotlar, M., On quasi-homogeneous Bessel potential operators. In: *Singular integrals* (Proc. Sympos. Pure Math., Chicago, Ill., 1966), 275–287. AMS, Providence, R.I., 1967.

[32] Simonenko, I. B., A new general method of investigating linear operator equations of the type of singular integral equations, *Soviet Math. Dokl.*, **5** (1964), 1323–1326.

[33] Speck, F.-O., Mixed boundary value problems of the type of Sommerfeld's half-plane problem, *Proc. R. Soc. Edinburgh, Sect. A*, **104** (1986), 261–277.

[34] Speck, F.-O. and Duduchava, R., Bessel potential operators for the quarter-plane, *Appl. Anal.*, **45** (1992), 49–68.

[35] Teixeira, F. S., *Wiener-Hopf Operators in Sobolev Spaces and Applications to Diffraction Theory*, Ph.D. Thesis in Mathematics, I.S.T., Technical University of Lisbon, 1989 (in Portuguese).

[36] Triebel, H., *Theory of Function Spaces II*, Monographs in Mathematics, **84**. Birkhäuser Verlag, Basel, 1992.

[37] Wendland, W. L., Stephan, E., and Hsiao, G. C., On the integral equation method for the plane mixed boundary value problem of the Laplacian, *Math. Methods Appl. Sci.*, **1** (1979), 265–321.

[38] Wloka, J., *Partial Differential Equations*. Cambridge University Press, Cambridge, 1987.

FACTORIZATION OF SOME CLASSES OF MATRIX FUNCTIONS AND THE RESOLVENT OF A HANKEL OPERATOR

Ana C. Conceição
Área Departamental de Matemática, F.C.T., Universidade do Algarve, Portugal
aconcei@ualg.pt

Viktor G. Kravchenko
Área Departamental de Matemática, F.C.T., Universidade do Algarve, Portugal
vkravch@ualg.pt

Francisco S. Teixeira
Departamento de Matemática, Instituto Superior Técnico, Portugal
fteix@math.ist.utl.pt

Dedicated to Professor Georgii S. Litvinchuk on the occasion of his 70th birthday

Abstract The factorization of some classes of matrix-valued functions is obtained, which yields some new results for a special class of Hankel integral operators in L_2^+. For each of its elements, it is shown that the resolvent operator can be explicitly determined through a matrix factorization obtained by solving two non-homogeneous equations.

Keywords: factorization, Hankel operators, resolvent operator, singular integral operators

1. Introduction

The explicit factorization problem for matrix-valued functions has applications in different areas, such as the theory of singular integral operators, boundary value problems, and the theory of non-linear differential equations (see, for instance, [1, 3, 6, 5]). However, only for some restricted classes of matrix functions (such as rational matrix functions) there exists an algorithm that allows us to determine explicit factorizations. In this work we intend to obtain factorizations of some particular classes of matrix-valued functions, by solving two non-homogeneous equations. And, as an application, we are able to get an explicit representation for the resolvents of certain Hankel integral operators. Some of the new results that we are going to present came out by the use of some of the ideas developed in [2], and [7]).

In [2] we introduced a class of matrix-valued functions, of the form

$$A_\gamma(b) = \begin{pmatrix} e & b \\ b^* & b^*b + \gamma e \end{pmatrix},$$

where $b \in [L_\infty(\mathbb{T})]_{n\times n}$, that is, b is a matrix-valued function whose components are essentially bounded functions on the unit circle, b^* is the Hermitian adjoint of b, e represents the identity matrix function, and γ is an arbitrary complex constant, for which it turned out to be possible to compute a canonical factorization (see Theorem 4.4 in [2]) when the function b is in a certain decomposing algebra $\mathcal{A}(\mathbb{T})$ of continuous functions (see, for instance, [6, 4]) and satisfies some additional conditions. The method used therein was based on the construction of the resolvent of the operator $N_+ = P_+bP_-b^*P_+$, where $P_\pm = (I \pm S)/2$ denote the Cauchy projectors associated with the singular integral operator S,

$$S\varphi(t) = \frac{1}{\pi i}\int_{\mathbb{T}} \frac{\varphi(\tau)}{\tau - t}\,d\tau,\ t \in \mathbb{T},$$

where the integral is understood in the sense of its principal value and I represents the identity operator. For the case $n = 1$ (i. e., for two by two matrices), we are able now to generalize our previous result, simplifying some of the conditions imposed before, and using the solutions of two related non-homogeneous equations, φ_1 and φ_2, such that

$$(N_+ + \gamma I)\varphi_1 = 1 \quad \textit{and} \quad (N_+ + \gamma I)\varphi_2 = b. \tag{1.1}$$

So, the canonical factorization of the matrix-valued function $A_\gamma(b)$ can be constructed if

$$1, b \in \operatorname{im}(N_+ + \gamma I). \tag{1.2}$$

For the case $n = 1$, the condition (1.2) leads to the existence of the inverse operator $(N_+ + \gamma I)^{-1}$, i.e., $-\gamma$ belongs to the resolvent set of the operator N_+.

Let us note that the relation between the spectral characteristics of an operator of the form N_+ and a factorization of a hermitian matrix can be easily established using Theorem 12, chapter 4, in [5]. So, the proof of our Theorem 1 has some overlaps with the proof of Theorem 12 from [5] mentioned above. Nevertheless, we still decided to give a detailed proof of Theorem 1 because a similar reasoning applies in the setting of Theorem 3.

2. Factorization of a class of hermitian matrix functions

Let $\mathbb{T}$ denote the unit circle. We will consider the space $L_2(\mathbb{T})$. As usual,

$$L_2^+(\mathbb{T}) = \mathrm{im}P_+,\ L_2^{-,0}(\mathbb{T}) = \mathrm{im}P_-,\ L_2^-(\mathbb{T}) = L_2^{-,0}(\mathbb{T}) \oplus \mathbb{C}.$$

Let us consider a matrix-valued function $A \in G[L_\infty(\mathbb{T})]_{2\times 2}$. We say that A admits a left (right) factorization in $L_2(\mathbb{T})$ if it can be represented as

$$A = A_+ \Lambda A_- \quad (A_- \Lambda A_+),$$

where $A_+^{\pm 1} \in [L_2^+(\mathbb{T})]_{2\times 2}$, $A_-^{\pm 1} \in [L_2^-(\mathbb{T})]_{2\times 2}$, and $\Lambda = \mathrm{diag}\,\{t^{\mu_1}, t^{\mu_2}\}$, $\mu_1 \geq \mu_2$ being integers. If $\mu_1 = \mu_2 = 0$, then A is said to admit a left (right) canonical factorization.

Let $\mathcal{A}(\mathbb{T})$ be a decomposing algebra of continuous functions, and let

$$\mathcal{A}^\pm(\mathbb{T}) = \mathcal{A}(\mathbb{T}) \cap \mathcal{C}^\pm(\mathbb{T}),$$

where $\mathcal{C}^\pm(\mathbb{T})$ refer to the classes of continuous functions on $\mathbb{T}$ which have an holomorphic extension to $\mathbb{T}^\pm$ and are continuous on $\mathbb{T}^\pm \cup \mathbb{T}$. Here $\mathbb{T}^+$ denotes the open unit disk and $\mathbb{T}^-$ stands for the exterior region of the unit circle.

Now let us introduce the matrix-valued functions

$$A_\gamma(b) = \begin{pmatrix} 1 & b \\ \bar{b} & |b|^2 + \gamma \end{pmatrix}, \tag{2.1}$$

where $b \in L_\infty(\mathbb{T})$ and $\gamma \in \mathbb{C}$.

For those matrix-valued functions, a canonical factorization (when it exists) can be obtained. To this end, we notice that, if we consider the function b represented as

$$b = b_+ + b_-,$$

where $b_+ \in \mathcal{A}^+(\mathbb{T})$ and $b_- \in \mathcal{A}^-(\mathbb{T})$, then $A_\gamma(b)$ can be expressed in the form

$$A_\gamma(b) = \begin{pmatrix} 1 & 0 \\ \overline{b_-} & 1 \end{pmatrix} A_\gamma(b_+) \begin{pmatrix} 1 & b_- \\ 0 & 1 \end{pmatrix}.$$

So, the factorization problem of the matrix-valued function $A_\gamma(b)$,

$$A_\gamma(b) = A_\gamma^+ \Lambda A_\gamma^-, \tag{2.2}$$

is equivalent to the problem of factorizing the matrix-valued function $A_\gamma(b_+)$ (see, for $\gamma = -1$, [5]),

$$A_\gamma(b_+) = C_\gamma^+ \Lambda C_\gamma^-,$$

where, obviously,

$$A_\gamma^+ = \begin{pmatrix} 1 & 0 \\ \overline{b_-} & 1 \end{pmatrix} C_\gamma^+ \quad \text{and} \quad A_\gamma^- = C_\gamma^- \begin{pmatrix} 1 & b_- \\ 0 & 1 \end{pmatrix}.$$

It turns out that a factorization of $A_\gamma(b_+)$ always leads to a factorization of $A_\gamma(b)$, independently of how complicated b_- is.

Let us assume, from now on, that $b \in \mathcal{A}^+(\mathbb{T})$.

With the purpose of obtaining a factorization of the matrix-valued function $A_\gamma(b)$, we start considering the operators

$$N_+ = P_+bP_-\bar{b}P_+ \quad \text{and} \quad N_- = P_-\bar{b}P_+bP_-.$$

We know that the spectra of the operators N_+ and N_- coincide (see [2]), i.e.,

$$\sigma(N_+) = \sigma(N_-).$$

If $-\gamma$ does not belong to the spectrum of the operator N_+, that is, if

$$-\gamma \in \rho(N_+),$$

then, in particular,

$$\gamma \neq 0,$$

since it is easy to see that zero belongs to the spectrum of N_+. In that case, we have that

$$\det A_\gamma(b) = \gamma \neq 0,$$

and, therefore, $A_\gamma(b)$ admits a factorization in $[\mathcal{A}(\mathbb{T})]_{2\times 2}$. Besides, we have that

$$\text{ind}\,\det A_\gamma(b) = 0.$$

Thus, the matrix-valued function $A_\gamma(b)$ admits the factorization of the form (2.2) (see, for instance, [1]) where the diagonal matrix function Λ has the following representation

$$\Lambda(t) = \begin{pmatrix} t^\mu & 0 \\ 0 & t^{-\mu} \end{pmatrix}.$$

It follows that the Riemann-Hilbert problem

$$\overrightarrow{\varphi}_+ = A_\gamma(b)\,\overrightarrow{\varphi}_-, \quad \overrightarrow{\varphi}_\pm \in [L_2^\pm(\mathbb{T})]_2, \tag{2.3}$$

has μ linear independent solutions, and, therefore, if the problem has only the trivial solution, then $\mu = 0$. Besides that, considering directly the equations

$$\begin{pmatrix} \varphi_1^+ \\ \varphi_2^+ \end{pmatrix} = \begin{pmatrix} 1 & b \\ \bar{b} & |b|^2 + \gamma \end{pmatrix} \begin{pmatrix} \varphi_1^- \\ \varphi_2^- \end{pmatrix}$$

it is straightforward to prove that

$$\varphi_2^- \in Ker(N_- + \gamma I).$$

Since $-\gamma \in \rho(N_+) = \rho(N_-)$, then $\varphi_2^- = 0$. Further easy computations show that the trivial solution is the only solution of the Riemann-Hilbert problem,

$$\overrightarrow{\varphi}_- = \overrightarrow{0}.$$

Therefore $\mu = 0$ and the factorization of the matrix-valued function $A_\gamma(b)$ is a canonical factorization:

$$A_\gamma(b) = A_\gamma^+ A_\gamma^-.$$

Moreover, analyzing the Riemann-Hilbert problem (2.3), it can also be proved that

$$\mu = \dim \operatorname{Ker}(N_- + \gamma I).$$

On the other hand, if $A_\gamma(b)$ admits a canonical factorization, $\mu = 0$, and

$$0 = \dim \operatorname{Ker}(N_- + \gamma I),$$

from which its follows that

$$-\gamma \in \rho(N_-) = \rho(N_+).$$

Following [2], the canonical factorization of $A_\gamma(b)$ can be determined explicitly through the solutions φ_1 and φ_2 of the two non-homogeneous equations (1.1),

$$\varphi_1 = (N_+ + \gamma I)^{-1} 1 = \mathcal{R}(1) \quad \text{and} \quad \varphi_2 = (N_+ + \gamma I)^{-1} b = \mathcal{R}(b).$$

That is, the canonical factorization of the matrix-valued function $A_\gamma(b)$ can be constructed if

$$1, b \in \operatorname{im}(N_+ + \gamma I).$$

So, we have obtained the following result:

Theorem 2.1 *If $b \in \mathcal{A}^+(\mathbb{T})$, then the matrix function $A_\gamma(b)$ admits a canonical factorization if and only if $-\gamma \in \rho(N_+)$. And, in that case,*

$$A_\gamma(b) = A_\gamma^+ A_\gamma^-,$$

where

$$A_\gamma^+ = \Psi_+ \quad \textit{and} \quad A_\gamma^- = (E + \Psi_-)^{-1},$$

with

$$\Psi_+ = \gamma \begin{pmatrix} \mathcal{R}(1) & \mathcal{R}(b) \\ P_+ \bar{b} \mathcal{R}(1) & 1 + P_+ \bar{b} \mathcal{R}(b) \end{pmatrix},$$

and

$$\Psi_- = \begin{pmatrix} P_- b P_- \bar{b} \mathcal{R}(1) & P_- b P_- \bar{b} \mathcal{R}(b) \\ -P_- \bar{b} \mathcal{R}(1) & -P_- \bar{b} \mathcal{R}(b) \end{pmatrix}.$$

3. The resolvent of a Hankel operator

As usual, let $L_2(\mathbb{R})$ denote the Hilbert space formed by all measurable functions φ on $\mathbb{R}$ for which $|\varphi|^2$ is Lebesgue integrable. The space $L_2(\mathbb{R})$, henceforth denoted by L_2, can be decomposed into the direct sum of two closed subspaces, $L_2^{\pm}$, formed, respectively, by the functions in L_2 whose supports are contained in $\overline{\mathbb{R}^{\pm}}$, i. e., we have $L_2 = L_2^+ \oplus L_2^-$. Let $P^{\pm} = (I \pm S_{\mathbb{R}})/2$ denote the Cauchy projectors associated with the singular integral operator on $\mathbb{R}$.

Let us consider the Hankel integral operator

$$\mathcal{K} : L_2^+ \longrightarrow L_2^+$$

defined by

$$\mathcal{K}\varphi(t) = \int_o^{+\infty} k(t+\tau)\varphi(\tau)d\tau, \ t > o \tag{3.1}$$

where the kernel-function k is such that its Fourier transform K,

$$K(\omega) = \int_{-\infty}^{+\infty} k(t)e^{i\omega t}dt, \omega \in \mathbb{R}, \tag{3.2}$$

belongs to $L_\infty(\mathbb{R})$.

As is well known, using the Fourier transformation $\mathcal{F}$ on L_2, the Hankel operator (3.1) can be written in the form

$$\mathcal{K} : L_2^+ \longrightarrow L_2^+,$$

$$\mathcal{K} = \mathcal{F}^{-1}P^+KJ\,|_{\mathrm{Im}P^+}\,\mathcal{F},$$

where J is the reflection operator given by

$$J : L_2 \longrightarrow L_2,$$

$$J\varphi(t) = \varphi(-t) \ \ a.e..$$

We will consider the integral operator $\mathcal{K}_\lambda$,

$$\mathcal{K}_\lambda = \lambda I - \mathcal{K}, \ \ \lambda \in \mathbb{C}\setminus\{0\}, \tag{3.3}$$

characterizing, as much as possible, its invertibility.

With $\mathcal{K}_\lambda$ we associate a Wiener-Hopf operator T,

$$T : [L_2^+]_2 \longrightarrow [L_2^+]_2,$$

$$T = \mathcal{F}^{-1}P^+G|_{\mathrm{Im}P^+}\mathcal{F}, \tag{3.4}$$

with a matrix-valued presymbol G,

$$G = -\lambda^{-1} \begin{pmatrix} -JK & 1 \\ \lambda^2 - KJK & K \end{pmatrix}, \tag{3.5}$$

such that ([7], Theorem 3.2): If T is an invertible operator, then the operator $\mathcal{K}_\lambda$ is also an invertible operator.

Since T is a Wiener-Hopf operator in $[L_2^+]_2$ with presymbol G in $[L_\infty(\mathbb{R})]_{2\times 2}$, T is an invertible operator if and only if G admits a canonical generalized factorization with respect to L_2.

It was shown in [7] that a canonical generalized factorization of G,

$$G = G_- G_+, \tag{3.6}$$

yields an expression for the inverse of $\mathcal{K}_\lambda$.

Let us assume that λ has been fixed, and that the matrix function G admits a canonical generalized factorization. Then the operator T is invertible with an inverse given by the expression

$$T^{-1} = \mathcal{F}^{-1} G_+^{-1} P^+ G_-^{-1}|_{\mathrm{Im} P^+} \mathcal{F}.$$

In that case, the operator $\mathcal{K}_\lambda$ is also an invertible operator, that is,

$$\forall f^+ \in L_2^+,$$

the equation

$$\mathcal{K}_\lambda \varphi^+ = f^+$$

has the unique solution φ^+,

$$\varphi^+ = \hat{\pi} T^{-1} \breve{\pi} f^+,$$

where

$$\hat{\pi} : [L_2]_2 \longrightarrow L_2, \quad \hat{\pi} \begin{pmatrix} \varphi_1 \\ \varphi_2 \end{pmatrix} = \varphi_1,$$

and

$$\breve{\pi} : L_2^+ \longrightarrow [L_2^+]_2, \quad \breve{\pi} f^+ = \lambda^{-1} \begin{pmatrix} 0 \\ \mathcal{K}_{-\lambda} f^+ \end{pmatrix}.$$

So, the inverse of the operator $\mathcal{K}_\lambda$ is obtained through the inverse operator of T ([7], Theorem 3.3):

Theorem 3.1 *Let T be the Wiener-Hopf operator associated with $\mathcal{K}_\lambda$, as in (3.4). If the operator T is invertible then $\mathcal{K}_\lambda$ is invertible, with inverse given by*

$$\mathcal{K}_\lambda^{-1} = \hat{\pi} T^{-1} \breve{\pi}.$$

Under adequate assumptions which will be stated in the sequence, let us calculate now a canonical factorization of the matrix-valued function G, to make it possible the explicit determination of the resolvent of the Hankel integral operator $\mathcal{K}$. The matrix-valued function G can be represented in the form

$$G = -\lambda^{-1} B_\lambda^t \begin{pmatrix} 0 & 1 \\ 1 & 0 \end{pmatrix}, \tag{3.7}$$

where

$$B_\lambda = \begin{pmatrix} 1 & K \\ -JK & \lambda^2 - KJK \end{pmatrix}, \tag{3.8}$$

and B_λ^t is its transpose matrix function. If B_λ has a canonical factorization,

$$B_\lambda = B_\lambda^+ B_\lambda^-,$$

then we can easily obtain a canonical factorization of the matrix function G, given by (3.6), where the factors G_- and G_+ are given by

$$G_- = (B_\lambda^-)^t$$

and

$$G_+ = -\lambda^{-1} (B_\lambda^+)^t \begin{pmatrix} 0 & 1 \\ 1 & 0 \end{pmatrix}.$$

Let $\mathcal{A}$ be a decomposing algebra of continuous functions on the real line, with finite and equal limits at $\pm\infty$, such that $\mathcal{A}$ can be written as a direct sum of the subalgebras $\mathcal{A}^+$ and $\mathcal{A}^-$, that is,

$$\mathcal{A} = \mathcal{A}^+ \oplus \mathcal{A}^-.$$

Let us introduce now the operator $N^+ = -P^+KP^-(JK)P^+$, and the function $K \in \mathcal{A}$ given by (3.2), represented as

$$K = K_+ + K_-,$$

where $K_+ = P^+K$, and $K_- = P^-K$.

It is worthwhile to notice that the matrix-valued function B_λ has a similar structure to that of $A_\gamma(b)$ (see (2.1)). As a consequence, using the solutions $R(1)$ and $R(K)$ of the non-homogeneous equations

$$(N^+ + \lambda^2 I)R(1) = 1 \quad \text{and} \quad (N^+ + \lambda^2 I)R(K) = K,$$

that is,

$$R(1) = (N^+ + \lambda^2 I)^{-1}1 \quad \text{and} \quad R(K) = (N^+ + \lambda^2 I)^{-1}K,$$

we are able to determine a factorization of the matrix function B_λ, following a reasoning similar to that used for $A_\gamma(b)$.

Theorem 3.2 *If $K \in \mathcal{A}$, then the matrix function B_λ admits a canonical factorization if and only if $-\lambda^2 \in \rho(N_+)$. And, in that case,*

$$B_\lambda = B_\lambda^+ B_\lambda^-,$$

where

$$B_\lambda^+ = \begin{pmatrix} 1 & 0 \\ -JK_- & 1 \end{pmatrix} \Psi_+ \quad \textit{and} \quad B_\lambda^- = (E + \Psi_-)^{-1} \begin{pmatrix} 1 & K_- \\ 0 & 1 \end{pmatrix},$$

$$\Psi_+ = \lambda^2 \begin{pmatrix} R(1) & R(K) \\ -P^+(JK)R(1) & 1 - P^+(JK)R(K) \end{pmatrix},$$

and

$$\Psi_- = \begin{pmatrix} -P^-KP^-(JK)R(1) & -P^-KP^-(JK)R(K) \\ P^-(JK)R(1) & P^-(JK)R(K) \end{pmatrix}.$$

Hence, if $K \in \mathcal{A}$ and $-\lambda^2 \in \rho(N^+)$, we can obtain a canonical factorization of the matrix-valued function G. In fact, if G has the form (3.5) and $-\lambda^2 \in \rho(N^+)$, then it admits a canonical factorization

$$G = G_- G_+,$$

where the factors G_- and G_+ are given by (see (3.7), (3.8), and the above theorem),

$$G_- = \begin{pmatrix} 1 - P^-KP^-(JK)R(1) & P^-(JK)R(1) \\ -P^-K - P^-KP^-(JK)R(K) & 1 + P^-(JK)R(K) \end{pmatrix}^{-1} \tag{3.9}$$

and

$$G_+ = -\lambda \begin{pmatrix} -P^+(JK)R(1) & R(1) \\ 1 - P^+(JK)R(K) & R(K) \end{pmatrix}. \tag{3.10}$$

Consequently, we have obtained the following result:

Theorem 3.3 *Let $\mathcal{K}$ be the Hankel operator defined by (3.1). If $-\lambda^2 \in \rho(N^+)$, then*

$$\mathcal{K}_\lambda = \lambda I - \mathcal{K}$$

is invertible, with inverse given by

$$\mathcal{K}_\lambda^{-1} = \hat{\pi}\mathcal{F}^{-1}G_+^{-1}P^+G_-^{-1}|_{\mathrm{Im}P^+}\mathcal{F}\check{\pi},$$

where G_+ and G_- are the matrix-valued functions given by (3.10) and (3.9), respectively.

Acknowledgements

This research was partially supported by Centro de Matemática Aplicada at Instituto Superior Técnico (Portugal) and by Fundação para a Ciência e Tecnologia (Portugal), Project PRAXIS/2/2.1/FIS/286/94.

References

[1] Clancey, K. and Gohberg, I., *Factorization of Matrix Functions and Singular Integral Operators*, Operator Theory: Advances and Applications, **3**. Birkhäuser Verlag, Basel, 1981.

[2] Conceição, A. C., Kravchenko, V. G., and Teixeira, F. S., Factorization of matrix functions and the resolvents of certain operators. To appear in: Operator Theory: Advances and Applications, **142**. Birkhäuser Verlag, 2003.

[3] Faddeev, L. D. and Takhtajan, L. A., *Hamiltonian Methods in the Theory of Solitons*. Springer-Verlag, Berlin, 1987.

[4] Gohberg, I., and Krupnik, N., *One-Dimensional Linear Singular Integral Operators*, Vols. I–II, Operator Theory: Advances and Applications, **53, 54**. Birkhäuser Verlag, Basel, 1992.

[5] Litvinchuk, G. S., *Solvability Theory of Boundary Value Problems and Singular Integral Equations with Shift*, Mathematics and its Applications, **523**. Kluwer Academic Publishers, Dordrecht, 2000.

[6] Litvinchuk, G. S. and Spitkovskii, I. M., *Factorization of Measurable Matrix Functions*, Operator Theory: Advances and Applications, **25**. Birkhäuser Verlag, Basel, 1987.

[7] Teixeira, F. S., On a class of Hankel operators: Fredholm properties and invertibility, *Integr. Equat. Oper. Theor.*, **12** (1989), no. 4, 592–613.

COMPACTNESS OF COMMUTATORS ARISING IN THE FREDHOLM THEORY OF SINGULAR INTEGRAL OPERATORS WITH SHIFTS

Alexei Karlovich
Instituto Superior Técnico, Portugal
akarlov@math.ist.utl.pt

Yuri Karlovich
Universidad Autónoma del Estado de Morelos, México
karlovich@buzon.uaem.mx

To Professor G. S. Litvinchuk on the occasion of his 70th birthday

Abstract The paper is devoted to the compactness of the commutators $aS_\Gamma - S_\Gamma aI$ and $W_\alpha S_\Gamma - S_\Gamma W_\alpha$, where S_Γ is the Cauchy singular integral operator, a is a bounded measurable function, W_α is the shift operator given by $W_\alpha f = f \circ \alpha$, and α is a bi-Lipschitz homeomorphism (shift). The cases of the unit circle and the unit interval are considered. We prove that these commutators are compact on rearrangement-invariant spaces with nontrivial Boyd indices if and only if the function a or, respectively, the derivative of the shift α has vanishing mean oscillation.

Keywords: Cauchy singular integral operator, shift operator, commutator, compact operator, rearrangement-invariant space, Boyd indices, interpolation of compactness.

1. Introduction

Let Γ be the unit interval $\mathbb{I} := (0, 1)$, or an arc on the unit circle $\mathbb{T}$, or the unit circle $\mathbb{T}$. For a function $f \in L^1(\Gamma)$, the Cauchy singular integral of f is given by

$$(S_\Gamma f)(t) := \frac{1}{\pi i}\, v.p. \int_\Gamma \frac{f(\tau)}{\tau - t}\, d\tau, \quad t \in \Gamma.$$

Let α be a homeomorphism (*shift*) of $\overline{\Gamma}$ onto itself which preserves the orientation on $\overline{\Gamma}$. Consider the shift operator W_α given by

$$(W_\alpha f)(t) := f[\alpha(t)], \quad t \in \Gamma. \tag{1.1}$$

Under appropriate assumptions, guaranteeing the boundedness of the operators S_Γ, W_α, and the operator aI of multiplication by a measurable complex-valued function a, we study the compactness of the commutators

$$aS_\Gamma - S_\Gamma aI, \quad W_\alpha S_\Gamma - S_\Gamma W_\alpha \tag{1.2}$$

on rearrangement-invariant spaces $X(\Gamma)$. These commutators arise naturally in the Fredholm theory of singular integral operators with shifts [14, 20, 24, 25]. Moreover, Fredholm criteria for singular integral operators with shifts depend essentially on whether the commutators (1.2) are compact or not.

Commutators of multiplication operators and singular integral operators are closely connected with operators of Hankel type. The compactness of both these types of operators on Lebesgue spaces was studied by many authors. We mention here [12, 13, 41] and also a survey [33]. Sufficient conditions for the compactness of commutators of multiplication operators and singular integral operators with kernels of Calderón-Zygmund type on $L^p(X)$, where X is a space of homogeneous type, are obtained in [18, Part II, Theorem 1.1], see also the references therein.

The compactness of the commutator of the Cauchy singular integral operator S_Γ and the shift operator W_α was studied on $L^p(\Gamma)$ by D. Kveselava under assumptions that Γ is a Lyapunov curve and α' is a Hölder continuous function (see [24, Ch. 1, Theorem 3.1]). That theorem was extended to the case of smooth curves and C^1-diffeomorphisms in [11], its detailed proof is in [10]. All mentioned results give only sufficient conditions for the compactness of $W_\alpha S_\Gamma - S_\Gamma W_\alpha$. A compactness criterion for $W_\alpha S_\mathbb{T} - S_\mathbb{T} W_\alpha$ on $L^p(\mathbb{T}), p \in (1,\infty)$, where $\alpha : \mathbb{T} \to \mathbb{T}$ is an orientation-preserving bi-Lipschitz homeomorphism, was obtained in [30].

In this paper we extend the known compactness criteria for the commutators (1.2) to the case of arbitrary rearrangement-invariant spaces $X(\mathbb{T})$ on the unit circle $\mathbb{T}$. Further, we prove compactness criteria for the commutators (1.2) on arbitrary rearrangement-invariant spaces $X(\mathbb{I})$ on the unit interval $\mathbb{I}$. Note that these criteria are new even in the case of Lebesgue spaces $L^p(\mathbb{I}), p \in (1,\infty)$, on the unit interval $\mathbb{I}$.

The paper is organized as follows. In Section 2 we give necessary definitions of rearrangement-invariant spaces and their interpolation characteristics (the Boyd indices). Further we formulate two interpolation theorems for compact operators which allow us to reduce the problem to the simplest case of the Lebesgue space $L^2(\Gamma)$. We conclude this section with boundedness conditions for the operators aI, W_α, and S_Γ on rearrangement-invariant spaces $X(\Gamma)$.

Section 3 is the heart of this paper. Here we study the compactness of the commutators (1.2) on the Lebesgue space $L^2(\Gamma)$. First we formulate the known results in the case of the unit circle. Further we "cut" the unit circle and study the compactness of the commutator $aS_{\mathbb{T}_+} - S_{\mathbb{T}_+}aI$ where $\mathbb{T}_+$ is the upper half-circle of the unit circle $\mathbb{T}$. The necessity of compactness for this commutator is a difficult part. Its proof is based on the fact saying that the compactness of $aS_{\mathbb{T}_+} - S_{\mathbb{T}_+}aI$ implies the compactness of $aN_{\mathbb{T}_+} - N_{\mathbb{T}_+}aI$ where $N_{\mathbb{T}_+}$ is a singular integral operator with two fixed singularities at -1 and 1. This allows us to reduce studying the compactness of $aS_{\mathbb{T}_+} - S_{\mathbb{T}_+}aI$ to studying the compactness of $bS_{\mathbb{T}} - S_{\mathbb{T}}bI$. After that we prove a compactness criterion for the commutator $aS_{\mathbb{I}} - S_{\mathbb{I}}aI$. Arguing similarly, we reduce studying the compactness of $W_\alpha S_{\mathbb{I}} - S_{\mathbb{I}}W_\alpha$ on the unit interval to the same problem but on the unit circle. As a result, we obtain a compactness criterion for the commutator $W_\alpha S_{\mathbb{I}} - S_{\mathbb{I}}W_\alpha$ on the Lebesgue space $L^2(\mathbb{I})$.

In Section 4 we state the main results of this paper: compactness criteria for the commutators (1.2) on rearrangement-invariant spaces $X(\Gamma)$ with nontrivial Boyd indices, where Γ is either the unit circle or the unit interval.

In Section 5 we obtain some corollaries concerning the commutators (1.2) with slowly oscillating data and concerning the commutator of the Cauchy singular integral operator and the isometric shift operator on Lebesgue spaces.

2. Preliminaries

2.1 Rearrangement-invariant spaces

For a general discussion of rearrangement-invariant spaces, see [1, 21, 23]. In this subsection we collect necessary facts in the abstract setting of (finite) measure spaces.

Let $(\mathcal{R}, \mu)$ be a nonatomic finite measure space. Denote by $\mathcal{M} = \mathcal{M}(\mathcal{R}, \mu)$ the set of all μ-measurable complex-valued functions on $\mathcal{R}$, and let $\mathcal{M}^+$ be the subset of functions in $\mathcal{M}$ whose values lie in $[0, \infty]$. The characteristic function of a μ-measurable set $E \subset \mathcal{R}$ will be denoted by χ_E. A mapping $\rho : \mathcal{M}^+ \to [0, \infty]$ is called a *function norm* if for all functions $f, g, f_n \in \mathcal{M}^+$ $(n \in \mathbb{N})$, for all constants $a \geq 0$, and for all μ-measurable subsets E of $\mathcal{R}$, the following properties hold:

(a) $\rho(f) = 0 \Leftrightarrow f = 0$ μ-a.e., $\rho(af) = a\rho(f)$, $\rho(f+g) \leq \rho(f) + \rho(g)$,

(b) $0 \leq g \leq f$ μ-a.e. $\Rightarrow$ $\rho(g) \leq \rho(f)$ (the lattice property),

(c) $0 \leq f_n \uparrow f$ μ-a.e. $\Rightarrow$ $\rho(f_n) \uparrow \rho(f)$ (the Fatou property),

(d) $\rho(\chi_E) < \infty$, $\displaystyle\int_E f\,d\mu \leq C_E \rho(f)$

with $C_E \in (0, \infty)$ depending on E and ρ but independent of f. The collection $X(\mathcal{R}, \mu)$ of all functions $f \in \mathcal{M}$ for which $\rho(|f|) < \infty$ is called a Banach function space. For each $f \in X$, the norm of f is defined by

$$\|f\|_X := \rho(|f|).$$

If ρ is a function norm, its associate norm ρ' is defined on $\mathcal{M}^+$ by

$$\rho'(g) := \sup\left\{\int_{\mathcal{R}} fg\,d\mu \ :\ f \in \mathcal{M}^+,\ \rho(f) \le 1\right\}, \quad g \in \mathcal{M}^+.$$

The Banach function space $X'(\mathcal{R},\mu)$ determined by the function norm ρ' is called the associate space of $X(\mathcal{R},\mu)$. The associate space $X'(\mathcal{R},\mu)$ is a subspace of the dual space $X^*(\mathcal{R},\mu)$.

Let $\mathcal{M}_0$ and $\mathcal{M}_0^+$ be the classes of μ-a.e. finite functions in $\mathcal{M}$ and $\mathcal{M}^+$, respectively. Two functions $f, g \in \mathcal{M}_0$ are said to be equimeasurable if

$$\mu\{x \in \mathcal{R} : |f(x)| > \lambda\} = \mu\{x \in \mathcal{R} : |g(x)| > \lambda\} \quad \text{for all} \quad \lambda \ge 0.$$

A function norm $\rho : \mathcal{M}^+ \to [0,\infty]$ is said to be rearrangement-invariant if $\rho(f) = \rho(g)$ for every pair of equimeasurable functions $f, g \in \mathcal{M}_0^+$. In that case, the Banach function space $X(\mathcal{R},\mu)$ generated by ρ is called a *rearrangement-invariant space.* The Lebesgue space $L^p(\mathcal{R},\mu), 1 \le p \le \infty$, is the simplest example of a rearrangement-invariant space. Orlicz and Lorentz spaces are other important classical examples of rearrangement-invariant spaces.

2.2 Interpolation of compact operators on rearrangement-invariant spaces

In this subsection we state two interpolation theorems for compact operators which allow us to reduce studying the compactness on rearrangement-invariant spaces to studying the compactness on the Lebesgue space $L^2(\mathcal{R},\mu)$.

Theorem 2.1 *Let $X(\mathcal{R},\mu)$ be a rearrangement-invariant space and $X'(\mathcal{R},\mu)$ its associate space. Suppose a linear operator A is bounded and compact on $X(\mathcal{R},\mu)$ and is bounded on $X'(\mathcal{R},\mu)$. Then A is bounded and compact on the Lebesgue space $L^2(\mathcal{R},\mu)$.*

This result follows from Berezhnoi's interpolation theorem for compact operators on Calderón-Lozanovskii spaces $\varphi(X_0, X_1)$ [2] and Lozanovskii's result saying that $L^2(\mathcal{R},\mu)$ is isometrically isomorphic to the Calderón-Lozanovskii space $\varphi(X(\mathcal{R},\mu), X'(\mathcal{R},\mu))$ with $\varphi(t) = \sqrt{t}$, $t \in \mathbb{R}_+ := (0,\infty)$ [26, Theorem 5], see also [8]. One can find more information on Calderón-Lozanovskii spaces in [7] and [28, Ch. 15].

We need a notion of the Boyd indices of rearrangement-invariant spaces to state the second interpolation theorem. By the Luxemburg representation theorem [1, Ch. 2, Theorem 4.10], there is a unique rearrangement-invariant function norm $\overline{\rho}$ over $[0,\mu(\mathcal{R})]$ with the Lebesgue measure m such that $\rho(f) = \overline{\rho}(f^*)$ for all $f \in \mathcal{M}_0^+$, where f^* is the non-increasing rearrangement of f (see, e.g., [1, p. 39]). The rearrangement-invariant space over $([0,\mu(\mathcal{R})], m)$ generated by $\overline{\rho}$ is called the Luxemburg representation of $X(\mathcal{R},\mu)$ and is denoted by $\overline{X}$. For each $s \in \mathbb{R}_+$, let E_s denote the dilation operator defined on $\mathcal{M}_0([0,\mu(\mathcal{R})], m)$

by

$$(E_s f)(t) := \begin{cases} f(st), & st \in [0, \mu(\mathcal{R})] \\ 0, & st \notin [0, \mu(\mathcal{R})] \end{cases}, \quad t \in [0, \mu(\mathcal{R})].$$

For every $s \in \mathbb{R}_+$, the operator $E_{1/s}$ is bounded on $\overline{X}$ [1, p. 165], its norm is denoted by $h_X(s)$. The function $h_X : \mathbb{R}_+ \to \mathbb{R}_+$ is submultiplicative and non-decreasing. From [21, Ch. 2, Theorem 1.3] it follows that the limits

$$\underline{\sigma}_X := \lim_{s\to 0} \frac{\log h_X(s)}{\log s}, \quad \overline{\sigma}_X := \lim_{s\to\infty} \frac{\log h_X(s)}{\log s}$$

exist and $\underline{\sigma}_X \leq \overline{\sigma}_X$. The numbers $\underline{\sigma}_X$ and $\overline{\sigma}_X$ are called the *lower and upper Boyd indices* of the rearrangement-invariant space $X(\mathcal{R}, \mu)$, respectively. For an arbitrary rearrangement-invariant space, its Boyd indices belong to $[0, 1]$. The Boyd indices of the rearrangement-invariant spaces $X(\mathcal{R}, \mu)$ and $X'(\mathcal{R}, \mu)$ are connected via the following identity:

$$\underline{\sigma}_X + \overline{\sigma}_{X'} = \underline{\sigma}_{X'} + \overline{\sigma}_X = 1. \tag{2.1}$$

One can find the proofs of these and other properties of the Boyd indices in [1, 23, 27].

For the Lebesgue spaces $L^p(\mathcal{R}, \mu), 1 \leq p \leq \infty$, the Boyd indices coincide and equal $1/p$. We will say that the Boyd indices are nontrivial if $\underline{\sigma}_X, \overline{\sigma}_X \in (0, 1)$. In the case of Orlicz spaces the latter condition is equivalent to the reflexivity of the space (see, e.g., [27]). An example of Young functions which generate reflexive Orlicz spaces whose Boyd indices do not coincide is given, for instance, in [28, p. 93].

Theorem 2.2 *Let $1 < q < p < \infty$. If a linear operator A is bounded on the Lebesgue spaces $L^p(\mathcal{R}, \mu)$ and $L^q(\mathcal{R}, \mu)$, then it is bounded on every rearrangement-invariant space $X(\mathcal{R}, \mu)$ whose Boyd indices satisfy*

$$1/p < \underline{\sigma}_X \leq \overline{\sigma}_X < 1/q. \tag{2.2}$$

Moreover, if in addition A is compact on $L^p(\mathcal{R}, \mu)$ or on $L^q(\mathcal{R}, \mu)$, then A is compact on $X(\mathcal{R}, \mu)$.

The boundedness part was proved in [6] (see also [1, Ch. 3, Theorem 5.16], [23, Theorem 2.b.11]). The compactness part follows from [19] and [39, Corollary 1].

Corollary 2.3 *Suppose a linear operator A is bounded on the Lebesgue space $L^p(\mathcal{R}, \mu)$ for every $p \in (1, \infty)$. Then A is compact on a rearrangement-invariant space $X(\mathcal{R}, \mu)$ with nontrivial Boyd indices $\underline{\sigma}_X, \overline{\sigma}_X$ if and only if it is compact on the Lebesgue space $L^{p_0}(\mathcal{R}, \mu)$ for some fixed $p_0 \in (1, \infty)$.*

Proof. In view of [19], since A is bounded on $L^p(\mathcal{R},\mu)$ for every $p \in (1,\infty)$, the operator A is compact on $L^{p_0}(\mathcal{R},\mu)$ if and only if A is compact on $L^p(\mathcal{R},\mu)$ for every $p \in (1,\infty)$.

Necessity. From the boundedness part of Theorem 2.2 and (2.1) it follows that the operator A is bounded on the rearrangement-invariant spaces $X(\mathcal{R},\mu)$ and $X'(\mathcal{R},\mu)$. Since A is compact on $X(\mathcal{R},\mu)$, it is compact on the Lebesgue space $L^2(\mathcal{R},\mu)$, by Theorem 2.1. Therefore, A is compact on every $L^p(\mathcal{R},\mu)$ and, hence, on $L^{p_0}(\mathcal{R},\mu)$.

Sufficiency. If A is compact on the Lebesgue space $L^{p_0}(\mathcal{R},\mu)$, then it is compact on the Lebesgue spaces $L^p(\mathcal{R},\mu)$ and $L^q(\mathcal{R},\mu)$, where p and q satisfy (2.2) (such numbers p and q exist because the Boyd indices $\underline{\sigma}_X$ and $\overline{\sigma}_X$ are nontrivial). By the compactness part of Theorem 2.2, A is compact on the rearrangement-invariant space $X(\mathcal{R},\mu)$. □

2.3 Boundedness of the multiplication operator, the shift operator, and the Cauchy singular integral operator

Let Γ be the unit interval $\mathbb{I}$, or an arc on the unit circle $\mathbb{T}$, or the unit circle $\mathbb{T}$, equipped with the Lebesgue length measure $|d\tau|$. In the following we will consider all rearrangement-invariant spaces over $(\Gamma, |d\tau|)$. For the brevity we will omit $|d\tau|$, so we will use the notation $X(\Gamma)$ for rearrangement-invariant spaces on Γ.

Let I be the identity operator on $X(\Gamma)$ and let $a : \Gamma \to \mathbb{C}$ be a measurable function. The operator aI of multiplication by a is bounded on a rearrangement-invariant space $X(\Gamma)$ if and only if $a \in L^\infty(\Gamma)$ [29, Theorem 1].

Let α be a homeomorphism of $\overline{\Gamma}$ onto itself and let α_{-1} be its inverse. Assume that α is a bi-Lipschitz homeomorphism, i.e., both α and α_{-1} satisfy the Lipschitz condition. As is well known (see, e.g., [31, Ch. 9]), a function $f : \overline{\Gamma} \to \mathbb{C}$ is Lipschitz if and only if f is absolutely continuous and $f' \in L^\infty(\Gamma)$. Hence, if α is a bi-Lipschitz homeomorphism, then $\alpha' \in \mathcal{G}L^\infty(\Gamma)$, where $\mathcal{G}L^\infty(\Gamma)$ denotes the class of functions $f \in L^\infty(\Gamma)$ for which $\operatorname{ess\,inf}\{|f(t)| : t \in \Gamma\} > 0$. For such α, the shift operator W_α given by (1.1) is bounded and continuously invertible on every Lebesgue space $L^p(\Gamma), 1 \le p \le \infty$. Its inverse equals $W_\alpha^{-1} = W_{\alpha_{-1}}$. Therefore, by the Calderón-Mitjagin interpolation theorem (see, e.g., [23, Theorem 2.a.10]), the operator W_α is bounded and continuously invertible on an arbitrary rearrangement-invariant space $X(\Gamma)$ whenever α and α_{-1} are absolutely continuous and $\alpha' \in \mathcal{G}L^\infty(\Gamma)$.

The following theorem gives a criterion for the boundedness of the Cauchy singular integral operator on rearrangement invariant spaces.

Theorem 2.4 *Let Γ be the unit circle $\mathbb{T}$ or the unit interval $\mathbb{I}$. The Cauchy singular integral operator S_Γ is bounded on a rearrangement-invariant space $X(\Gamma)$ if and only if $X(\Gamma)$ has nontrivial Boyd indices.*

The idea of the proof of Theorem 2.4 goes back to D. Boyd, who considered the case of the real line (for the proof, see, e.g., [1, Ch. 3, Theorem 5.18]). The proofs for the unit interval and for the unit circle are contained in [21, Ch. 2, Section 8.6].

Finally, if $a \in L^\infty(\Gamma)$, α is a bi-Lipschitz homeomorphism of $\overline{\Gamma}$ onto itself, and the Boyd indices of $X(\Gamma)$ are nontrivial, then the commutators (1.2) are bounded on $X(\Gamma)$. In particular, they are bounded on every Lebesgue space $L^p(\Gamma)$ for $p \in (1,\infty)$. Thus, we can apply Corollary 2.3 to study the compactness of the commutators (1.2).

2.4 Vanishing mean oscillation

In this subsection we mainly follow [37].

The length of an interval $I \subset \mathbb{R}$ will be denoted by $|I|$. For f in $L^1(I)$, the average of f over I will be denoted by $I(f)$:

$$I(f) := \frac{1}{|I|}\int_I f(t)dt.$$

Considering all subintervals J of I, one can introduce the quantity

$$M(f,I) := \lim_{s\to 0+} \sup_{J\subset I,\,|J|\le s} \frac{1}{|J|}\int_J |f(t) - J(f)|dt.$$

The space $VMO(I)$ is defined to be the class of functions f in $L^1(I)$ that satisfy $M(f,I) = 0$ (VMO means *"vanishing mean oscillation"*).

For f an integrable function defined in an open interval containing the point λ of $\mathbb{R}$, the integral gap of f is given by

$$\gamma_\lambda(f) := \limsup_{\delta\to 0}\left|\frac{1}{\delta}\int_\lambda^{\lambda+\delta} f(t)dt - \frac{1}{\delta}\int_{\lambda-\delta}^{\lambda} f(t)dt\right|.$$

An elementary estimate shows that if $f \in VMO(I)$, then $\gamma_\lambda(f) = 0$ for each interior point λ of I.

Analogous definitions can obviously be made for functions given on an arc of the unit circle $\mathbb{T}$ or on the whole unit circle $\mathbb{T}$. We suppose that the unit circle is counter-clockwise oriented. For $t_1, t_2 \in \mathbb{T}$ we denote by (t_1, t_2) the arc of $\mathbb{T}$ with the starting point t_1 and the terminating point t_2.

Lemma 2.5 *Let $t_1, t_2 \in \mathbb{T}$ and $t_1 \ne t_2$. Suppose a function $f : \mathbb{T} \to \mathbb{C}$ satisfies the properties $f|_{(t_1,t_2)} \in VMO(t_1,t_2)$, $f|_{(t_2,t_1)} \in VMO(t_2,t_1)$, and $\gamma_{t_j}(f) = 0$ for $j \in \{1,2\}$. Then $f \in VMO(\mathbb{T})$.*

This statement is proved by the repetition of the proof of [37, Lemma 2] with minor modifications.

We remark that

$$QC(\mathbb{T}) := (C(\mathbb{T}) + H^\infty) \cap (C(\mathbb{T}) + \overline{H^\infty}) = L^\infty(\mathbb{T}) \cap VMO(\mathbb{T}) \qquad (2.3)$$

is a nonseparable C^*-subalgebra of $L^\infty(\mathbb{T})$ properly containing the C^*-algebra $C(\mathbb{T})$ of all continuous functions on $\mathbb{T}$ (see [36, Section 4]). We refer to [5], [32], and [38] for these and other properties of the algebra $QC(\mathbb{T})$ of quasicontinuous functions on $\mathbb{T}$.

The following statement is probably known (cf. [40, Ch. VIII, Problem 6.13], where a BMO-analogue was stated). For convenience of the readers we give its proof.

Lemma 2.6 *Suppose I is a finite interval of $\mathbb{R}$, or an arc of $\mathbb{T}$, or $\mathbb{T}$. If $f \in VMO(I)$, then $|f|^\alpha$ belongs to $VMO(I)$ for every $\alpha \in (0,1]$.*

Proof. Assume for definiteness that I is a finite interval of $\mathbb{R}$. Other cases are considered analogously.

By [40, Ch. VIII, Corollary 1.2], for any $F \in L^1(I)$ and every $J \subset I$,

$$\frac{1}{|J|}\int_J |F(t) - J(F)|dt \le 2\inf_{c\in\mathbb{C}} \frac{1}{|J|}\int_J |F(t) - c|dt. \tag{2.4}$$

Applying (2.4), the well-known inequality

$$||a|^\alpha - |b|^\alpha| \le |a-b|^\alpha, \quad \alpha \in (0,1], \quad a,b \in \mathbb{C},$$

and Hölder's inequality, we get

$$\begin{aligned}
&\frac{1}{|J|}\int_J ||f(t)|^\alpha - J(|f|^\alpha)|dt \le 2\inf_{c\in\mathbb{C}} \frac{1}{|J|}\int_J ||f(t)|^\alpha - |c|^\alpha|dt \\
&\le 2\inf_{c\in\mathbb{C}} \frac{1}{|J|}\int_J |f(t) - c|^\alpha dt \le 2\left(\inf_{c\in\mathbb{C}} \frac{1}{|J|}\int_J |f(t) - c|dt\right)^\alpha \\
&\le 2\left(\frac{1}{|J|}\int_J |f(t) - J(f)|dt\right)^\alpha.
\end{aligned}$$

So, $M(|f|^\alpha, I) \le 2(M(f,I))^\alpha$, which implies the required statement. □

3. Compactness of commutators on L^2

3.1 Case of the unit circle

For a Banach space X, let $\mathfrak{B}(X)$ be the Banach algebra of all bounded linear operators on X and $\mathcal{K}(X)$ the ideal of all compact operators on X. For $C, D \in \mathfrak{B}(X)$, put $[C,D] := CD - DC$.

Combining the compactness criteria from [12] and Sarason's representations of quasicontinuous functions (2.3) immediately gives the following criterion (see also [33, Section 2]).

Theorem 3.1 *Suppose $a \in L^\infty(\mathbb{T})$. The operator $aS_\mathbb{T} - S_\mathbb{T}aI$ is compact on $L^2(\mathbb{T})$ if and only if $a \in VMO(\mathbb{T})$.*

The next statement follows from [30, Theorem 4.1 and Proposition 4.5].

Theorem 3.2 *Let α be an orientation-preserving bi-Lipschitz homeomorphism of $\mathbb{T}$ onto itself. The operator $W_\alpha S_\mathbb{T} - S_\mathbb{T} W_\alpha$ is compact on $L^2(\mathbb{T})$ if and only if $\alpha' \in VMO(\mathbb{T})$.*

3.2 Operators with fixed singularities on the upper half-circle

Let $\mathbb{T}_+ := \{\tau \in \mathbb{T} : \arg \tau \in (0, \pi)\}$ be the upper half-circle of the unit circle $\mathbb{T}$. Using standard methods we prove in this subsection that the operator

$$(N_{\mathbb{T}_+} f)(t) := \frac{1}{\pi i} \int_{\mathbb{T}_+} \frac{f(\tau)}{1 - \tau t} \, d\tau, \quad t \in \mathbb{T}_+,$$

having fixed singularities at -1 and 1, belongs to $\operatorname{alg}(I, S_{\mathbb{T}_+})$, the smallest Banach subalgebra of $\mathfrak{B}(L^2(\mathbb{T}_+))$ which contains the operators I and $S_{\mathbb{T}_+}$.

Consider a Möbius transformation of $\overline{\mathbb{T}_+}$ onto $\overline{\mathbb{R}_+}$ which sends the points 1 and -1 to 0 and $+\infty$, respectively. Namely, let

$$\psi : \overline{\mathbb{T}_+} \to \overline{\mathbb{R}_+}, \quad \psi(t) = i \frac{1-t}{1+t}.$$

Obviously, its inverse is

$$\xi : \overline{\mathbb{R}_+} \to \overline{\mathbb{T}_+}, \quad \xi(x) = \frac{i - x}{i + x}.$$

The map B given by

$$(Bf)(x) = \frac{\sqrt{2}}{i + x} f(\xi(x)), \quad x \in \mathbb{R}_+, \tag{3.1}$$

is an isometric isomorphism of $L^2(\mathbb{T}_+)$ onto $L^2(\mathbb{R}_+)$ with the inverse

$$(B^{-1}\varphi)(t) = \frac{i\sqrt{2}}{1 + t} \varphi(\psi(t)), \quad t \in \mathbb{T}_+. \tag{3.2}$$

Lemma 3.3 *The operator $N_{\mathbb{T}_+}$ belongs to the algebra $\operatorname{alg}(I, S_{\mathbb{T}_+})$.*

Proof. For $f \in L^2(\mathbb{R}_+)$, put

$$(N_{\mathbb{R}_+} f)(x) := \frac{1}{\pi i} \int_{\mathbb{R}_+} \frac{f(y)}{x + y} \, dy, \quad x \in \mathbb{R}_+.$$

It is well known (see, e.g., [22, Theorem 2.3] or [35, Proposition 2.4]), that this operator belongs to $\operatorname{alg}(I, S_{\mathbb{R}_+}) \subset \mathfrak{B}(L^2(\mathbb{R}_+))$. Using (3.1) and (3.2), one can straightforwardly check that $B S_{\mathbb{T}_+} B^{-1} = S_{\mathbb{R}_+}$ and $B N_{\mathbb{T}_+} B^{-1} = -N_{\mathbb{R}_+}$. Thus, $N_{\mathbb{T}_+}$ is bounded on $L^2(\mathbb{T}_+)$ and, moreover, it belongs to $\operatorname{alg}(I, S_{\mathbb{T}_+})$. □

This lemma immediately implies the following.

Corollary 3.4 *Suppose $a \in L^\infty(\mathbb{T}_+)$. If $aS_{\mathbb{T}_+} - S_{\mathbb{T}_+} aI$ is compact on $L^2(\mathbb{T}_+)$, then $aN_{\mathbb{T}_+} - N_{\mathbb{T}_+} aI$ is compact on $L^2(\mathbb{T}_+)$ too.*

3.3 Compactness of commutators with multiplication operators. Case of the upper half-circle

Let $\mathbb{T}_- := \{\tau \in \mathbb{T} : \arg\tau \in (\pi, 2\pi)\}$ be the lower half-circle of the unit circle $\mathbb{T}$. For a function $a \in L^\infty(\mathbb{T}_+)$, we consider the function

$$A(\tau) := \begin{cases} a(\tau), & \tau \in \mathbb{T}_+, \\ a(\overline{\tau}), & \tau \in \mathbb{T}_- \end{cases} \tag{3.3}$$

given on the whole circle $\mathbb{T}$. The functions a and A have an analogous behavior in the neighborhoods of ± 1.

Lemma 3.5 *A function a belongs to $L^\infty(\mathbb{T}_+) \cap VMO(\mathbb{T}_+)$ if and only if the function A given by* (3.3) *belongs to $L^\infty(\mathbb{T}) \cap VMO(\mathbb{T})$.*

Proof. Sufficiency is obvious because $A|_{\mathbb{T}_+} = a$.

Necessity. Since $a \in L^\infty(\mathbb{T}_+)$, we have $A|_{\mathbb{T}_\pm} \in L^\infty(\mathbb{T}_\pm)$. Hence, $A \in L^\infty(\mathbb{T})$. On the other hand, $A|_{\mathbb{T}_+} = a \in VMO(\mathbb{T}_+)$ if and only if $A|_{\mathbb{T}_-} \in VMO(\mathbb{T}_-)$.

For $\delta \in (0, \pi)$, we set

$$\begin{aligned} \mathbb{T}_+(1,\delta) &:= \Big\{\tau \in \mathbb{T} : \arg\tau \in (0,\pi),\ |(1,\tau)| < \delta\Big\}, \\ \mathbb{T}_-(1,\delta) &:= \Big\{\tau \in \mathbb{T} : \arg\tau \in (\pi,2\pi),\ |(\tau,1)| < \delta\Big\}, \\ \mathbb{T}_+(-1,\delta) &:= \Big\{\tau \in \mathbb{T} : \arg\tau \in (0,\pi),\ |(\tau,-1)| < \delta\Big\}, \\ \mathbb{T}_-(-1,\delta) &:= \Big\{\tau \in \mathbb{T} : \arg\tau \in (\pi,2\pi),\ |(-1,\tau)| < \delta\Big\}, \end{aligned}$$

where $|J|$ denotes the length of an interval $J \subset \mathbb{T}$. Then for $t \in \{-1, 1\}$ we get

$$\int_{\mathbb{T}_+(t,\delta)} A(\tau)|d\tau| = \int_{\mathbb{T}_+(t,\delta)} a(\tau)|d\tau| = \int_{\mathbb{T}_-(t,\delta)} a(\overline{\tau})|d\tau| = \int_{\mathbb{T}_-(t,\delta)} A(\tau)|d\tau|.$$

Hence, $\gamma_t(A) = 0$ for $t \in \{-1, 1\}$. Thus, $A \in VMO(\mathbb{T})$ due to Lemma 2.5. □

Consider the Hilbert space $L_2^2(\mathbb{T}_+)$ of vector functions $\varphi = \{\varphi_j\}_{j=1}^2$ with the entries $\varphi_j \in L^2(\mathbb{T}_+)$ which is equipped with the norm

$$\|\varphi\|_{L_2^2(\mathbb{T}_+)} = \left(\|\varphi_1\|^2_{L^2(\mathbb{T}_+)} + \|\varphi_2\|^2_{L^2(\mathbb{T}_+)}\right)^{1/2}.$$

It is easy to see that the operator $\sigma : L^2(\mathbb{T}) \to L_2^2(\mathbb{T}_+)$ defined by

$$(\sigma\varphi)(\tau) := \begin{Bmatrix} \varphi(\tau) \\ \overline{\tau}\,\varphi(\overline{\tau}) \end{Bmatrix}, \quad \tau \in \mathbb{T}_+, \tag{3.4}$$

is bounded and continuously invertible, and its inverse is given by

$$\left(\sigma^{-1}\begin{Bmatrix} \varphi_1 \\ \varphi_2 \end{Bmatrix}\right)(\tau) = \begin{cases} \varphi_1(\tau), & \tau \in \mathbb{T}_+, \\ \overline{\tau}\,\varphi_2(\overline{\tau}), & \tau \in \mathbb{T}_-. \end{cases} \tag{3.5}$$

Lemma 3.6 *Suppose $a \in L^\infty(\mathbb{T}_+)$ and A is given by* (3.3). *Then*

$$\sigma(AI)\sigma^{-1} = \begin{pmatrix} aI & 0 \\ 0 & aI \end{pmatrix}, \quad \sigma S_{\mathbb{T}}\sigma^{-1} = \begin{pmatrix} S_{\mathbb{T}_+} & N_{\mathbb{T}_+} \\ -N_{\mathbb{T}_+} & -S_{\mathbb{T}_+} \end{pmatrix}.$$

Proof. The first equality is obvious. Let us prove the second one. Assume that $\varphi_1, \varphi_2 \in L^2(\mathbb{T}_+)$ and $t \in \mathbb{T}$. Then from (3.5) we get

$$\left(S_{\mathbb{T}}\sigma^{-1}\left\{\begin{array}{c}\varphi_1\\ \varphi_2\end{array}\right\}\right)(t) = \frac{1}{\pi i}\int_{\mathbb{T}_+}\frac{\varphi_1(\tau)}{\tau - t}\,d\tau + \frac{1}{\pi i}\int_{\mathbb{T}_-}\frac{\overline{\tau}\,\varphi_2(\overline{\tau})}{\tau - t}\,d\tau.$$

Changing variables in the second integral, we infer in view of $\overline{\tau} = 1/\tau$ that

$$\left(S_{\mathbb{T}}\sigma^{-1}\left\{\begin{array}{c}\varphi_1\\ \varphi_2\end{array}\right\}\right)(t) = (S_{\mathbb{T}_+}\varphi_1)(t) + (N_{\mathbb{T}_+}\varphi_2)(t), \quad t \in \mathbb{T}. \tag{3.6}$$

Again taking into account that $\overline{t} = 1/t$, we obtain for $t \in \mathbb{T}_+$,

$$\overline{t}\,(S_{\mathbb{T}_+}\varphi_1)(\overline{t}) = -(N_{\mathbb{T}_+}\varphi_1)(t), \quad \overline{t}\,(N_{\mathbb{T}_+}\varphi_2)(\overline{t}) = -(S_{\mathbb{T}_+}\varphi_2)(t). \tag{3.7}$$

From (3.4) and (3.6)–(3.7) we get the second equality. □

Theorem 3.7 *Suppose $a \in L^\infty(\mathbb{T}_+)$. The operator $aS_{\mathbb{T}_+} - S_{\mathbb{T}_+}aI$ is compact on $L^2(\mathbb{T}_+)$ if and only if $a \in VMO(\mathbb{T}_+)$.*

Proof. Suppose $a \in L^\infty(\mathbb{T}_+)$ and A is given by (3.3). From Lemma 3.6 we deduce that

$$\sigma[AI, S_{\mathbb{T}}]\sigma^{-1} = \begin{pmatrix} [aI, S_{\mathbb{T}_+}] & [aI, N_{\mathbb{T}_+}] \\ [N_{\mathbb{T}_+}, aI] & [S_{\mathbb{T}_+}, aI] \end{pmatrix}. \tag{3.8}$$

Necessity. If $[aI, S_{\mathbb{T}_+}] \in \mathcal{K}(L^2(\mathbb{T}_+))$, then $[aI, N_{\mathbb{T}_+}] \in \mathcal{K}(L^2(\mathbb{T}_+))$ due to Corollary 3.4. Hence, (3.8) implies that $[AI, S_{\mathbb{T}}] \in \mathcal{K}(L^2(\mathbb{T}))$. By Theorem 3.1, $A \in VMO(\mathbb{T})$. Thus, $a = A|_{\mathbb{T}_+} \in VMO(\mathbb{T}_+)$. Necessity is proved.

Sufficiency. If $a \in VMO(\mathbb{T}_+)$, then the function A defined by (3.3) belongs to $L^\infty(\mathbb{T}) \cap VMO(\mathbb{T})$, due to Lemma 3.5. By Theorem 3.1, $[AI, S_{\mathbb{T}}] \in \mathcal{K}(L^2(\mathbb{T}))$. Taking into account (3.8), we get $[aI, S_{\mathbb{T}_+}] \in \mathcal{K}(L^2(\mathbb{T}_+))$. □

3.4 Compactness of commutators with multiplication operators. Case of the unit interval

In this subsection we "transplant" the commutator $aS_{\mathbb{I}} - S_{\mathbb{I}}aI$ from $\mathbb{I}$ to $\mathbb{T}_+$ using the C^∞-diffeomorphism

$$\theta : \mathbb{I} \to \mathbb{T}_+, \quad \theta(x) := e^{\pi i x}, \quad x \in \mathbb{I}.$$

This transformation preserves the compactness of operators and the class of coefficients. This allows us to apply Theorem 3.7.

The inverse for θ is given by

$$\theta_{-1} : \mathbb{T}_+ \to \mathbb{I}, \quad \theta_{-1}(\tau) = \frac{\log \tau}{\pi i} = \frac{\arg \tau}{\pi}, \quad \tau \in \mathbb{T}_+.$$

Lemma 3.8 *A function f belongs to $L^\infty(\mathbb{I}) \cap VMO(\mathbb{I})$ if and only if the function $f \circ \theta_{-1}$ belongs to $L^\infty(\mathbb{T}_+) \cap VMO(\mathbb{T}_+)$.*

Lemma 3.8 is proved by analogy with [4, Proposition 2.5].

Clearly, the bounded linear operator $G : L^2(\mathbb{T}_+) \to L^2(\mathbb{I})$ given by

$$(G\varphi)(x) := \varphi(\theta(x)), \quad x \in \mathbb{I},$$

is continuously invertible and its inverse is given by

$$(G^{-1}f)(\tau) = f(\theta_{-1}(\tau)), \quad \tau \in \mathbb{T}_+.$$

Lemma 3.9 *The operator $GS_{\mathbb{T}_+}G^{-1} - S_{\mathbb{I}}$ is compact on $L^2(\mathbb{I})$.*

This statement is proved by analogy with [24, Ch. 1, Theorem 3.1] using that the kernel of the integral operator $GS_{\mathbb{T}_+}G^{-1} - S_{\mathbb{I}}$ has weak singularities on $\mathbb{I}$. But this is not the case if to take arbitrary smooth curves instead of the Lyapunov curves $\mathbb{I}$ and $\mathbb{T}_+$. Generalizations of Lemma 3.9 to the case of arbitrary smooth curves can be found in [10] and in [17].

Theorem 3.10 *Suppose $a \in L^\infty(\mathbb{I})$. The operator $aS_{\mathbb{I}} - S_{\mathbb{I}}aI$ is compact on $L^2(\mathbb{I})$ if and only if $a \in VMO(\mathbb{I})$.*

Proof. Let $a \in L^\infty(\mathbb{I})$. From the obvious equality $aI = G(a \circ \theta_{-1})G^{-1}$ and Lemma 3.9 we get

$$[aI, S_{\mathbb{I}}] - G[(a \circ \theta_{-1})I, S_{\mathbb{T}_+}]G^{-1} = [aI, S_{\mathbb{I}} - GS_{\mathbb{T}_+}G^{-1}] \in \mathcal{K}(L^2(\mathbb{I})). \tag{3.9}$$

Necessity. Suppose $[aI, S_{\mathbb{I}}] \in \mathcal{K}(L^2(\mathbb{I}))$. Then from (3.9) we deduce that $[(a \circ \theta_{-1})I, S_{\mathbb{T}_+}] \in \mathcal{K}(L^2(\mathbb{T}_+))$. By Theorem 3.7, $a \circ \theta_{-1} \in VMO(\mathbb{T}_+)$. Hence, in view of Lemma 3.8, $a \in VMO(\mathbb{I})$. Necessity is proved.

Sufficiency. If $a \in VMO(\mathbb{I})$, then $a \circ \theta_{-1} \in VMO(\mathbb{T}_+)$, due to Lemma 3.8. Therefore, by Theorem 3.7, $[(a \circ \theta_{-1})I, S_{\mathbb{T}_+}] \in \mathcal{K}(L^2(\mathbb{T}_+))$. Hence, taking into account (3.9), we get $[aI, S_{\mathbb{I}}] \in \mathcal{K}(L^2(\mathbb{I}))$. □

3.5 Compactness of commutators with the shift operator

In this subsection we prove a compactness criterion for the commutator $W_\alpha S_{\mathbb{I}} - S_{\mathbb{I}} W_\alpha$, where α is a bi-Lipschitz homeomorphism of $[0, 1]$ onto itself. The main step of the proof is constructing a shift $\beta : \mathbb{T} \to \mathbb{T}$ with such behavior in neighborhoods of 1 and -1 as α has in neighborhoods of 0 and 1, respectively.

Lemma 3.11 *Let α be an orientation-preserving bi-Lipschitz homeomorphism of $[0,1]$ onto itself such that $\alpha(0)=0$ and $\alpha(1)=1$. Then*

$$\beta(\tau) := \begin{cases} (\theta\circ\alpha\circ\theta_{-1})(\tau), & \tau\in\mathbb{T}_+, \\ \overline{(\theta\circ\alpha\circ\theta_{-1})(\overline{\tau})}, & \tau\in\mathbb{T}_-, \\ \tau, & \tau\in\{-1,1\} \end{cases} \tag{3.10}$$

is an orientation-preserving bi-Lipschitz homeomorphism of $\mathbb{T}$ onto itself. Moreover, $\alpha'\in VMO(\mathbb{I})$ if and only if $\beta'\in VMO(\mathbb{T})$.

Proof. Obviously, β is a homeomorphism of $\mathbb{T}$ onto itself which preserves the orientation on $\mathbb{T}$. Since θ is a C^∞-diffeomorphism, θ and θ_{-1} are Lipschitz functions. Therefore $\alpha:\mathbb{I}\to\mathbb{I}$ and $\beta:\mathbb{T}\to\mathbb{T}$ are bi-Lipschitz homeomorphisms only simultaneously. Hence $\alpha'\in\mathcal{G}L^\infty(\mathbb{I})$ and $\beta'\in\mathcal{G}L^\infty(\mathbb{T})$.

Since $\theta'(x)=\pi i e^{\pi i x}$ whenever $x\in\mathbb{I}$, for $\tau\in\mathbb{T}_+$ we obtain

$$\beta'(\tau)=\frac{\theta'[(\alpha\circ\theta_{-1})(\tau)]}{\theta'[\theta_{-1}(\tau)]}\alpha'[\theta_{-1}(\tau)] = e^{\pi i[(\alpha\circ\theta_{-1})(\tau)-\theta_{-1}(\tau)]}\alpha'[\theta_{-1}(\tau)]. \tag{3.11}$$

If $\tau\in\mathbb{T}_-$, then in view of $\beta'(\tau)=\overline{\beta'(\overline{\tau})}$ and (3.11), we get

$$\beta'(\tau)=e^{-\pi i[(\alpha\circ\theta_{-1})(\overline{\tau})-\theta_{-1}(\overline{\tau})]}\alpha'[\theta_{-1}(\overline{\tau})]. \tag{3.12}$$

From (3.11) and (3.12) we see that

$$\beta'(\tau)=\varphi_\alpha(\tau)\psi_\alpha(\tau), \quad \tau\in\mathbb{T}, \tag{3.13}$$

where

$$\begin{aligned}\varphi_\alpha(\tau) \;&:=\; \begin{cases} e^{\pi i[(\alpha\circ\theta_{-1})(\tau)-\theta_{-1}(\tau)]}, & \tau\in\mathbb{T}_+, \\ e^{-\pi i[(\alpha\circ\theta_{-1})(\overline{\tau})-\theta_{-1}(\overline{\tau})]}, & \tau\in\mathbb{T}_-, \\ 1, & \tau\in\{-1,1\} \end{cases} \\ &= \begin{cases} e^{\pi i\left[\alpha\left(\frac{\arg\tau}{\pi}\right)-\frac{\arg\tau}{\pi}\right]}, & \tau\in\mathbb{T}_+, \\ e^{\pi i\left[\frac{2\pi-\arg\tau}{\pi}-\alpha\left(\frac{2\pi-\arg\tau}{\pi}\right)\right]}, & \tau\in\mathbb{T}_-, \\ 1, & \tau\in\{-1,1\} \end{cases}\end{aligned}$$

and

$$\psi_\alpha(\tau):=\begin{cases} \alpha'[\theta_{-1}(\tau)], & \tau\in\mathbb{T}_+, \\ \alpha'[\theta_{-1}(\overline{\tau})], & \tau\in\mathbb{T}_-. \end{cases}$$

Clearly, φ_α is a continuous function on $\mathbb{T}$, which satisfies $|\varphi_\alpha(\tau)|=1$. Thus,

$$\varphi_\alpha, 1/\varphi_\alpha\in C(\mathbb{T})\subset QC(\mathbb{T})=L^\infty(\mathbb{T})\cap VMO(\mathbb{T}). \tag{3.14}$$

Taking into account that $QC(\mathbb{T})$ is an algebra, we infer from (3.13) and (3.14) that $\beta' \in QC(\mathbb{T})$ if and only if $\psi_\alpha \in QC(\mathbb{T})$. By Lemma 3.5, $\psi_\alpha \in QC(\mathbb{T})$ if and only if $\alpha' \circ \theta_{-1} \in L^\infty(\mathbb{T}_+) \cap VMO(\mathbb{T}_+)$. On the other hand, the latter inclusion is equivalent to $\alpha' \in L^\infty(\mathbb{I}) \cap VMO(\mathbb{I})$, in view of Lemma 3.8. Thus, $\beta' \in \mathcal{G}L^\infty(\mathbb{T}) \cap VMO(\mathbb{T})$ if and only if $\alpha' \in \mathcal{G}L^\infty(\mathbb{I}) \cap VMO(\mathbb{I})$. □

Theorem 3.12 *Let $\alpha : [0,1] \to [0,1]$ be an orientation-preserving bi-Lipschitz homeomorphism such that $\alpha(0) = 0$ and $\alpha(1) = 1$. The operator $W_\alpha S_\mathbb{I} - S_\mathbb{I} W_\alpha$ is compact on $L^2(\mathbb{I})$ if and only if $\alpha' \in VMO(\mathbb{I})$.*

Proof. Since α is an orientation-preserving bi-Lipschitz homeomorphism, α' belongs to $\mathcal{G}L^\infty(\mathbb{I})$.

Necessity. The idea of the proof of necessity part is borrowed from [30, Proposition 4.5]. Since $W_\alpha^{-1} = W_{\alpha_{-1}}$, we conclude that $[W_\alpha, S_\mathbb{I}] \in \mathcal{K}(L^2(\mathbb{I}))$ if and only if

$$W_{\alpha_{-1}}[W_\alpha, S_\mathbb{I}]W_{\alpha_{-1}} = [S_\mathbb{I}, W_{\alpha_{-1}}] \in \mathcal{K}(L^2(\mathbb{I})).$$

On the other hand, since $S_\mathbb{I}$ is self-adjoint (see, e.g., [9, Ch. 1, Theorem 7.1] or [20, Ch. 1, Corollary 4]) and since $(W_{\alpha_{-1}})^* = \alpha' W_\alpha$, we have

$$[\alpha' W_\alpha, S_\mathbb{I}] = ([S_\mathbb{I}, W_{\alpha_{-1}}])^* \in \mathcal{K}(L^2(\mathbb{I})).$$

Taking into account that, by assumption, W_α and $S_\mathbb{I}$ commute modulo compact operators, we infer that

$$[\alpha' I, S_\mathbb{I}]W_\alpha = [\alpha' W_\alpha, S_\mathbb{I}] - \alpha'[W_\alpha, S_\mathbb{I}] \in \mathcal{K}(L^2(\mathbb{I})).$$

Thus, $[\alpha' I, S_\mathbb{I}] \in \mathcal{K}(L^2(\mathbb{I}))$. Finally, by Theorem 3.10, $\alpha' \in VMO(\mathbb{I})$. Necessity is proved.

Sufficiency. Sufficiency is proved by analogy with Theorem 3.10.

If α is a bi-Lipschitz homeomorphism and $\alpha' \in VMO(\mathbb{I})$, then, in view of Lemma 3.11, the function β defined by (3.10) is an orientation-preserving bi-Lipschitz homeomorphism of $\mathbb{T}$ onto itself such that $\beta' \in VMO(\mathbb{T})$. Hence, by Theorem 3.2, $[W_\beta, S_\mathbb{T}] \in \mathcal{K}(L^2(\mathbb{T}))$. Therefore, $\chi_{\mathbb{T}_+}[W_\beta, S_\mathbb{T}]\chi_{\mathbb{T}_+} I \in \mathcal{K}(L^2(\mathbb{T}))$. Since $\beta(\pm 1) = \pm 1$ and $\beta|_{\mathbb{T}_+} = \theta \circ \alpha \circ \theta_{-1}$ maps $\mathbb{T}_+$ onto itself, the latter inclusion is equivalent to

$$W_{\theta\circ\alpha\circ\theta_{-1}} S_{\mathbb{T}_+} - S_{\mathbb{T}_+} W_{\theta\circ\alpha\circ\theta_{-1}} \in \mathcal{K}(L^2(\mathbb{T}_+)). \tag{3.15}$$

Taking into account the obvious equality $G^{-1} W_\alpha G = W_{\theta\circ\alpha\circ\theta_{-1}}$, from (3.15) we get

$$[G W_{\theta\circ\alpha\circ\theta_{-1}} G^{-1}, G S_{\mathbb{T}_+} G^{-1}] = [W_\alpha, G S_{\mathbb{T}_+} G^{-1}] \in \mathcal{K}(L^2(\mathbb{I})). \tag{3.16}$$

From (3.16) and Lemma 3.9 we get $W_\alpha S_\mathbb{I} - S_\mathbb{I} W_\alpha \in \mathcal{K}(L^2(\mathbb{I}))$. □

4. Compactness of commutators on rearrangement-invariant spaces

4.1 Case of the unit circle

Let $X(\mathbb{T})$ be a rearrangement-invariant space with nontrivial Boyd indices over the unit circle $\mathbb{T}$.

Theorem 4.1 *Suppose $a \in L^\infty(\mathbb{T})$. The operator $aS_\mathbb{T} - S_\mathbb{T} aI$ is compact on $X(\mathbb{T})$ if and only if $a \in VMO(\mathbb{T})$.*

Theorem 4.2 *Let α be an orientation-preserving bi-Lipschitz homeomorphism of $\mathbb{T}$ onto itself. Then the operator $W_\alpha S_\mathbb{T} - S_\mathbb{T} W_\alpha$ is compact on $X(\mathbb{T})$ if and only if $\alpha' \in VMO(\mathbb{T})$.*

These theorems follow from Corollary 2.3 with $p_0 = 2$ and Theorems 3.1, 3.2, respectively. Note that the corresponding results for Lebesgue spaces $L^p(\mathbb{T})$, $1 < p < \infty$, were obtained in [30, Theorems 3.2 and 4.1] by using some modification of the interpolation theorem from [19].

4.2 Case of the unit interval

Let $\mathbb{I} = (0, 1)$ and let $X(\mathbb{I})$ be a rearrangement-invariant space with nontrivial Boyd indices.

Theorem 4.3 *Suppose $a \in L^\infty(\mathbb{I})$. The operator $aS_\mathbb{I} - S_\mathbb{I} aI$ is compact on $X(\mathbb{I})$ if and only if $a \in VMO(\mathbb{I})$.*

Theorem 4.4 *Let $\alpha : [0, 1] \to [0, 1]$ be an orientation-preserving bi-Lipschitz homeomorphism such that $\alpha(0) = 0$ and $\alpha(1) = 1$. The operator $W_\alpha S_\mathbb{I} - S_\mathbb{I} W_\alpha$ is compact on $X(\mathbb{I})$ if and only if $\alpha' \in VMO(\mathbb{I})$.*

These theorems follow from Corollary 2.3 and Theorems 3.10, 3.12, respectively.

5. Some corollaries

5.1 Commutators having slowly oscillating data

In this subsection we formulate corollaries on the compactness of commutators having slowly oscillating data. Note that singular integral operators and functional operators with shifts which have slowly oscillating data are intensively studied during the last time (see, e.g., [3], [15], [16], and the references therein).

For a function defined on an open open interval $J := (u, v) \subset \mathbb{R}$, we let $\omega(f, J)$ denote the oscillation of f over J:

$$\omega(f, J) := \sup_{s,t \in J} |f(s) - f(t)|.$$

Following [37], we say that the function f on J slowly oscillates at u if f is bounded and continuous on J and if, for each $\eta \in (0,1)$,

$$\lim_{\delta \to 0+} \omega(f, [u + \eta\delta, u + \delta]) = 0.$$

The notion of a slowly oscillating function at v is defined analogously. If a function f slowly oscillates at both endpoints u and v, then we will write $f \in SO(J)$.

Lemma 5.1 *We have $SO(J) \subset L^\infty(J) \cap VMO(J)$.*

This lemma was obtained in [37, Lemma 3], see also [34, Lemma A4, p. 80].

Suppose $X(\mathbb{I})$ is a rearrangement-invariant space with nontrivial Boyd indices over $\mathbb{I}$.

Corollary 5.2 *If $a \in SO(\mathbb{I})$, then the operator $aS_{\mathbb{I}} - S_{\mathbb{I}}aI$ is compact on $X(\mathbb{I})$.*

Corollary 5.3 *Let α be an orientation-preserving homeomorphism of $[0,1]$ onto itself such that $\alpha(0) = 0$ and $\alpha(1) = 1$. If $\alpha' \in \mathcal{G}L^\infty(\mathbb{I}) \cap SO(\mathbb{I})$, then the operator $W_\alpha S_{\mathbb{I}} - S_{\mathbb{I}} W_\alpha$ is compact on $X(\mathbb{I})$.*

Corollaries 5.2 and 5.3 follow from Lemma 5.1 and Theorems 4.3, 4.4, respectively, because α is a bi-Lipschitz homeomorphism if $\alpha' \in \mathcal{G}L^\infty(\mathbb{I}) \cap SO(\mathbb{I})$.

5.2 Commutators with isometric shift operators on Lebesgue spaces

Suppose $1 < p < \infty$ and consider the isometric shift operator on Lebesgue spaces $L^p(\Gamma)$ which is given by

$$(U_\alpha f)(t) := |\alpha'(t)|^{1/p} f[\alpha(t)], \quad t \in \Gamma,$$

where Γ is either the unit circle $\mathbb{T}$ or the unit interval $\mathbb{I}$.

Theorem 5.4 *Let α be an orientation-preserving bi-Lipschitz homeomorphism of $\mathbb{T}$ onto itself. If $\alpha' \in VMO(\mathbb{T})$, then the operator $U_\alpha S_{\mathbb{T}} - S_{\mathbb{T}} U_\alpha$ is compact on $L^p(\mathbb{T}), 1 < p < \infty$.*

Proof. Clearly,

$$U_\alpha S_{\mathbb{T}} - S_{\mathbb{T}} U_\alpha = M_\alpha (W_\alpha S_{\mathbb{T}} - S_{\mathbb{T}} W_\alpha) + (M_\alpha S_{\mathbb{T}} - S_{\mathbb{T}} M_\alpha) W_\alpha, \tag{5.1}$$

where $(M_\alpha f)(t) := |\alpha'(t)|^{1/p} f(t)$ for $t \in \mathbb{T}$. From Lemma 2.6 we infer that $\alpha' \in VMO(\mathbb{T})$ implies $|\alpha'|^{1/p} \in VMO(\mathbb{T})$ for every $p \in (1,\infty)$. Since α is bi-Lipschitz, $|\alpha'|^{1/p} \in L^\infty(\mathbb{T})$. Hence, the operator $M_\alpha S_{\mathbb{T}} - S_{\mathbb{T}} M_\alpha$ is compact on $L^p(\mathbb{T})$, due to Theorem 4.1. On the other hand, $W_\alpha S_{\mathbb{T}} - S_{\mathbb{T}} W_\alpha$ is compact on $L^p(\mathbb{T})$, in view of Theorem 4.2. Thus, taking into account (5.1), we deduce that $U_\alpha S_{\mathbb{T}} - S_{\mathbb{T}} U_\alpha$ is also compact on $L^p(\mathbb{T})$. □

Theorem 5.5 *Let α be an orientation-preserving bi-Lipschitz homeomorphism of* $[0,1]$ *onto itself such that* $\alpha(0) = 0$, $\alpha(1) = 1$, *and* $\alpha' \in VMO(\mathbb{I})$. *Then the operator* $U_\alpha S_\mathbb{I} - S_\mathbb{I} U_\alpha$ *is compact on* $L^p(\mathbb{I}), 1 < p < \infty$.

This theorem is proved by analogy with Theorem 5.4. Here we have to use Theorems 4.3 and 4.4 instead of Theorems 4.1 and 4.2, respectively.

Corollary 5.6 *Let α be an orientation-preserving homeomorphism of* $[0,1]$ *onto itself such that* $\alpha(0) = 0$ *and* $\alpha(1) = 1$. *If* $\alpha' \in \mathcal{G}L^\infty(\mathbb{I}) \cap SO(\mathbb{I})$, *then the operator* $U_\alpha S_\mathbb{I} - S_\mathbb{I} U_\alpha$ *is compact on* $L^p(\mathbb{I}), 1 < p < \infty$.

This statement is an immediate consequence of Lemma 5.1 and Theorem 5.5.

Acknowledgements

The authors are partially supported by F.C.T. (Portugal) grant POCTI 34222/MAT/2000. The first author is also supported by F.C.T. (Portugal) grant PRAXIS XXI/BPD/22006/99, the second author is also supported by PROMEP (México).

References

[1] Bennett, C. and Sharpley, R., *Interpolation of Operators*, Pure and Applied Mathematics, **129**. Academic Press Inc., Boston, MA, 1988.

[2] Berezhnoi, E. I., Interpolation of linear and compact operators in the spaces $\varphi(X_0, X_1)$. In: *Qualitative and approximate methods for the investigation of operator equations*, 19–29. Yaroslav. Gos. Univ., Yaroslavl', 1980 (in Russian).

[3] Böttcher, A., Karlovich, Yu. I., and Rabinovich, V. S., The method of limit operators for one-dimensional singular integrals with slowly oscillating data, *J. Operator Theory*, **43** (2000), 171–198.

[4] Böttcher, A., Karlovich, Yu. I., and Silbermann, B., Singular integral equations with *PQC* coefficients and freely transformed argument, *Math. Nachr.*, **166** (1994), 113–133.

[5] Böttcher, A. and Silbermann, B., *Analysis of Toeplitz Operators*. Springer-Verlag, Berlin, 1990.

[6] Boyd, D. W., Indices of function spaces and their relationship to interpolation, *Canad. J. Math.*, **21** (1969), 1245–1254.

[7] Brudnyi, Yu. A. and Krugljak, N. Ya., *Interpolation Functors and Interpolation Spaces*, Vol. I, North-Holland Mathematical Library, **47**. North-Holland Publishing Co., Amsterdam, 1991.

[8] Calderón, A.-P., Intermediate spaces and interpolation, the complex method, *Studia Math.*, **24** (1964), 113–190.

[9] Gohberg, I. and Krupnik, N., *One-Dimensional Linear Singular Integral Equations*, Vol. I, Operator Theory: Advances and Applications, **53**. Birkhäuser Verlag, Basel, 1992.

[10] Grudskii, S. M., On compactness of an integral operator, Deposited in VINITI, No. 4856-80 Dep., VINITI, 1980 (in Russian).

[11] Grudskii, S. M. and Dybin, V. B., A Riemann boundary value problem with discontinuities of almost periodic type in its coefficient, *Soviet Math. Dokl.*, **18** (1978), 1383–1387.

[12] Hartman, P., On completely continuous Hankel matrices, *Proc. Amer. Math. Soc.*, **9** (1958), 862–866.

[13] Janson, S., Mean oscillation and commutators of singular integral operators, *Ark. Mat.*, **16** (1978), 263–270.

[14] Karapetiants, N. and Samko, S., *Equations with Involutive Operators.* Birkhäuser Boston, Inc., Boston, MA, 2001.

[15] Karlovich, A. Yu., Karlovich, Yu. I., and Lebre, A. B., Invertibility of functional operators with slowly oscillating non-Carleman shifts. To appear in: Operator Theory: Advances and Applications, **142**. Birkhäuser Verlag, 2003.

[16] Karlovich, Yu. I. and Lebre, A. B., Algebra of singular integral operators with a Carleman backward slowly oscillating shift, *Integr. Equat. Oper. Theor.*, **41** (2001), 288–323.

[17] Karlovich, Yu. I. and Ramírez de Arellano, E., Singular integral operators with fixed singularities on weighted Lebesgue spaces, *Integr. Equat. Oper. Theor.*, to appear.

[18] Krantz, S. and Song-Ying Li, Boundedness and compactness of integral operators on spaces of homogeneous type and applications. I-II, *J. Math. Anal. Appl.*, **258** (2001), 629-641 (I), 642-657 (II).

[19] Krasnoselskii, M. A., On a theorem of M. Riesz, *Soviet Math. Dokl.*, **1** (1960), 229–231.

[20] Kravchenko, V. G. and Litvinchuk, G. S., *Introduction to the Theory of Singular Integral Operators with Shift*, Mathematics and its Applications, **289**. Kluwer Academic Publishers, Dordrecht, 1994.

[21] Krein, S. G., Petunin, Ju. I., and Semenov, E. M., *Interpolation of Linear Operators*, Translations of Mathematical Monographs, **54**. AMS, Providence, R.I., 1982.

[22] Krupnik, N. Ya. and Nyaga, V. I., Singular integral operators with a shift along a piecewise Lyapunov contour, *Soviet Math. (Izv. VUZ)*, **19** (1975), no. 6, 49–59.

[23] Lindenstrauss, J. and Tzafriri, L., *Classical Banach Spaces. Function Spaces.* Springer Verlag, New York, Berlin, 1979.

[24] Litvinchuk, G. S., *Boundary Value Problems and Singular Integral Equations with Shift.* Nauka, Moscow, 1977 (in Russian).

[25] Litvinchuk, G. S., *Solvability Theory of Boundary Value Problems and Singular Integral Equations with Shift*, Mathematics and its Applications, **523**. Kluwer Academic Publishers, Dordrecht, 2000.

[26] Lozanovskii, G. Ja., Certain Banach lattices, *Sibirsk. Mat. Zh.*, **10** (1969), 584–599 (in Russian).

[27] Maligranda, L., Indices and interpolation, *Dissertationes Math. (Rozprawy Mat.)*, **234** (1985), 1–49.

[28] Maligranda, L., *Orlicz Spaces and Interpolation*, Seminars in Mathematics, **5**. Univ. Estadual de Campinas, Campinas SP, Brazil, 1989.

[29] Maligranda, L. and Persson, L.-E., Generalized duality of some Banach function spaces, *Nederl. Akad. Wetensch. Indag. Math.*, **51** (1989), 323–338.

[30] Muhly, P. and Xia, J., Calderón-Zygmund operators, local mean oscillation and certain automorphisms of the Toeplitz algebra, *Amer. J. Math.*, **117** (1995), 1157–1201.

[31] Natanson, I. P., *Theory of Real Variable Functions.* Nauka, Moscow, 1974 (in Russian).

[32] Nikol'skii, N. K., *Treatise on the Shift Operator.* Springer-Verlag, Berlin, 1986.

[33] Power, S. C., Hankel operators on Hilbert space, *Bull. London Math. Soc.*, **12** (1980), 422–442.

[34] Power, S. C., *Hankel Operators on Hilbert Space*, Research Notes in Mathematics, **64**. Pitman Advanced Publishing Program, Boston, London, Melbourne, 1982.

[35] Roch, S. and Silbermann, B., *Algebras of Convolution Operators and their Image in the Calkin Algebra*, Report MATH, 90-05. Akademie der Wissenschaften der DDR, Karl-Weierstrass-Institut für Mathematik, Berlin, 1990.

[36] Sarason, D., Functions of vanishing mean oscillation, *Trans. Amer. Math. Soc.*, **207** (1975), 391–405.

[37] Sarason, D., Toeplitz operators with piecewise quasicontinuous symbols, *Indiana Univ. Math. J.*, **26** (1977), 817–838.

[38] Sarason, D., *Function Theory on the Unit Circle*, Virginia Polytechnic Institute and State University, Department of Mathematics, Blacksburg, Va., 1978.

[39] Sharpley, R., Interpolation theorems for compact operators, *Indiana Univ. Math. J.*, **22** (1972/73), 965–984.

[40] Torchinsky, A., *Real-Variable Methods in Harmonic Analysis*, Pure and Applied Mathematics, **123**. Academic Press Inc., Orlando, FL, 1986.

[41] Uchiyama, A., On the compactness of operators of Hankel type, *Tôhoku Math. J.*, **30** (1978), 163–171.

SOME PROBLEMS IN THE THEORY OF INTEGRAL AND DIFFERENTIAL EQUATIONS OF FRACTIONAL ORDER

Anatoly A. Kilbas
Belarusian State University, Belarus
kilbas@bsu.by

To Professor G. S. Litvinchuk on the occasion of his 70th birthday

Abstract The paper is devoted to some aspects of the so-called integral and differential equations of fractional order in which an unknown function is contained under the operation of integration and differentiation of fractional order. Unified methods based on the Laplace transform, operational calculus and compositional relations are presented to study such linear integral and differential equations. Problems and new trends of research are discussed.

Keywords: Fractional calculus, integral and differential equations of fractional order, Cauchy-type problem, Volterra integral equation, Laplace transform, operational calculus and compositional methods, Mittag-Leffler functions

1. Introduction

The fractional integral and ordinary differential equations have the following general form

$$F\left(x, \varphi(x), \mathcal{J}^{\alpha_1}\varphi(x), \mathcal{J}^{\alpha_2}\varphi(x), \ldots, \mathcal{J}^{\alpha_n}\varphi(x)\right) = g(x), \tag{1.1}$$

where $F(x, y_1, y_2, \ldots, y_n)$ and $g(x)$ are given functions and $\mathcal{J}^{\alpha_k} = I^{\alpha_k}$ and $\mathcal{J}^{\alpha_k} = D^{\alpha_k}$ are the operators of fractional integration and differentiation, respectively, with real $\alpha_k > 0$ or complex α_k, $\mathrm{Re}(\alpha_k) > 0$ $(k = 1, 2, \ldots, n)$. Among them linear integral and differential equations of the form

$$\sum_{k=1}^{n} c_k(x)\mathcal{J}^{\alpha_k}\varphi(x) + c_0(x)\varphi(x) = f(x) \tag{1.2}$$

which arise in practical applications, were studied more extensively.

The fractional integration and differentiation operators in (1.1)-(1.2) may have different forms - see the book by S. Samko, A. Kilbas and O. Marichev [44]; however, the equations with the Riemann-Liouville fractional integral $\mathcal{J}^{\alpha}\varphi = I_{a+}^{\alpha}\varphi$ and fractional derivative $\mathcal{J}^{\alpha}\varphi = D_{a+}^{\alpha}\varphi$, $a \in \mathbf{R}$, were studied more than equations with other forms. For complex $\alpha \in \mathbf{C}$, $\operatorname{Re}(\alpha) > 0$, such fractional integrals and derivatives are defined by

$$(I_{a+}^{\alpha}\varphi)(x) = \frac{1}{\Gamma(\alpha)} \int_a^x \frac{\varphi(t)dt}{(x-t)^{1-\alpha}} \tag{1.3}$$

and

$$(D_{a+}^{\alpha}y)(x) = \left(\frac{d}{dx}\right)^n (I_{a+}^{n-\alpha}y)(x), \ n = [\operatorname{Re}(\alpha)] + 1, \tag{1.4}$$

respectively, where $[\operatorname{Re}(\alpha)]$ means the integral part of $\operatorname{Re}(\alpha)$.

Integral equations of fractional order are integral equations of Volterra type. Therefore, the methods known for Volterra integral equations can be applied to study these equations. The integral equations of the first kind containing the Riemann-Liouville integrals (1.3) with an unknown function $\varphi(x)$ are called Abel integral equations. In the book of R. Gorenflo and S. Vessela [17] there were given methods of investigation of such equations and some results on existence and uniqueness theorems for Abel integral equations of the first and second kind, on stabilization of the corresponding ill-posed problems together with presentation of numerical methods and applications.

Fractional ordinary differential equations are less studied. A survey of the results in this direction was given in the paper by A. Kilbas and J. Trujillo [27].

The problem of the closed form solvability for fractional integral and differential equations is important in applications. In this paper we present a unified approach to find the closed form solution for both integral and differential equations of fractional order. We discuss the unified methods based on the Laplace transform, operational calculus and composition relations, paying more attention to ideas and results which were not reflected in existing monographs.

The paper is organized as follows. The Laplace transform method is discussed in Section 2. Section 3 deals with the operational calculus approach. Section 4 is devoted to the method of composition relations. Some problems and new trends of research are discussed in Section 5.

2. Laplace transform method

We consider the linear integro-differential equation of fractional order with constant coefficients

$$\sum_{k=1}^{m} c_k (\mathcal{J}_{a+}^{\alpha_k}\varphi)(x) + c_0\varphi(x) = f(x) \ (\operatorname{Re}(\alpha_i) > 0; \ i = 1, \ldots, m), \tag{2.1}$$

where $m \geq 1$, $\mathcal{J}_{a+}^{\alpha_k}\varphi$ $(k = 1, 2, \ldots, m)$ are the Riemann-Liouville fractional integrals $I_{a+}^{\alpha_k}\varphi$ or the Riemann-Liouville fractional derivatives $D_{a+}^{\alpha_k}\varphi$, given by (1.3) and (1.4), and $c_k \neq 0$ $(k = 0, 1, \ldots, m)$ are real or complex constants. When $\mathcal{J}_{a+}^{\alpha_k}\varphi = I_{a+}^{\alpha_k}\varphi$ $(k = 1, 2, \ldots, m)$, equation (2.1) is a fractional integral equation, while for $\mathcal{J}_{a+}^{\alpha_k}\varphi = D_{a+}^{\alpha_k}\varphi$ $(k = 1, 2, \ldots, m)$ (2.1) is the fractional differential equation.

Without loss of generality we consider fractional integro-differential equation (2.1) with $a = 0$. The scheme to solve the nonhomogeneous fractional integro-differential equation (2.1) $(f(x) \neq 0)$ is based on the relation

$$(\mathcal{L}\mathcal{J}_{0+}^{\alpha}\varphi)(p) = p^{-\alpha}(\mathcal{L}\varphi)(p), \tag{2.2}$$

for the Laplace transform $(\mathcal{L}\varphi)(p) = \int_0^\infty \varphi(t)e^{-pt}dt$ of the Riemann-Liouville fractional integral $\mathcal{J}_{0+}^{\alpha}\varphi = I_{0+}^{\alpha}\varphi$ $(\mathrm{Re}(\alpha) > 0)$ or derivative $\mathcal{J}_{0+}^{\alpha}\varphi = D_{0+}^{-\alpha}\varphi$ $(\mathrm{Re}(\alpha) < 0)$ for "sufficiently good function" φ (see S. Samko, A. Kilbas and O. Marichev, [44, Section 7.2]). Applying the Laplace transform to both sides of (2.1) and making use of (2.2) we have

$$\left[\sum_{k=1}^{m} c_k p^{-\alpha_k} + c_0\right] (\mathcal{L}\varphi)(p) = (\mathcal{L}f)(p).$$

Hence we obtain a particular solution of the nonhomogeneous integral equation (2.1) in the form

$$\varphi(x) = \left(\mathcal{L}^{-1}\left[\frac{(\mathcal{L}f)(p)}{\sum_{k=1}^{m} c_k p^{-\alpha_k} + c_0}\right]\right)(x) \tag{2.3}$$

in terms of the Laplace transform $\mathcal{L}$ and the inverse Laplace transform $\mathcal{L}^{-1}$ - see, e.g., the book by V. Ditkin and A. Prudnikov [8]. When $\mathrm{Re_k}(\alpha) > 0$ $(k = 1, 2, \ldots, m)$ and $\mathrm{Re}(\alpha_k) < 0)$ $(k = 1, 2, \ldots, m)$, (2.3) yields the explicit solution of the fractional integral and differential equations, respectively. Solution (2.3) was obtained formally, so that conditions for its validity must be studied.

The Laplace transform method to solve fractional integral equations of Abel type was first used by E. Hille and J. Tamarkin [19]. Using this approach, they obtained the explicit solution of the integral equation

$$\varphi(x) - \frac{\lambda}{\Gamma(\alpha)} \int_0^x \frac{\varphi(t)dt}{(x-t)^{1-\alpha}} = f(x) \ (x > 0) \tag{2.4}$$

in the form

$$\varphi(x) = \frac{d}{dx} \int_0^x E_\alpha \left[\lambda(x-t)^\alpha\right] f(t)dt,$$

where

$$E_\alpha(z) = \sum_{k=0}^{\infty} \frac{z^k}{\Gamma(\alpha k + 1)} \ (\alpha > 0) \tag{2.5}$$

is the Mittag-Leffler function, see A. Erdelyi, etc. [11, Section 18.1] or M. Dzhrbashyan [9]. One can find results on integral equations (2.1) and (2.4) in S. Samko, A. Kilbas and O. Marichev [44, Sections 2.4 and 42.4] and R. Gorenflo and S. Vessela [17].

D. Maravall [34] was probably the first who used a formal Laplace transform approach to obtain the explicit solution of another particular case of the fractional differential equation of form (2.1), but his paper was practically unknown.

The Laplace transform was used by K. Oldham and J. Spanier [38, Section 8.5] to solve the homogeneous equation ($f(x) = 0$)

$$y'(x) + b(D_{0+}^{\alpha}y)(x) + cy(x) = f(x), \ (0 < \alpha < 1; \ b \in \mathbf{R}, \ c \in \mathbf{R}) \tag{2.6}$$

with $\alpha = 1/2$, $b = 1$ and $c = -2$, and by D. Seitkazieva [45] to construct a solution of equation (2.6) with $0 < \alpha < 1$.

K. Miller and B. Ross [37] applied the Laplace transform method to solve the Cauchy problem for the particular case of the differential equation of form (2.1) with $\alpha_k = k\alpha$ and $1/\alpha = 1, 2, ...$:

$$\textstyle\sum_{k=1}^{m} c_k(D_{0+}^{k\alpha}y)(x) + c_0 y(x) = f(x), \ \frac{1}{\alpha} = 1, 2, 3, \ldots, \tag{2.7}$$

$$y(0) = y'(0) = \cdots y^{(m-1)}(0) = 0. \tag{2.8}$$

They introduced a fractional analogue of the Green function $G_\alpha(x)$ defined via the inverse Laplace transform $\mathcal{L}^{-1}$ by

$$G_\alpha(x) = \left(\mathcal{L}^{-1}\left[\frac{1}{P(x^\alpha)}\right]\right)(x), \ P(x) = \sum_{k=1}^{m} c_k x^k + c_0,$$

and proved that the unique solution $y(x)$ of (2.7)-(2.8) has the form of the Laplace convolution of $G_\alpha(x)$ and $f(x)$:

$$y(x) = \int_0^x G_\alpha(x-t)f(t)dt. \tag{2.9}$$

These investigations were developed by I. Podlubny [39, Chapter 5] who defined a similar fractional analogue of the Green function $G_\alpha(x)$ for an equation more general than the equation of form (2.1). He showed that the solution of the corresponding Cauchy-type problem can be expressed in terms of $G_\alpha(x)$, and in particular, found the explicit formula for $G_\alpha(x)$ as a multiple series containing the Mittag-Leffler function

$$E_{\alpha,\beta}(z) = \sum_{k=0}^{\infty} \frac{z^k}{\Gamma(\alpha k + \beta)} \ (\alpha > 0, \ \beta > 0). \tag{2.10}$$

The latter generalizes the function $E_\alpha(z) = E_{\alpha,1}(z)$, defined in (2.5), see A. Erdelyi, etc. [11, Section 18.1] and M. Dzhrbashyan [9].

As we have seen, finding a particular solution is, in a sense, the same both for the nonhomogeneous integral or nonhomogeneous differential equation of form (2.1). It is not the same for the corresponding homogeneous equations with $f(x) = 0$. The homogeneous differential equation of form (2.1)

$$\sum_{k=1}^{m} c_k(D_{a+}^{\alpha_k}y)(x) + c_0 y(x) = 0 \ (0 < \mathrm{Re}(\alpha_1) < \mathrm{Re}(\alpha_2) < \cdots < \mathrm{Re}(\alpha_m)), \quad (2.11)$$

may have nontrivial linear independent solutions similarly to ordinary differential equations of order m which are particular cases of (2.11) corresponding to $\alpha_k = k = 1, 2, \ldots, m$. I. Leskovskii [28] was probably the first who constructed some linear independent solutions of homogeneous equation (2.11) with specific distinct real exponents $0 < \alpha_k < 1$ $(k = 1, 2, \ldots, m)$ in the form of the Mittag-Leffler function (2.10).

As for the homogeneous integral equation of form (2.11)

$$\sum_{k=1}^{m} c_k(I_{a+}^{\alpha_k}y)(x) + c_0 y(x) = 0 \ (0 < \mathrm{Re}(\alpha_1) < \mathrm{Re}(\alpha_2) < \cdots < \mathrm{Re}(\alpha_m)), \quad (2.12)$$

it is a Volterra integral equation of the second kind, and therefore (2.12) has only a trivial solution $\varphi(x) = 0$. The situation may only change if we consider this equation on the infinite interval in which the Riemann-Liouville fractional integrals $I_{a+}^{\alpha_k}\varphi$ are replaced by the Liouville fractional integrals $I_{-}^{\alpha_k}\varphi$ defined by

$$(I_{-}^{\alpha}\varphi)(x) = \frac{1}{\Gamma(\alpha)}\int_x^\infty \frac{\varphi(t)dt}{(t-x)^{1-\alpha}} \ (\alpha \in \mathbf{C}, \ \mathrm{Re}(\alpha) > 0). \quad (2.13)$$

J. Alonso [2] proved that the equation of the form (2.12), with $I_{a+}^{\alpha_k}$ replaced by the Liouville fractional integrals $I_{-}^{\alpha_k}$, considered on $\mathbf{R}_+ = (0, \infty)$, has the solution in the form $\varphi(x) = e^{-\lambda x}$ if and only if the parameter λ satisfies the condition

$$\mathrm{Re}(\lambda) > 0, \ P_{-\alpha}(\lambda) \equiv c_m + \sum_{k=0}^{m-1} c_k \lambda^{-\alpha_k} = 0.$$

This result, as well as the known statements for the linear ordinary differential equations, yield an idea that solutions of homogeneous differential equation (2.11) can be connected with the roots λ of the characteristic quasipolynomial: $P_\alpha(\lambda) \equiv \sum_{k=1}^{m} c_k\lambda^{\alpha_k} + c_0 = 0$. Such an approach for the homogeneous differential equations (2.7) $(f(x) = 0)$ was developed by K. Miller and B. Ross [37]. L. Campos [4] presented this method for solving particular cases of equations of the form (2.11) with the Riemann-Liouville fractional derivatives $D_{0+}^{\alpha_k}y$

replaced by the fractional derivatives $y^{(\alpha_k)}$ of complex order $\alpha_k \in \mathbf{C}$ in the complex plane $\mathbf{C}$ defined as a generalization of the formula for differentiating the Cauchy type integral - see S. Samko, A. Kilbas and O. Marichev [44, Section 22.1].

K. Miller and B. Ross [37, Section VI.3] indicated that the Laplace transform can be applied to solve some homogeneous differential equations of form (2.7) ($f(x) \equiv 0$) with polynomial coefficients $c_k(x)$ and illustrated such an approach by solving the equation $(D_{0+}^{1/2}y)(x) = y/x$. Note that this equation was the first fractional differential equation discussed in mathematical literature, see S. Samko, A. Kilbas and O. Marichev [44, Section 43.1] and A. Kilbas and J. Trujillo [27, pp. 158-159] in this connection.

K. Miller and B. Ross [37] also solved the so-called sequential fractional differential equations of the form (2.7):

$$\left(P_m(D_{0+}^{\alpha})y\right)(x) \equiv \sum_{k=1}^{m} c_k \left((D_{0+}^{\alpha})^k y\right)(x) + c_0 y(x) = f(x), \ \frac{1}{\alpha} = 1, 2, \ldots, \qquad (2.14)$$

by using the roots $\lambda = \lambda_i$ of the characteristic polynomial $P_m(\lambda) = \sum_{k=0}^{m} c_k \lambda^k$.

Examples of linear fractional differential equations of form (2.7) and (2.1), solved via the Laplace transform method and analogues of Green function, were presented in the books by K. Miller and B. Ross [37, Chapters V and VI] and by I. Podlubny [39, Sections 4.1.1 and 4.2.1].

R. Gorenflo and F. Mainardi [15]-[16] applied the Laplace transform to solve the fractional differential equation

$$(D_*^{\alpha}y)(x) = -\rho^{\alpha}y(x) + f(x) \ (\alpha > 0; \ \rho > 0, \ x > 0),$$

and discussed the key role of the Mittag-Leffler function (2.10) in the cases $1 < \alpha < 2$ and $2 < \alpha < 3$. Here

$$\begin{aligned}(D_*^{\alpha}y)(x) &\equiv \left(D_{0+}^{\alpha}\left[y(t) - \sum_{k=0}^{n-1}\frac{t^k}{k!}y^{(k)}(0)\right]\right)(x) \\ &= \frac{1}{\Gamma(n-\alpha)}\int_0^x \frac{y^{(n)}(t)dt}{(x-t)^{n-\alpha-1}} \qquad (2.15)\end{aligned}$$

with $\alpha > 0$ $(n-1 < \alpha \leq n)$. This expression for the fractional derivative, called sometimes Caputo derivative, coincides with the Riemann-Liouville fractional derivative $D_{0+}^{\alpha}y$

$$(D_{0+}^{\alpha}y)(x) = \sum_{k=0}^{n-1}\frac{y^{(k)}(0)}{\Gamma(1+k-\alpha)}x^{k-\alpha} + \frac{1}{\Gamma(n-\alpha)}\int_0^x \frac{y^{(n)}(t)dt}{(x-t)^{\alpha-n+1}},$$

of order $\alpha > 0$, in the case $y^{(k)}(0) = 0, \ k = 0, 1, \ldots, n-1$, where $n-1 < \alpha \leq n$.

3. Operational calculus method

The usefulness of operational calculus for solving ordinary differential equations is well known - see, for example, V. Ditkin and A. Prudnikov [8]. The basics of the operational calculus for the operators of differentiation was developed by J. Mikusinski [36]. It is based on the interpretation of the Laplace convolution

$$(f * g)(x) = \int_0^x f(x-t)g(t)dt$$

as a multiplication of elements f and g in the ring of functions continuous on the half-axis $\mathbf{R}_+$. J. Mikusinski applied his operational calculus to solve ordinary differential equations with constant coefficients. We also mention that the Mikusinski's scheme was used by V. Ditkin [6], V. Ditkin and A. Prudnikov [7], N. Meller [35] and J. Rodriguez [40] to develop the operational calculus for a Bessel-type differential operators with nonconstant coefficients. Transform approach to the development of operational calculus was considered by I. Dimovski [5]. J. Rodriguez, J. Trujillo and M. Rivero [41] were probably the first who applied operational calculus to solve a Bessel-type differential equation of fractional order.

A series of papers were devoted to develop operational calculus for the fractional operators with applications to solution of integral and differential equations of fractional order. Yu. Luchko and H. Srivastava [31] constructed the operational calculus for the Riemann-Liouville fractional derivative $D_{0+}^{\alpha}y$ given in (1.4), in the space $\mathcal{C}_{-1}$ of functions $y(x)$ such as $x^{-p}(D_{0+}^{\alpha})^k y(x) \in C[0,\infty)$ $(k = 1, 2, \dots, m)$ for some $p > -1$. They proved that the operation $*_\lambda$ defined for $\lambda \geq 1$ by

$$(f *_\lambda g)(x) = (I_{0+}^{\lambda-1} f * g)(x) = \int_0^x (I_{0+}^{\lambda-1} f)(x-t)g(t)dt \ (\lambda \geq 1),$$

$I_{0+}^{\lambda-1}$ being the operation of Riemann-Liouville fractional integration (1.3), is the convolution (without divisors) of the linear operator I_{0+}^{α} for $\alpha > 0$ in the space $\mathcal{C}_{-1}$, and that the Riemann-Liouville operator I_{0+}^{α} has the convolution representation

$$(I_{0+}^{\alpha} f)(x) = (h *_\lambda f)(x) \left(1 \leq \lambda < \alpha + 1, \ h(x) := \frac{x^{\alpha-\lambda}}{\Gamma(\alpha - \lambda - 1)}\right). \tag{3.1}$$

They showed that the space $\mathcal{C}_{-1}$ with the operations $*_\lambda$ and $+$, having the property of distributivity

$$(f *_\lambda (g+h))(x) = (f *_\lambda g)(x) + (f *_\lambda h)(x) \ (f, g, h \in \mathcal{C}_{-1}),$$

becomes a commutative ring without divisors of zero, and therefore, following J. Mikusinski, $\mathcal{C}_{-1}$ can be extended to the quotient field $\mathcal{M}$. Luchko and

Srivastava indicated that the elements of the field $\mathcal{M}$ can be considered as convolution quotients f/g with the operations

$$\frac{f}{g}+\frac{f_1}{g_1}=\frac{(f*_\lambda g_1)+(g*_\lambda f_1)}{(g*_\lambda g_1)},\ \left(\frac{f}{g}\right)\left(\frac{f_1}{g_1}\right)=\frac{(f*_\lambda f_1)}{(g*_\lambda g_1)},$$

which means that the ring $\mathcal{C}_1$ can be embedded in the field $\mathcal{M}$ by the mapping

$$f(x)\rightarrow\frac{(h*_\lambda f)(x)}{h(x)}$$

with $h(x)$ in (3.1). On the basis of these facts they defined the algebraic inverse of the operator I^α_{0+} as an element S of the field $\mathcal{M}$ which is reciprocal to the element $h(x)$ in the field $\mathcal{M}$:

$$S=\frac{I}{h}\equiv\frac{h}{(h*_\lambda h)}\equiv\frac{h}{h^2},$$

where $I=h/h$ denotes the identity element of the field $\mathcal{M}$ with respect to the operation of multiplication.

Introducing the space

$$\Omega^m_\alpha(\mathcal{C}_{-1})=\left\{f(x)\in\mathcal{C}_{-1}:\ (D^{m\alpha}_{0+})^k f)(x)\in\mathcal{C}_{-1}\ (k=1,2,\dots,m)\right\}$$

with $m\in\mathbf{N}$ and $\alpha>0$, Luchko and Srivastava proved the relation for $f(x)\in\Omega^m_\alpha(\mathcal{C}_{-1})$ in the field $\mathcal{M}$

$$\left((D^\alpha_{0+})^m f\right)(x)=S^m f-\sum_{k=0}^{m-1}S^{m-k}F(D^\alpha_{0+})^k f,$$

where the operator $F=E-I^\alpha_{0+}D^\alpha_{0+}$ is given by

$$(Ff)(x):=\left((E-I^{m\alpha}_{0+}D^\alpha_{0+})f\right)(x)=\sum_{k=0}^{-[-\alpha]}\frac{x^{\alpha-k}}{\Gamma(\alpha-k+1)}\lim_{x\to0}(D^{\alpha-k}_{0+}f)(x)$$

and E is the identity operator. This result means that the Riemann-Liouville fractional differentiation operator D^α_{0+} is reduced to the operator of multiplication in the field $\mathcal{M}$.

Such an operational calculus was applied by Luchko and Srivastava to solve the following Cauchy-type problem for the equation of the form (2.14):

$$\left(P_m(D^\alpha_{0+})y\right)(x)=f(x),\ P_m(z)=\sum_{k=1}^{m}c_k z^k, \tag{3.2}$$

with any $\alpha>0$ and $f(x)\in\mathcal{C}_{-1}$ in the space $\Omega^m_\alpha(\mathcal{C}_{-1})$ with the initial conditions

$$\lim_{x\to0}\left(D^{\alpha-k}_{0+}(D^\alpha_{0+})^i y\right)(x)=b_{ik}\ (i=0,1,\dots m-1;\ k=1,\dots,n). \tag{3.3}$$

They reduced this problem to the following algebraic equation in the field $\mathcal{M}$

$$P_m(S) = f + \sum_{i=0}^{m-1} P_i(S)\gamma_i, \; P_i(S) = \sum_{j=1}^{m-i} c_{i+j}S^j \; (i = 0, 1, \ldots, m-1),$$

which enabled them to find its unique solution

$$y = \frac{I}{P_m(S)} + \sum_{i=0}^{m-1} \frac{P_i(S)}{P_m(S)}\gamma_i$$

and give the explicit solution of the Cauchy-type problem (3.2)-(3.3) in terms of the special function

$$E^{\varrho}_{\alpha,\beta}(z) = \sum_{k=0}^{\infty} \frac{(\varrho)_k z^k}{k!\Gamma(\alpha k + \beta)}, \tag{3.4}$$

where $(\varrho)_k$ is the Pochhammer symbol defined by

$$(\varrho)_k = 1, \; (\varrho)_k = \varrho(\varrho+1)\cdots(\varrho+k-1) \; (k = 1, 2, \ldots),$$

see A. Erdelyi, etc. [11, Section 2.1.1]. Obviously, (3.4) is a generalization of the Mittag-Leffler function (2.10), the latter is obtained from (3.4) when $\varrho = 1$.

Such an operational method was applied by Yu. Luchko and S. Yakubovich [32]-[33], M. Al-Bassam and Yu. Luchko [1] and S. Hadid and Yu. Luchko [18] to solve the Cauchy-type problems for fractional differential equations with constant coefficients involving the Erdelyi-Kober-type fractional derivatives - see S. Samko, A. Kilbas and O. Marichev [44, Section 18.1]. The explicit solutions in these cases are expressed via the Mittag-Leffler type function

$$E_{\varrho}\left((\alpha,\beta)_n; z\right) = \sum_{k=0}^{\infty} \frac{(\varrho)_k z^k}{k! \prod_{i=1}^{n} \Gamma(\alpha_i k + \beta_i)},$$

more general than (3.4).

The operational method similar to that in Yu. Luchko and H. Srivastava [31] was developed by R. Gorenflo and Yu. Luchko [13] to obtain explicit solutions to Abel type integral equations of second kind of form (2.1) in which the Riemann-Liouville fractional integration operators $I^{\alpha_k}_{0+}$ are replaced by more general constructions containing the generalized Mittag-Leffler function (3.4) in the kernel. In particular, Gorenflo and Luchko established the explicit solution of the integral equation of form (2.1) with $a = 0$. R. Gorenflo, Yu. Luchko and H. Srivastava [14] developed the operational method for an integral operator, more general than the Riemann-Liouville fractional integral (1.3), the kernel of which contains the Gauss hypergeometric function ${}_2F_1(a, b; c; z)$. They used this operational calculus to find the explicit solution of an integral equation of

Volterra type involving ${}_2F_1(a,b;c;z)$ in the kernel, in terms of integral operators with the generalized Mittag-Leffler type function and the generalized Wright function as a kernel.

Yu. Luchko and R. Gorenflo [30] used the operational method to prove that the Cauchy problem for the fractional differential equations with the Caputo fractional derivative (2.15)

$$(D_*^\alpha y)(x) - \lambda y(x) = f(x),$$
$$y^{(k)}(0) = b_k \ (k = 0, 1, \ldots n-1; \ n-1 < \alpha \leq n) \tag{3.5}$$

has the unique solution

$$y(x) = \sum_{k=0}^{n-1} b_k x^k E_{\alpha,k+1}(\lambda x^\alpha) + \int_0^x t^{\alpha-1} E_{\alpha,\alpha}(\lambda t^\alpha) f(x-t)dt$$

in a special space of functions. They also investigated the Cauchy problem for more general fractional differential equation

$$(D_*^\alpha y)(x) - \sum_{k=1}^{m} c_k (D_*^{\alpha_k} y)(x) = f(x) \ (\alpha > \alpha_1 > \cdots > \alpha_m \geq 0)$$

with initial conditions (3.5) and constructed its explicit solution via multivariate Mittag-Leffler function.

The above results were presented in a survey paper by Yu. Luchko [29].

We also note that D. Elizarrazaz and L. Verde-Star [10] obtained the explicit general solution of equation (3.2) and the explicit solution of Cauchy-type problem (3.2)-(3.3) by using linear algebra construction and classical methods of operational calculus. Their approach was based on introducing divided differences of fractional order, coinciding with the Riemann-Liouville fractional differential operators in a certain space of functions, and generalized exponential polynomials, which are connected with functions of Mittag-Leffler type.

4. Compositional method

The idea of the composition method is based on the known formulas for the Riemann-Liouville fractional integral (1.3) and derivative (1.4)

$$(I_{a+}^\alpha (t-a)^{\beta-1})(x) = \frac{\Gamma(\beta)}{\Gamma(\beta+\alpha)}(x-a)^{\beta+\alpha-1} \ (\mathrm{Re}(\alpha) > 0, \ \mathrm{Re}(\beta) > 0), \tag{4.1}$$

$$(D_{a+}^\alpha (t-a)^{\beta-1})(x) = \frac{\Gamma(\beta)}{\Gamma(\beta-\alpha)}(x-a)^{\beta-\alpha-1} \ (\mathrm{Re}(\beta) > \mathrm{Re}(\alpha) > 0) \tag{4.2}$$

(see S. Samko, A. Kilbas and O. Marichev [44, (2.44), (2.26) and (2.35)]). According to (1.3) and (1.4), formulas (4.1) and (4.2) mean that the compositions of the Riemann-Liouville fractional integral I_{a+}^α and fractional derivative D_{a+}^α

with the power function $(x-a)^{\beta-1}$ lead to the same function except for a certain constant factor. It means that $\varphi(x)=(x-a)^{\beta-1}$ is a solution of the homogeneous fractional integral equation

$$\varphi(x)=\frac{\Gamma(\alpha+\beta)(x-a)^{\alpha}}{\Gamma(\alpha)\Gamma(\beta)}\int_a^x\frac{\varphi(t)dt}{(x-t)^{1-\alpha}}\ (\mathrm{Re}(\alpha)>0;\ \mathrm{Re}(\beta)>0)$$

and the fractional differential equation

$$(D_{a+}^{\alpha}\varphi(x))(x)=\frac{\Gamma(\beta)(x-a)^{-\alpha}}{\Gamma(\beta-\alpha)}\varphi(x)(x)\ (\mathrm{Re}(\beta)>\mathrm{Re}(\alpha)>0).$$

These arguments lead us to the conjecture that compositions of fractional derivatives and integrals with elementary functions can give exact solutions of differential and integral equations of fractional order. Moreover, we arrive at a possibility to extend such results to compositions of fractional calculus operators with special functions. It allows us to find exact solutions of new classes of integral and differential equations of fractional order. Here we describe some results in this direction.

The investigations by A. Kilbas and M. Saigo [24]-[26] and M. Saigo and A. Kilbas [42] are based on compositions of the Riemann-Liouville fractional integrals and derivatives (1.3) and (1.4) with the special entire function of the form

$$E_{\alpha,m,l}(z)=\sum_{k=0}^{\infty}c_kz^k \tag{4.3}$$

with

$$c_0=1,\ c_k=\prod_{i=0}^{k-1}\frac{\Gamma[\alpha(im+l)+1]}{\Gamma[\alpha(im+l+1)+1]}\ (k=1,2,\ldots). \tag{4.4}$$

Such a function was introduced by A. Kilbas and M. Saigo [23] while studying asymptotic properties for solutions of linear integral equations of Abel-Volterra type. When $m=1$, this function coincides with the Mittag-Leffer function $E_{\alpha,\alpha l+1}(z)$, defined in (2.10), up to a constant factor:

$$E_{\alpha,1,l}(z)=\Gamma(\alpha l+1)E_{\alpha,\alpha l+1}(z).$$

A. Kilbas and M. Saigo [24] proved the relation

$$\left(I_{0+}^{\alpha}\left[t^{\alpha l}E_{\alpha,m,l}(at^{\alpha m})\right]\right)(x)=\frac{1}{a}x^{\alpha(l-m+1)}\left[E_{\alpha,m,l}(ax^{\alpha m})-1\right] \tag{4.5}$$

and applied this formula to obtain the solution of the Abel integral equation with a quasipolynomial free term

$$\varphi(x)=\frac{ax^{\beta}}{\Gamma(\alpha)}\int_0^x\frac{\varphi(t)dt}{(x-t)^{1-\alpha}}+\sum_{k=0}^{N}f_kx^{\mu_k}(0<x<d\leq\infty), \tag{4.6}$$

where f_k and μ_k $(k = 0, 1, \ldots)$ are constants, the solution being obtained in terms of the generalized Mittag-Leffler function (4.3)-(4.4) in the form

$$\varphi(x) = \sum_{k=0}^{N} f_k x^{\mu_k} E_{\alpha,1+\beta/\alpha,\mu_k/\alpha}(a x^{\alpha m}).$$

A. Kilbas and M. Saigo [24] also established a relation similar to (4.5), which connected the Liouville fractional integral (2.13) with the function $E_{\alpha,m,l}(at^{-\alpha m})$, and applied this result to finding the explicit solution of the integral equation of form (4.6)

$$\varphi(x) = \frac{a x^{-\beta}}{\Gamma(\alpha)} \int_x^{\infty} \frac{\varphi(t)dt}{(t-x)^{1-\alpha}} + \sum_{k=0}^{N} f_k x^{-\mu_k} (0 \leq d < x < \infty)$$

in terms of the generalized Mittag-Leffler function (4.3)-(4.4).

A. Kilbas and M. Saigo [25] and M. Saigo and A. Kilbas [42] proved the relation similar to (4.5) for the Riemann-Liouville fractional derivative (1.4)

$$\left(D_{0+}^{\alpha}\left[t^{\alpha(l-m+1)} E_{\alpha,m,l}(at^{\alpha m})\right]\right)(x) = = \frac{\Gamma[\alpha(l-m+1)+1]}{\Gamma[\alpha(l-m)+1]} x^{\alpha(l-m)} + a x^{\alpha l} E_{\alpha,m,l}(a x^{\alpha m}), \tag{4.7}$$

provided that $l > m - 1 - 1/\alpha$. In particular,

$$\left(D_{0+}^{\alpha}\left[t^{\alpha(l-m+1)} E_{\alpha,m,l}(at^{\alpha m})\right]\right)(x) = a x^{\alpha l} E_{\alpha,m,l}(a x^{\alpha m}), \tag{4.8}$$

when $\alpha(l-m) = -j$ for some $j = 1, 2, \ldots, -[-\alpha]$. Based on (4.8), they showed that the homogeneous fractional differential equation

$$(D_{0+}^{\alpha} y)(x) = a x^{\beta} y(x) \ (0 < x < d < \infty;\ \alpha > 0,\ a \neq 0,\ \beta > -\alpha) \tag{4.9}$$

with $\alpha \neq 1, 2, \ldots$ has $[\alpha] + 1$ solutions

$$y_j(x) = x^{\alpha-j} E_{\alpha,1+\beta/\alpha,1+(\beta-j)/\alpha}\left(a x^{\alpha+\beta}\right) \ (j = 1, 2, \ldots, [\alpha]+1).$$

They proved that these solutions are linearly independent for $\beta > \{\alpha\} \equiv \alpha - [\alpha]$, and obtained the solution of the corresponding Cauchy-type problem for (4.9) with the initial conditions

$$(D_{0+}^{\alpha-k} y)(x)|_{x=0} = b_k \ (k = 1, 2, \ldots, [\alpha]+1), \tag{4.10}$$

in the form

$$y(x) = \sum_{j=0}^{[\alpha]+1} b_j x^{\alpha-j} E_{\alpha,1+\beta/\alpha,1+(\beta-j)/\alpha}\left(a x^{\alpha+\beta}\right).$$

The result in (4.7) was also applied by A. Kilbas and M. Saigo [25] and M. Saigo and A. Kilbas [42] to obtain a particular solution $y_0(x)$ of the nonhomogeneous fractional differential equation with a quasipolynomial free term

$$(D_{0+}^{\alpha}y)(x) = ax^{\beta}y(x) + \sum_{i=0}^{n} f_i x^{\mu_i} \ (0 < x < d < \infty;\ \alpha > 0,\ a \neq 0,\ \beta \in \mathbf{R}), \quad (4.11)$$

where $\mu_i,\ f_i \in \mathbf{R}\ (i = 0, 1, \ldots, n)$, and the solution of the corresponding Cauchy-type problem for equation (4.11) with initial conditions (4.10). In particular, the exact solutions for the fractional differential equations of order 1/2, arising in the theory of voltametry at expanding electrodes (K. Oldham and J. Spanier [38, equation (8.6.1)]) and in the theory of polarography (K. Wiener [46]) were constructed.

A. Kilbas and M. Saigo [25] and M. Saigo and A. Kilbas [42] also considered a connection of the generalized Mittag-Leffler function (4.3)-(4.4) with the right-hand sided Liouville fractional derivative $D_{-}^{\alpha}y$ defined for $x > 0$ by

$$(D_{-}^{\alpha}y)(x) = \left(-\frac{d}{dx}\right)^{n} (I_{-}^{n-\alpha}y)(x),\ n = [\mathrm{Re}(\alpha)] + 1, \quad (4.12)$$

where $(I_{-}^{\alpha}y)(x)$ is the right-hand sided Liouville fractional integral (2.13). For such a fractional derivative they proved a relation similar to (4.7) and applied it to obtain a particular solution of the nonhomogeneous differential equation of form (4.11) with D_{0+}^{α} replaced by D_{-}^{α}, in terms of the generalized Mittag-Leffler type functions (4.3)-(4.4).

When $\alpha = n \in \mathbf{N}$ is an integer, the Liouville fractional derivative D_{0+}^{α} and D_{-}^{α} coincide with usual derivatives

$$(D_{0+}^{n}y)(x) = y^{(n)}(x),\ (D_{-}^{n}y)(x) = (-1)^{n}y^{(n)}(x).$$

The compositional relation (4.7) for D_{0+}^{α} and a similar formula for D_{-}^{α} remain valid for $\alpha = n \in \mathbf{N}$. The proofs of such relations were given by M. Saigo and A. Kilbas [43] with applications to the closed form solution of some classes of ordinary differential equations.

A. Kilbas, B. Bonilla, J. Rodriguez, J. Trujillo and M. Rivero [22] and B. Bonilla, A. Kilbas, M. Rivero, L. Rodriguez and J. Trujillo [3] studied compositions of the right-hand sided Liouville fractional integrals and derivatives (2.13) and (4.12) with the special functions

$$Z_{\rho}^{\nu}(z) = \int_{0}^{\infty} t^{\nu-1} \exp\left(-t^{\rho} - \frac{z}{t}\right) dt\ (z \in \mathbf{C},\ \mathrm{Re}(z) > 0;\ \rho > 0,\ \nu \in \mathbf{C}) \quad (4.13)$$

and

$$\lambda_{\gamma,\sigma}^{(\beta)}(z) = \frac{\beta}{\Gamma(\gamma + 1 - 1/\beta)} \int_{1}^{\infty} (t^{\beta} - 1)^{\gamma - 1/\beta} t^{\sigma} e^{-zt} dt\ (z \in \mathbf{C},\ \mathrm{Re}(z) > 0)\ (4.14)$$

$$(\beta > 0;\ \gamma \in \mathbf{C}, \mathrm{Re}(\gamma) > \tfrac{1}{\beta} - 1;\ \sigma \in \mathbf{R}).$$

These functions are modifications of Mcdonald function $K_\nu(z)$, analytic in z. Their important property is the fact that they are preserved by Liouville fractional derivatives and integrals (2.13) and (4.12) up to change of the indices:

$$(I_-^\alpha Z_\rho^\nu)(x) = Z_\rho^{\nu+\alpha}(x),\ (D_-^\alpha Z_\rho^\nu)(x) = Z_\rho^{\nu-\alpha}(x)$$

and

$$(I_-^\alpha \lambda_{\gamma,\sigma}^{(\beta)})(x) = \lambda_{\gamma,\sigma-\alpha}^{(\beta)}(x),\ (D_-^\alpha \lambda_{\gamma,\sigma}^{(\beta)})(x) = \lambda_{\gamma,\sigma+\alpha}^{(\beta)}(x).$$

These relations were applied in the above papers by A. Kilbas, B. Bonilla, J. Rodriguez, J. Trujillo and M. Rivero [22] and B. Bonilla, A. Kilbas, M. Rivero, L. Rodriguez and J. Trujillo [3] to obtain explicit solutions of certain Abel-Volterra integral equations of the third kind and fractional order differential equations in terms of functions (4.13) and (4.14).

Finally we mention one more statement on solvability of a fractional differential equations with the Liouville fractional derivative (4.12): the homogeneous equation

$$(D_-^\alpha y)(x) = ay(x)\ (0 < x < \infty,\ \alpha > 0,\ a > 0)$$

has the solution $y(x) = \exp\left(-a^{1/\alpha}x\right)$, see e.g. S. Samko, A. Kilbas and O. Marichev [44, Section 42.2], which is an obvious consequence of the relation

$$\left(D_-^\alpha (e^{-at})\right)(x) = a^\alpha e^{-ax}\ (a > 0). \tag{4.15}$$

5. Problems and new trends of research

In Sections 2-4 we presented a unified approach to both integral and differential equations of fractional order which allowed us to construct their solutions in closed form. We discussed some methods based on the Laplace transform, operational calculus and compositional relations, and formulated results which in main were not reflected in monographs. Now we characterize some problems related with those methods which can lead to new possible trends of research.

The Laplace transform and operational calculus methods can be used in a more suitable way to solve linear fractional integral and differential equations with constant coefficients. The main problem arising concerns function spaces for solutions. As we saw in Section 2, the formal procedure to derive a solution to nonhomogeneous linear fractional integral and differential equation of form (2.1) via the Laplace transform method is based on the relation in (2.2) which holds for the fractional integrals I_{0+}^α and derivatives D_{0+}^α given on an interval $(0,d)$ $(0 < \infty)$ or on the half-axis $\mathbf{R}_+$. Since the Laplace transform $\mathcal{L}$ is defined on $\mathbf{R}_+$, we must be careful with its justification in the case of equation (2.1) given on $(0,d)$. When (2.1) is considered on $\mathbf{R}_+$, we also need to investigate function spaces related to the Laplace transform $\mathcal{L}$ and its inversion $\mathcal{L}^{-1}$. Another problem is a representation of explicit solution (2.3) in a simpler form.

In some cases this solution can be represented via the fractional analog of the Green function G_α as a convolution of form (2.9).

We also note that the operator in the right-hand side of (2.3) is a special case of a more general composition $\mathcal{L}^{-1}a\mathcal{L}$ with a given function a. A development of the theory of such Laplace-type pseudo-differential operators, could clarify some problems in the theory of fractional integral and differential equations.

Section 3 shows that the operational calculus method allows us to solve nonhomogeneous linear fractional integral and differential equations with constant coefficients more general than (2.1). But their solutions are usually looked for in certain rather complicated function spaces. Of interest could be a justification of the procedure of their solution in simpler function spaces. Another problem of interest is studying properties of special functions of Mittag-Leffler type which generalize the Mittag-Leffler function given in (2.10).

The Laplace transform and operational calculus methods allow us to find closed form solutions for linear fractional integral and differential equations with constant coefficients. It is necessary to develop other methods to obtain explicit solutions for such equations with variable coefficients. One of these approaches presented in Section 4 and based on compositions of fractional integral and differential operators with elementary and special functions, provides a possibility to solve new classes of fractional order linear integral and differential equations with nonconstant coefficients. The results given in Section 4 were obtained by means of compositions of fractional integral and differential operators with special functions of Mittag-Leffler and Bessel type. We expect that it is possible to find new compositional formulas relating fractional operators to various special functions which will lead to explicit solutions of new classes of differential equations.

Solutions of homogeneous fractional differential equations can be obtained by above methods. As it was mentioned in Section 2, solutions of some linear equations with constant coefficients can be found in terms of the roots of the characteristic quasipolynomials. Relation (4.15) shows that in this way we can obtain explicit solutions of linear fractional differential equations of form (2.11) with the Riemann-Liouville fractional derivatives $D_{0+}^{\alpha_k}$ replaced by the right-hand sided Liouville fractional derivatives $D_{-}^{\alpha_k}$ as defined in (4.12). Such explicit solutions can be also proved for differential equations of form (2.11) with fractional derivatives $D_{0+}^{\alpha_k}$ replaced by the left-hand sided Liouville fractional derivatives $D_{+}^{\alpha_k}$ (S. Samko, A. Kilbas and O. Marichev [44, Section 5.1]).

Similarly to ordinary differential equations, one could expect that it is possible to obtain explicit solutions of nonhomogeneous linear fractional differential equations knowing solutions to the corresponding homogeneous equations.

An approach based on series expansions of known and unknown functions, can be also used to find explicit solutions of homogeneous and nonhomogeneous integral and differential equations of fractional order. Such an approach is closely connected with asymptotic behavior of solutions of fractional integral and differential equations near certain points. A development of asymptotic methods for such fractional equations is important in connection with applied problems. Asymptotic estimates for their solutions can be obtained from the explicit solution, but only a few fractional integral and differential equations can be solved in closed form. Therefore, a development of asymptotic methods in the theory of general fractional integral and differential equations has to be based on other approach.

Some results for the Abel-Volterra integral equations were obtained in this direction. A. Kilbas and M. Saigo [23] studied asymptotic behavior near $x = 0$ of the solution $\varphi(x)$ of the linear integral equation of form (2.4) with λ replaced by $a(x)$, in the case $0 < \alpha < 1$, provided that $a(x)$ and $f(x)$ have a special power asymptotic behavior at zero. They showed that solution of the equation under the consideration has the asymptotic expansion of the same form as $f(x)$ and considered the cases where the asymptotic expansions coincide with the exact solutions.

A. Kilbas and M. Saigo [26] considered a nonlinear Abel-Volterra integral equation of the form

$$\varphi^m(x) + \frac{a(x)}{\Gamma(\alpha)} \int_a^x \frac{\varphi(t)dt}{(x-t)^{1-\alpha}} = f(x) \ (0 < x < d \leq \infty) \tag{5.1}$$

where $\alpha > 0$ and $m \in \mathbf{R}$ $(m \neq 0, -1, -2, \ldots)$. Under some conditions on $a(x)$ and $f(x)$, including asymptotic estimates near the origin, they studied the asymptotic solvability and asymptotic behavior of the solution $\varphi(x)$ of equation (5.1) near the origin.

A. Kilbas and M. Saigo [26] also gave conditions under which equation (5.1) with $a(x) = ax^\beta$ and $fx) = bx^\gamma$ can be solved in closed form, and found its exact solution. The problem of uniqueness of solution of equations (5.1) was also discussed. Solvability of this nonlinear equation depends on that of some transcendental equations. The positive solutions of such transcendental equations were investigated by N. Karapetyants, A. Kilbas and M. Saigo [20] where a nonlinear integral equation, more general than (5.1), was investigated, see also N. Karapetyants, A. Kllbas, M. Saigo and S.G. Samko [21] in this connection.

As for asymptotic behavior of solutions of fractional order differential equations, we are aware only a result by M. Fujiwara [12] who considered the differential equation of fractional order $(\mathcal{D}_+^\alpha y)(x) = (\alpha/x)^\alpha y(x)$ with the Hadamard fractional derivative of order $\alpha > 0$ (see S. Samko, A. Kilbas and O.I. Marichev [44, (18.54)] on Hadamard fractional derivatives). Making use of the explicit solution to the above equation in terms of the Mellin-Barnes integral (A. Erdelyi, etc. [11, Section 1.19]), M. Fujiwara proved that if $y(x)$ has the power asymptotics at the origin of the form $y(x) \sim Ax^\lambda e^{-\mu/x}$, then $A = (2\pi)^{(\alpha-1)/2}/\sqrt{\alpha}$, $\lambda = (\alpha - 1)/2$ and $\mu = \alpha$. Probably, some approach to develop the asymptotic theory for solutions of fractional differential equations following Fujiwara can be based on asymptotic properties of fractional derivatives, while the other method can use asymptotic estimates of solutions for the corresponding Volterra integral equations.

References

[1] Al-Bassam, M. A. and Luchko, Yu. F., On generalized fractional calculus and its application to the solution of integro-differential equations, *J. Fract. Calc.*, **7** (1995), 69–88.

[2] Alonso, J., On differential equations of fractional order, *Diss. Abstrs.*, **26** (1964), no. 4, 947.

[3] Bonilla, B., Kilbas, A. A., Rivero, M., Rodriguez, L., and Trujillo, J. J., Modified Bessel-type function and solution of differential and integral equations, *Indian J. Pure Appl. Math.*, **31** (2000), no. 1, 93–109.

[4] Campos, L. M. B. C., On the solution of some simple fractional differential equations, *Internat. J. Math. Math. Sci.*, **13** (1990), no. 3, 481–496.

[5] Dimovski, I. H., Operational calculus for a class of differential operators, *C.R. Acad. Bulgar. Sci.*, **19** (1966), no. 12, 1111–1114.

[6] Ditkin, V. A., The theory of operator calculus, *Dokl. Akad. Nauk SSSR (N.S)*, **116(1)** (1957), 15–17 (in Russian).

[7] Ditkin, V. A. and Prudnikov, A. P., On the theory of operational calculus stemming from the Bessel equation, *USSR Comput. Math. Math. Phys.*, **3** (1963), 296–315.

[8] Ditkin, V. A. and Prudnikov, A. P., *Integral Transforms and Operational Calculus.* Pergamon Press, Oxford, 1965.

[9] Dzhrbashyan, M. M., *Harmonic Analysis and Boundary Value Problems in the Complex Domain*, Operator Theory: Advances and Applications, **65**. Birkhäuser Verlag, Basel, 1993.

[10] Elizarraraz, D. and Verde-Star, L., Fractional divided differences and the solution of differential equations of fractional order, *Adv. in Appl. Math.*, **24** (2000), no. 3, 260–283.

[11] Erdelyi, A., Magnus, W., Oberhettinger, F., and Tricomi, F. G., *Higher Transcendental Functions.* McGraw-Hill, New York, Toronto, London, 1953 (Vol. I), 1954 (Vol. II), 1955 (Vol. III).

[12] Fujiwara, M., On the integration and differentiation of an arbitrary order, *Tôhoku Math. J.*, **37** (1933), 110–121.

[13] Gorenflo, R. and Luchko, Yu. F., Operational method for solving generalized Abel integral equations of second kind, *Integral Transform. Spec. Funct.*, **5** (1997), no. 1-2, 47–58.

[14] Gorenflo, R., Luchko, Yu. F., and Srivastava, H. M., Operational method for solving integral equations with Gauss hypergeometric function as a kernel, *Int. J. Math. Stat. Sci.*, **6** (1997), no. 2, 179–200.

[15] Gorenflo, R. and Mainardi, F., Fractional oscillations and Mittag-Leffler functions, *Fachbereich Math. und Inf., Freie Univ. Berlin*, A14 (1996), 1–22.

[16] Gorenflo, R. and Mainardi, F., Fractional calculus and stable probability distributions. In: *Fourth Meeting on Current Ideas in Mechanics and Related Fields* (Krakow, 1997), *Arc. Mech. (Arch. Mech. Stos.)*, **50**, no. 3, 377–388, 1998.

[17] Gorenflo, R. and Vessela, S., *Abel Integral Equations. Analysis and Applications*, Lecture Notes in Mathematics, **1461**. Springer-Verlag, Berlin, 1991.

[18] Hadid, S. B. and Luchko, Yu. F., An operational method for solving fractional differential equations of an arbitrary real order, *Panamer. Math. J.*, **6** (1996), no. 1, 57–73.

[19] Hille, E. and Tamarkin, J. D., On the theory of linear integral equations, *Ann. Math.*, **31** (1930), 479–528.

[20] Karapetyants, N. K., Kilbas, A. A., and Saigo, M., On the solution of nonlinear Volterra convolution equation with power nonlinearity, *J. Integral Equations Appl.*, **8** (1996), no. 4, 429–445.

[21] Karapetyants, N. K., Kilbas, A. A., Saigo, M., and Samko, S. G., Upper and lower bounds for solution of nonlinear Volterra convolution equation with power nonlinearity, *J. Integral Equations Appl.*, **12** (2000), no. 4, 421–448.

[22] Kilbas, A. A., Bonilla, B., Rodriguez, J., Trujillo, J. J., and Rivero, M., Compositions of fractional integrals and derivatives with a Bessel-type function and the solution of differential and integral equations, *Dokl. Nats. Akad. Nauk Belarusi*, **42** (1998), no. 2, 25–29 (in Russian).

[23] Kilbas, A. A. and Saigo, M., On solution of integral equation of Abel-Volterra type, *Differential Integral Equations*, **8** (1995), no. 5, 993–1011.

[24] Kilbas, A. A. and Saigo, M., On Mittag-Leffler type function, fractional calculus operators and solutions of integral equations, *Integral Transform. Spec. Funct.*, **4** (1996), no. 4, 355–370.

[25] Kilbas, A. A. and Saigo, M., Solution in closed form of a class of linear differential equations of fractional order, *Differential Equations*, **33** (1997), no. 2, 194–204.

[26] Kilbas, A. A. and Saigo, M., On solution of a nonlinear Abel-Volterra integral equation, *J. Math. Anal. Appl.*, **229** (1999), no. 1, 41–60.

[27] Kilbas, A. A. and Trujillo, J. J., Differential equations of fractional order: methods, results and problems, I, *Applicable Analysis*, **78** (2001), no. 1-2, 153–192.

[28] Leskovskii, I. P., On the solution of the linear homogeneous differential equations with fractional derivatives and constant coefficients. In: *Some Questions of Diff. Equat. in the Solution of Appl. Problems*, 85–88. Tul'sk Politechn. Inst., Tula, 1980.

[29] Luchko, Yu. F., Operational method in fractional calculus, *Frac. Calc. Appl. Anal.*, **4** (1999), no. 2, 463–488.

[30] Luchko, Yu. F. and Gorenflo, R., An operational method for solving fractional differential equations with the Caputo derivatives, *Acta Math. Vietnam*, **24** (1999), no. 2, 207–233.

[31] Luchko, Yu. F. and Srivastava, H. M., The exact solution of certain differential equations of fractional order by using operational calculus, *Comput. Math. Appl.*, **29** (1995), no. 8, 73–85.

[32] Luchko, Yu. F. and Yakubovich, S. B., Operational calculus for the generalized fractional differential operator and applications, *Math. Balkanica (N.S.)*, **7** (1993), no. 2, 119–130.

[33] Luchko, Yu. F. and Yakubovich, S. B., An operational method for solving some classes of integro-differential equations, *Differential Equations*, **30** (1994), no. 2, 247–256.

[34] Maravall, D., Linear differential equations of non-integer order and fractional oscilations, *Rev. Ac. Ci. Madrid*, **65** (1971), 245–258 (in Spanish).

[35] Meller, N. A., Some aplications of operational calculus to problems of analysis, *J. Vychisl. Mat. i Mat. Fiz.*, **3** (1963), no. 1, 71–78 (in Russian).

[36] Mikusinski, J., *Operational Calculus*. Pergamon Press, New York, 1959.

[37] Miller, K. S. and Ross, B., *An Introduction to the Fractional Calculus and Fractional Differential Equations*. John Wiley and Sons, New York, 1993.

[38] Oldham, K. B. and Spanier, J., *The Fractional Calculus*. Academic Press, New York, London, 1974.

[39] Podlubny, I., *Fractional Differential Equations*, Mathematics in Sciences and Engineering, **198**. Academic Press, San-Diego, CA, 1999.

[40] Rodriguez, J., Operational calculus for the generalized Bessel operator, *Serdica*, **15** (1989), no. 3, 179–186.

[41] Rodríguez, J., Trujillo, J. J., and Rivero, M., Operational fractional calculus of Kratzel integral transformation. In: *Differential equations* (Xanthi, 1987), Lecture Notes in Pure and Appl. Math., **118**, 613–620. Dekker, New York, 1989.

[42] Saigo, M. and Kilbas, A. A., On Mittag-Leffler type function and applications, *Integral Transform. Spec. Funct.*, **7** (1998), no. 1-2, 97–112.

[43] Saigo, M. and Kilbas, A. A., Solution of a class of linear differential equations in terms of Mittag-Leffler type functions, *Differential Equations*, **36** (2000), no. 2, 193–200.

[44] Samko, S. G., Kilbas, A. A., and Marichev, O. I., *Fractional Integrals and Derivatives. Theory and Applications.* Gordon and Breach Science Publishers, Yverdon, 1993.

[45] Seitkazieva, D., Integro-differential equations with a kernel that has a weak singularity. In: *Studies in integro-differential equations*, **13**, 132–135. Ilim, Frunze, (in Russian).

[46] Wiener, K., On solutions of a differential equation of nonintegral order that occurs in the theory of polarography, *Wiss. Z. Martin-Luther-Univ. Halle-Wittenberg Math.-Natur. Reiche*, **32** (1983), no. 1, 41–46 (in German).

FRACTIONAL DIFFERENTIAL EQUATIONS: A EMERGENT FIELD IN APPLIED AND MATHEMATICAL SCIENCES

Anatoly A. Kilbas
Belarusian State University, Belarus
kilbas@bsu.by

Hari M. Srivastava
University of Victoria, Canada
harimsri@math.uvic.ca

Juan J. Trujillo
University of La Laguna, Spain
juan.trujillo@ull.es

To Professor G. S. Litvinchuk on the occasion of his 70th birthday

Abstract The paper is devoted to some aspects of differential equations of fractional order and their applications. It is explained a fact that the subject of fractional differential equations is an emergent topic as a very useful tool to model many anomalous phenomena in nature and in the theory of complexity systems. Various fractional integral and fractional derivative are presented together with some of their properties. Simple partial differential equations are discussed in connection with anomalous diffusion. A new model, useful in both sub-diffusion and super-fast diffusion processes, is introduced. Such a model, generalizing the clasical problem associated with the heat equation, involves the generalized Liouville fractional derivative over the time variable. The one-dimensional case is studied and explicitely solved, and its generalization to the multi-dimensional case is discussed.

Keywords: Fractional integrals and derivatives, fractional differential equations, fractional diffusion equations, subdiffusion and superdiffusion.

1. Introduction

Questions as to what we mean by, and where we could apply, the fractional calculus operators have fascinated us all ever since 1695 when the so-called fractional calculus was conceptualized in connection with the infinitesimal calculus [117, 101]. We should like to recall the famous correspondence between Leibniz and L'Hôspital. Many great mathematicians have been interested in these questions during last 170 years, and some remarkable applications of fractional calculus have emerged. A comprehensive presentation of fractional calculus can be found in [122] and some of its applications in [101, 112].

The interest in the study of the fractional differential equations (FDE) as a separate topic arose some 40 year ago. Questions about the existence of solutions of Cauchy-type problems involving the Riemann-Liouville operators D^{α}_{a+} ($\alpha \in \mathbb{C}$ and $a \in \mathbb{R}$) and some other problems have attracted the attention of several mathematicians. A systematic and rigourous study of some problems of this kind involving FDE can be found in the recent literature; see [57, 58] for a review.

Although numerous applications of the fractional calculus operators have been found during its long history, many mathematicians and applied researchers have tried to get models of real processes using FDE. However, except possibly in a handful of applications, the literature on fractional calculus does not seem to contain results of far-reaching consequences. This may be due, in part, to the fact that many of the useful properties of the ordinary derivative D are not known to carry over analogously to the case of the fractional derivative operator D^{α}. For example, a clear geometric or physical meaning, product rules, chain rules, and so on.

But then, what are the useful (and good) properties of these fractional calculus operators to help in the models of so many anomalous processes? From the point of view of the authors and from the known experimental results, most of the processes associated with complexity systems have a non-local dynamics involving long-memory in time, and the fractional integral and fractional derivative operators do have some of those characteristics. Perhaps this is one of the reasons why these fractional calculus operators lose the above-mentioned useful properties of the ordinary derivative D.

It is known that the classical calculus provides a power tool to explain and to model many important dynamical processes in most parts of applied areas of the sciences. But the experiments and reality teach us that there are many complexity systems in nature with anomalous dynamics, including biological and social sciences, transport of chemical contaminant through water around rocks, dynamics of viscoelastic materials as polymers, diffusion of pollution in the atmosphere, diffusion processes involving cells, traffic in a net, transmission of signals through strong magnetic fields and the effect of speculation on the profitability of stocks in the stock markets, and many more.

In most of the above-mentioned cases, this kind of anomalous processes have a macroscopic complex behaviour, and their dynamics can not be characterized by classical derivative models. Nevertheless, the heuristic explanation of the corresponding models of some of those processes frequently are not so complicated using tools from statistical physics. For such an explanation, one must use some generalized concepts from classical physics such as fractional Brownian motion, continuous time random walk (CTRW) method involving Lévi stable distributions (instead of Gaussian distributions), the generalized central limit theorem (instead of the classical limit theorem), and non-Markovian distributions which means non-local distributions (instead of the classical Markovian ones). From this approach it is also important to remark that the anomalous behaviour of many complex processes include the multi-scaling in the time and space variables and so also the fractal characteristics of the media.

The above-mentioned tools have been used extensively during last 30 years. But the connections of these statistical models with some FDE, involving the fractional integral and derivative operators (Riemann-Liouville, Caputo, Liouville or Weyl and Riesz) have been formally established during the last 10 years and more intensively during the last 5 years, as we will show later. The mentioned connections can be obtained in several ways. One of the more often used approaches is based on generalization of the propagator of the model, by the use of Fourier and/or Laplace transforms. In this approach the characteristic function of the corresponding probability density function (pdf), containing the exponential function e^x in the classical case, is replaced by the Mittag-Leffler, Wright or Fox functions which satisfy some of the main properties of e^x. Using the above idea, many researchers have arrived at nice and useful fractional dynamical models involving fractional derivative operators in both time and space variables, which can assume the long-memory and non-local dependence of the anomalous processes [13, 12, 78, 90, 67, 77, 21, 157]. At present, the more commonly studied models are connected with the anomalous diffusion processes; in this way there have been obtained the fractional versions of the diffusion and advection-diffusion equations as well as equations of Fokker-Planck, Kramer and some others; see [90] for a review.

Here we should remark that the above-mentioned fractional integral and fractional derivative operators have permitted us to model some special situations. But there are many other types of fractional calculus operators, fractional pseudo-differential operators and perhaps other nice singular operators, that could help us to remodel some of the anomalous processes. In this connection, we shall use a generalized Liouville operator to obtain a new model for super-diffusion processes where the fractional derivative operator only works over the time variable.

We may conclude from the foregoing presentation that it is important to work out a rigourous theory of FDE analogous to that for ordinary differential equations (ODE). In fact, there are a lot of similar and new open questions in this subject, answer to which can help in the development of the applied sciences. Let us mention here, as an illustrative historical example, that the delta function, introduced by Dirac at the end of 19th century, has been used by the physicists in many of their ordinary derivative models formally for a long time. And their investigations became rigourous only about 50 years later after using distribution theory presented by Laurent Schwartz in 1944; for example, see [128].

This paper is organized as follows. The first part given in Section 2 is dedicated to the explanation of the fact that FDE is an emergent topic in the development of modelling of complexity systems. The second part includes mathematical justification of new model for super-diffusion processes in terms of Liouville-type fractional derivatives in the time variable. First, in Section 3, some properties of several fractional integral and fractional derivative operators are presented. Next, in Section 4, examples of simple partial FDE with boundary conditions are discussed in connection with anomalous diffusion equations. Here a new model, which is useful in the simulation of both sub-diffusion and super-fast diffusion processes, is introduced. Such a model involves a generalized Liouville fractional derivative operator in the time variable. In this connection, we note also that some earlier authors have studied only the models of super-diffusion processes involving the Riesz fractional derivative in the space variables.

2. The Complexity Systems

The asymptotic approach of physical systems to equilibrium is captured by the second law of thermodynamics, which states that the entropy of an isolated system either remains fixed or increases over time. All physical states in equilibrium are statistically equivalent and this is the end point of all physical activity. There are many systems in the nature which apparently contradict the second law of thermodynamics. This apparent conflict with physical law arises, mainly, because there are no systems which are completely isolated [154, 8, 15, 155, 137, 157].

Schrödinger has discussed the idea that macroscopic laws, are the consequences of the interaction of a large number of particles. Therefore, such laws can only be described by means of statistical physics. For this reason, the Gaussian or normal distribution of probability, the Brownian motion and the classical central limit theorem are important and useful. These tools have permitted important advances to explain many natural processes stemming from the classical physical models which are based upon the use of the infinitesimal calculus and involve the ordinary and local derivative operator D. One relevant example of such utility is the classical diffusion and Fokker-Planck equations [12, 90, 10].

The complexity systems are very familiar and occur often in nature. Examples of this kind of dynamical processes appear in all sciences. In most of such systems, the classical models and analytic functions are often not sufficient to describe their basic features. One of the familiar cases is the so-called Brownian motion, which describes certain erratic way through space of a particle.

This Brownian motion in nature could be modelled as a random walk on a two-dimensional plane, where the size of the jumps and the waiting time before each jump are random and not uniform. This is the background of the so-called "continuous time random walk" (CTRW) method. Such a method is the basis for a heuristic explanation for the physical behaviour of the normal and anomalous diffusion processes.

For example, we can see macroscopic dynamical processes of a normal diffusion that are modelled by a diffusion equation or Fokker-Planck equation, with fixed parameters, by using the CTRW-approach. That is, we can see the above processes as the mean line of the CTRW-method applied to a set of microscopic particles whose motions are random in the time-waiting and in the space-jump, and these random motions are governed by characteristic distributions. When both random variables are independent and their pdf's are Gaussian and Markovian (local process in the sense that the entire past of the system does not influence on the future and only the present determines the immediate future), we can apply the classical central limit theorem. But when, for instance, the waiting-time follows Lévi stable distribution instead of the Gaussian one, by using the CTRW-method and the generalized central limit theorem, we can describe more complex processes such as sub-diffusion and super-diffusion.

The CTRW-method can be characterized by the moments of the random mean motion. If the process is non-local or with memory waiting-time, then the second moment of the random variable X of the jumps $< X^2 >$, is proportional to a power t^α of order α of the time when the time is sufficiently large. In this way, we can characterize the sub-diffusion, normal-diffusion, and super-diffusion processes in the cases when $0 < \alpha < 1$, $\alpha = 1$, and $\alpha > 1$, respectively.

On the other hand, when the random variables of the jumps in the space do not follow the Gaussian distribution, we have a relation with the so-called fractal media behaviour of many complex phenomena.

We would like to indicate that there are close relations between dynamics of many complexity systems including anomalous processes, CTRW-method, fractional Brownian motion, fractional diffusion, fractal media, and fractional differential equations.

The anomalous diffusion, perhaps, is the more frequently studied complex problem, until this moment. Richardson [113] was probably the first who considered the existence of such anomalous kinetics in connection with a turbulent diffusion in the atmosphere. Later on, other authors observed the same phenomena in other areas including chemical and plasma physics [148, 69]. The characterization of the Lévi stable distributions in terms of the Fourier transform of their probability density functions (pdf), the generalized central limit theorem and the fractional Brownian motion concept were introduced by Lévi [75, 76]. The last concept was developed by Mandelbrot and Van Ness [86]. Subsequently, Mandelbrot [84] found some connections between fractals and the geometry of complex systems in nature.

The CTRW-method was probably first introduced by Gnedenko and Kolmogorov in 1949 in the Russian version of their classical book [38]. Later on, such an approach was applied by Montroll and Weiss [94] to study the mechanical properties of lattices. We note that one can find good introduction to the above statistical tools in the book [30].

The book [101] was the first book dedicated specifically to the subject of fractional calculus and its applications. This book contains a very good history of the developments of fractional integration and fractional differentiation and their applications from 1695 to 1974 written by B. Ross, and also some of their applications to the chemical and physical sciences. In 1987 Samko, Kilbas and Marichev published in Russian [122], where one can find an encyclopedic, deep and rigourous study of the fractional calculus operators. The book [93] presents the first introduction to fractional differential equations (see also a more recent book [112]).

The first Ph.D. thesis devoted to applications of fractional differential equations to modelling the behaviour of viscoelastic materials was written by Bagley advised by Torvik [4], in the Air Force Institute of Technology and Material Laboratory of Ohio. After that many researchers paid attention to application of fractional differential equations to viscoelastic materials and other complex processes including anomalous diffusion phenomena.

We shall classify a short selection of relevant applied and review papers on the topic of this presentation, which will give to the reader an idea of the evolution of the mathematical modelling of some *Complexity Systems*. We begin from the articles published in the period 1970-1990. Perhaps Oldham and Spanier [100] (see also [101]) first introduced the fractional differential equation as a model for certain chemical processes. The *CTRW-method* was applied to some physical problems by [131], [134], [61], and [152]; the last paper contains a survey of the results in this approach. *Anomalous Processes in Amorphous and Disordered Media* were studied in [124], [109], and [110]. *Models of Viscoelastic Materials* in terms of fractional differential equations were discussed in [141], [5], [6], [62], and [7]. Problems involving *Kinetics of Anomalous Diffusion* were considered, by using random walk methods, in [45], [46], [59], [132], and [18]. A generalization of the entropy concept was presented in [144, 145], and this concept has a close connection with the behaviour of some complexity systems.

Fractional diffusion and fractional wave equations, generalizing the classical ones, were investigated in [126] and [33].

Now we continue a survey of publications over the topic of our paper from 1991. We present some articles devoted to anomalous models in different applied sciences. Some models in *Material Theory* were studied in [35], [80], [81], [138], [25], [130], [17], and [136]. *Transport Processes* were investigated in [55], [9], [16], [106], some *Fluid Flow Phenomena* were presented in [82], [1], [24], [105], and [127], while *Wave Propagation* was considered in [41], [123], and [44]. Some models of *Biology* were investigated in [37], [47], [3], [48], [143], [73], [149], [49], and [150]; of *Electromagnetic Theory* in [31], [26], [27], [28]; of *Thermodynamics* in [23], [88] and [139]; of *Mechanics* in [115], [118], and [111]; of *Geology* and *Astrophysics* in [79], [70], and [21]; of *Economics* in [8], [78] and [157], of *Chemistry* in [142] and of *Signal Theory* in [95]. Anomalous models connected with *Chaos and Fractals* were studied in [159], [102], [161], [56], [71], [72], [103], and [146].

The models of some *anomalous processes in terms of fractional differential equations* were discussed in [34], [32], [36], [89], [22], [119], [13], [39], [40], [116], [11], [12], [91], [77], [140], [147], [2], [42], [43], [54], and [92].

We also indicate that one can find the applications of the described fractional calculus methods to modelling anomalous processes in the above and other sciences in the following books and survey papers [108], [158], [96], [151], [52], [87], [98], [133], [53], [68], [19], [20], [83], [13], [160], [85], [107], [129], [153], [14], [51], [90], [99], [104] and [50]. See also [125], [135] and [60], where one may find introduction to anomalous phenomena.

We have mentioned a series of papers and books devoted to the modelling of some complexity systems by using different tools including fractional differential equations. We choose only some of many publications in this field appeared basically last years. In Table I, we show the evolution of the number of these publications. We have used for this only those papers in scientific journals which are included in the lists of Science Citation Index and Social Science Citation Index of the Institute for Scientific Information (ISI) from 1945 to 2001.

We have not included here those papers that are devoted to the applications of the CTRW-method, although they have close relations with the fractional derivative models which we have considered in this paper.

We can conclude from this study that the number of publications, devoted to the modelling of anomalous processes, has considerably increased during the last 10 years. We also note that the use of the fractional derivative operators has an important role to obtain the corresponding ordinary derivative models of such processes. These models include non-local and long-memory properties and connections with the fractional Brownian motion and with the CTRW-method.

Table 1. **Evolution of the Number of Publications in Connection with Models of Complexity Systems Considered by ISI**

Descriptive Phrases	45-80	81-90	91-96	97-01	Total
Fractional Brownian Motion	3	6	158	323	490
Anomalous Diffusion	116	108	520	672	1416
Anomalous Relaxation	10	14	23	38	85
Super-Diffusion or Sub-Diffusion	0	18	30	82	130
Anomalous Dynamics or Anomalous Processes or Fractional Models or Fractional Relaxation or Fractional Kinetics	18	28	79	87	212
Fractional Diffusion Equation or Fractional Fokker-Planck Equation	0	2	19	61	82

We remark that, during the last few years, there was a special interest in the modelling of anomalous diffusion processes involving the ultra-slow diffusion. But the anomalous phenomena, such as super-diffusion, super-conductivity, wave propagation and others, have not been studied thoroughly.

3. Fractional Integral and Fractional Derivative Operators

In this section, we present some properties of the known fractional integral and fractional derivative operators [122]. Let f be a real function, $\alpha \in \mathbb{C}$ and $n \in \mathbb{N} = \{1, 2, 3, \ldots\}$. The classical Liouville (or Weyl) fractional integral $I_{-}^{\alpha} f$ of order α with $\Re(\alpha) > 0$ is defined by

$$(I_{-}^{\alpha} f)(x) = \frac{1}{\Gamma(\alpha)} \int_{x}^{\infty} \frac{f(t)dt}{(t-x)^{1-\alpha}} \quad (x > a \geqq -\infty), \tag{3.1}$$

where $\Gamma(\alpha)$ is the Euler Gamma-function. The corresponding fractional derivative $D_{-}^{\alpha} f$ of order α ($\Re(\alpha) \geqq 0,\ \alpha \neq 0$) has the form:

$$(D_{-}^{\alpha} f)(x) = (-D)^{n} (I_{-}^{n-\alpha} f)(x) \quad (x > a \geqq -\infty), \tag{3.2}$$

with $D^n = (d/dx)^n$ and $n = [\Re(\alpha)] + 1$, where $[\Re(\alpha)]$ means the integer part of $\Re(\alpha)$.

The Liouville fractional calculus operators I_{-}^{α} and D_{-}^{α} of exponential function yield the same exponential function, both apart from a constant multiplication factor:

Lemma 3.1 *Let* $\alpha, \lambda \in \mathbb{C}$ $(\Re(\alpha) > 0;\ \Re(\lambda) > 0)$. *Then*

$$(I_-^\alpha e^{-\lambda t})(x) = \lambda^{-\alpha} e^{-\lambda x} \quad \textit{and} \quad (D_-^\alpha e^{-\lambda t})(x) = \lambda^\alpha e^{-\lambda x}. \tag{3.3}$$

Now we present the so-called fractional integrals and fractional derivatives of a function f with respect to another function g; see Section 18.2 of the book [122]. Such operators play a main role in the new model of sub- and super-diffusion processes. We shall introduce these processes in the next section and hope that their rigourous study will be done in the near future.

Let $g(x)$ be an increasing and positive monotone function on (a, ∞), having continuous derivative $g'(x)$ on (a, ∞). We define the general fractional integral operator $I_{-;g}^\alpha f$ of a function f with respect to a function g, of order α $(\Re(\alpha) > 0)$ by

$$(I_{-;g}^\alpha f)(x) = \frac{1}{\Gamma(\alpha)} \int_x^\infty \frac{g'(t) f(t) dt}{[g(t) - g(x)]^{1-\alpha}} \quad (x > a \geqq -\infty). \tag{3.4}$$

If $g'(x) \neq 0$ $(a < x < \infty)$, then the operator $I_{-;g}^\alpha$ can be expressed *via* the fractional integral I_-^α by

$$I_{-;g}^\alpha f = Q_g I_-^\alpha Q_g^{-1} f, \tag{3.5}$$

where Q_g is the substitution operator $(Q_g f)(x) = f[g(x)]$ and Q_g^{-1} is its inverse operator.

When $g'(x) \neq 0$ $(a < x < b)$, the corresponding fractional derivative $D_{-;g}^\alpha f$ of a function f with respect to g of order α $(\Re(\alpha) \geqq 0;\ \alpha \neq 0)$ has the form:

$$(D_{-;g}^\alpha f)(x) = \left(-\frac{1}{g'(x)} D\right)^n (I_{-;g}^{n-\alpha} f)(x) =$$

$$= \frac{1}{\Gamma(n-\alpha)} \left(-\frac{1}{g'(x)} D\right)^n \int_x^\infty \frac{g'(t) f(t) dt}{[g(t) - g(x)]^{\alpha-n+1}} \quad (x > a \geqq -\infty), \tag{3.6}$$

where $D = d/dx$ and $n = [\Re(\alpha)] + 1$.

When $\alpha = n \in \mathbb{N}$, the Liouville derivative (3.2) and the generalized derivative (3.6) have the following forms:

$$(D_-^n f)(x) = (-1)^n f^{(n)}(x) \text{ and } (D_{-;g}^n f)(x) = \left(-\frac{1}{g'(x)} \frac{d}{dx}\right)^n f(x). \tag{3.7}$$

In particular, $(D_-^1 f)(x) = -f'(x)$ and $(D_{-;g}^1 f)(x) = -f'(x)/g'(x)$.

Similarly to Lemma 3.1, there exist the corresponding properties for the Liouville generalized operators (3.4) and (3.6) acting on the exponential function $e^{-\lambda g(x)}$. These properties will be important in the development of new dynamical models of super-diffusion and sub-diffusion.

Lemma 3.2 *Let* $\alpha \in \mathbb{C}$ $(\Re(\alpha) > 0)$ *and* $\lambda > 0$. *Then*

$$(I^{\alpha}_{-;g} e^{-\lambda g(t)})(x) = \lambda^{-\alpha} e^{-\lambda g(x)} \tag{3.8}$$

and

$$(D^{\alpha}_{-;g} e^{-\lambda g(t)})(x) = \lambda^{\alpha} e^{-\lambda g(x)}. \tag{3.9}$$

Proof. Using the relation (3.3) and taking (3.4) into account, we have

$$\left(I^{\alpha}_{-;g} e^{-\lambda g(t)}\right)(x) = Q_g\left((I^{\alpha}_{-} e^{-\lambda t})(x)\right) = Q_g\left(\lambda^{-\alpha} e^{-\lambda x}\right) = \lambda^{-\alpha} e^{-\lambda g(x)},$$

and (3.8) is proved. Applying (3.6) and (3.8) with α replaced by $n - \alpha$, we deduce (3.9) as follows:

$$\left(D^{\alpha}_{-;g} e^{-\lambda g(t)}\right)(x) = Q_g\left((D^{\alpha}_{-} e^{-\lambda t})(x)\right) = Q_g\left(\lambda^{\alpha} e^{-\lambda x}\right) = \lambda^{\alpha} e^{-\lambda g(x)}.$$

□

In particular, when $a = 0$ and $g(x) = x^{\sigma}$ $(\sigma > 0)$, (3.4) and (3.6) yield the following special cases of the fractional integral and fractional derivative operators of order α with respect to the function x^{σ}:

$$(I^{\alpha}_{-;x^{\sigma}} f)(x) = \frac{\sigma}{\Gamma(\alpha)} \int_x^{\infty} \frac{t^{\sigma-1} f(t)dt}{(t^{\sigma} - x^{\sigma})^{1-\alpha}} \quad (x > 0), \tag{3.10}$$

when $\Re(\alpha) > 0$, and

$$(D^{\alpha}_{-;x^{\sigma}} f)(x) = \frac{\sigma^{1-n}}{\Gamma(n-\alpha)} \left(-x^{1-\sigma} D\right)^n \int_x^{\infty} \frac{t^{\sigma-1} f(t)dt}{(t^{\sigma} - x^{\sigma})^{\alpha-n+1}} \quad (x > 0), \tag{3.11}$$

when $\Re(\alpha) \geqq 0$ $(\alpha \neq 0)$ *with* $n = [\Re(\alpha)] + 1$.

We note that $I^{\alpha}_{-;x^{\sigma}} f$ is connected with the fractional integral $I^{\alpha}_{-} f$ by

$$(I^{\alpha}_{-;x^{\sigma}} f)(x) = \left(I^{\alpha}_{-} f(t^{1/\sigma})\right)(x^{\sigma}) \quad (x > 0), \tag{3.12}$$

and also with the so-called Erdelyi-Kober type operator apart from a power multiplier factor, see (18.39) in [122].

As a particular case of Lemma 3.2, we have

Lemma 3.3 *Let* $\sigma > 0$, $\lambda > 0$, *and* $\alpha \in \mathbb{C}$ $(\Re(\alpha) > 0)$. *Then*

$$(I^{\alpha}_{-;x^{\sigma}} e^{-\lambda t^{\sigma}})(x) = \lambda^{-\alpha} e^{-\lambda x^{\sigma}} \quad \text{and} \quad (D^{\alpha}_{-;x^{\sigma}} e^{-\lambda t^{\sigma}})(x) = \lambda^{\alpha} e^{-\lambda x^{\sigma}}. \tag{3.13}$$

Remark. The formula (3.9) and the second relations in (3.3) and (3.13) are valid for $\alpha \in \mathbb{C}$ with $\Re(\alpha) = 0$ and $\alpha \neq 0$.

We give a special attention to the Riesz integral and Riesz derivative operators I^α and D^α of order $\alpha \in \mathbb{C}$ $(\Re(\alpha) > 0)$ in the n-dimensional Euclidean space $\mathbb{R}^n$. These operators are defined as negative $(-\Delta)^{-\alpha/2}$ and positive $(-\Delta)^{\alpha/2}$ powers of the Laplace operator

$$\Delta = \frac{\partial^2}{\partial x_1^2} + \frac{\partial^2}{\partial x_2^2} + \cdots + \frac{\partial^2}{\partial x_n^2}. \tag{3.14}$$

They can be represented in terms of the direct $\mathcal{F}$ and inverse $\mathcal{F}^{-1}$ Fourier transforms by

$$I^\alpha f \equiv (-\Delta)^{-\alpha/2} f = \mathcal{F}^{-1}|\bar{x}|^{-\alpha}\mathcal{F}f; \quad D^\alpha f \equiv (-\Delta)^{\alpha/2} f = \mathcal{F}^{-1}|\bar{x}|^{\alpha}\mathcal{F}f, \tag{3.15}$$

where $\bar{x} = (x_1, \ldots, x_n) \in \mathbb{R}^n$ and $|\bar{x}| = \left(x_1^2 + \cdots + x_n^2\right)^{1/2}$. When $0 < \alpha < n$, the Riesz fractional integration I^α can be realized as the Riesz potential, given (for $\bar{x} \in \mathbb{R}^n$) by

$$(I^\alpha f)(\bar{x}) = \gamma(n,\alpha) \int_{\mathbb{R}^n} \frac{f(\bar{y})}{|\bar{x} - \bar{y}|^{n-\alpha}} d\bar{y} \quad \left(\gamma(n,\alpha) = \frac{\Gamma[(n-\alpha)/2]}{2^\alpha \pi^{n/2}\Gamma(\alpha/2]}\right). \tag{3.16}$$

We note that the Riesz potential (3.16) and the Riesz derivative operator D^α, developed explicitly as a hypersingular integral by P. I. Lizorkin, are studied sufficiently well; for example, see [114], [120], [121] and Sections 25-27 in [122]. In connection with the fractional diffusion processes such a multidimensional Riesz fractional differentiation was used by Kochubei [63]-[67]. We also note that the Riesz derivative operator was used by many applied researchers, but really they have used only its characterization in terms of Fourier transforms.

In conclusion, we remark that there are a great number of other fractional calculus operators of one and several variables, which preserve different memory properties. Some fractional calculus operators, such as the Riemann-Liouville, Grünwald-Letnikov, Marchaud, Erdélyi-Kober, Hadamard, etc., can be found in [122]. We also mention another smooth fractional calculus operator, known as the Caputo derivative (see, for example, [39, 40]), which is often used in many applied fractional calculus models, explicitly or implicitly. Note that more general fractional derivative of such a kind was introduced by Dzhrbashyan and Nersesyan; for example, see the relation (4.10) in [122]. Furthermore, it is possible to introduce or define new singular operators with long memory and non-local properties.

4. A New Model for the Super-Diffusion Processes

As we have mentioned in the first part of this presentation, there are many papers published during the last ten years, presenting fractional calculus models for kinetics of natural anomalous processes into the complexity systems, which keep long-memory and non-local properties of the corresponding dynamics. Special interest has been paid to the anomalous diffusion processes, that is, super-slow diffusion (or sub-diffusion) and super-fast diffusion (or super-diffusion) processes. Nigmatullin [97] is probably the first who considered the following *diffusion equation with a memory*:

$$\frac{\partial U(\bar{x},t)}{\partial t} = \int_a^t K(t-\tau)\Delta U(\bar{x},\tau)d\tau \quad (t>0;\ \bar{x}\in\mathbb{R}^n), \tag{4.1}$$

where Δ is the Laplace operator (3.14). In the particular case when $K(t) = \rho^2\delta(t-a)$, $\delta(t-a)$ being the Dirac delta function, this relation represents the classical diffusion equation:

$$\frac{\partial U(\bar{x},t)}{\partial t} = \rho^2\Delta U(\bar{x},t) \quad (t>0;\ \bar{x}\in\mathbb{R}^n). \tag{4.2}$$

As we have seen in the first part of this presentation, there are many nice fractional calculus models for the sub-diffusion processes, without and with external force field, such as the *Fractional Diffusion Equation* and the *Advection-Diffusion* or the *Fokker-Planck Equations*, respectively. All these models include the fractional derivative operators of Riemann-Liouville and of Caputo, acting on the time variable, and the fractional derivative operators of Liouville (or Weyl) $D_{\pm}^{\alpha}$ and of Riesz D^{α}, acting on the space variables, respectively.

However, the super-diffusion processes are studied less frequently. Furthermore, there are only few papers including fractional calculus models, where the authors introduce the Riesz derivative operator D^{α} acting on the space variables by using its definition, which is given in (3.15) in terms of the direct and inverse Fourier transforms. For example, see [29], [156], [74], and [21].

We propose here a new fractional derivative model, including the generalized Liouville fractional derivative operator $D_{-;g}^{\alpha}$, which acts on the time variable. This model allows us to have a strong control of both the sub- and super-diffusion processes. It also has a great flexibility due to the free function g, and there are many other advantages from the purely technical point of view in the sense that we can use the same methods as in the ordinary case (that is, those based upon integral transforms and separation of variables) to solve boundary-value problems associated with the fractional calculus models.

First we consider our new fractional calculus model in the one-dimensional case. We begin with a scheme for the solution of the classical problem associated with the one-dimensional heat equation:

$$\frac{\partial U(x,t)}{\partial t} = \rho^2\frac{\partial^2}{\partial x^2}U(x,t) \quad (\rho>0;\ \ t>0;\ \ 0<x<l), \tag{4.3}$$

with the boundary conditions:

$$U(0,t) = U(l,t) = 0 \quad (t>0), \tag{4.4}$$

(f being a *sufficiently smooth* function on $[0,l]$) and with the initial condition:

$$U(x,0) = f(x) \quad (0<x<l). \tag{4.5}$$

We note that the parameter ρ^2 can characterize the thermal or diffusion constant of the media.

Using the method of separation of variables, we seek the solution of the problem (4.3)-(4.5) in the form $U(x,t) = X(x)T(t)$. Then

$$\frac{X''(x)}{X(x)} = \frac{T'(t)}{\rho^2 T(t)} = -\lambda^2. \tag{4.6}$$

Therefore, $X(x)$ is a solution of the Sturm-Liouville problem:

$$X''(x) + \lambda X(x) = 0, \quad X(0) = X(l) = 0, \tag{4.7}$$

for which the eigenvalues and the corresponding eigenfunctions are given by

$$\lambda_n = \mathbf{a}^2; \; X_n(x) = \sin(\mathbf{a}x); \; \mathbf{a} = \frac{n\pi}{l} \quad (n \in \mathbb{N}). \tag{4.8}$$

On the other hand, $T(t)$ is a solution of the differential equation:

$$T'(t) + (\mathbf{a}\rho)^2 T(t) = 0, \tag{4.9}$$

given by

$$T_n(t) = \exp\left[-(\mathbf{a}\rho)^2 t\right] \quad (n \in \mathbb{N}). \tag{4.10}$$

Finally, the solution of the problem modelled by (4.3) to (4.5) is obtained as a series solution whose coefficients are found from the following Fourier development of the function f involved in the initial condition (4.5):

$$U(x,t) = \sum_{n=1}^{\infty} c_n X_n(x) T_n(t), \tag{4.11}$$

where

$$c_n = \frac{2}{l} \int_0^l \sin(\mathbf{a}x) f(x) dx \quad (n \in \mathbb{N}), \tag{4.12}$$

and $X_n(x)$ and $T_n(t)$ are given in (4.8) and (4.10), respectively. Thus we need only to prove the convergence on $[0,l] \times [0,\infty)$ of the double series for $U(x,t)$ in (4.11) and of the series for $U_{xx}(x,t)$ and $U_t(x,t)$ obtained from (4.11) by term-by-term differentiation with respect to x and t, respectively.

Now we can see more clearly that a sub-diffusive process, associated with the above problem, can be obtained by replacing the factor $T_n(t)$ in the solution (4.11) by some other function $T_n(t,\alpha)$ depending on the time t and a new parameter α. Such a new parameter gives us the possibility to have control over the decay of the process in the time variable; for instance, we have

$$T_n(t,\alpha) = \exp\left[-(\mathbf{a}\rho)^{2/\alpha} g(t,\alpha)\right] \quad (n \in \mathbb{N}). \tag{4.13}$$

In this case, (4.11) is extended to the form:

$$U(x,t) = \sum_{n=1}^{\infty} c_n X_n(x) T_n(t,\alpha), \tag{4.14}$$

where $X_n(t)$ and $T_n(t,\alpha)$ are given in (4.8) and (4.13), respectively. Thus we shall have the control of a sub- or super-diffusive process by means of the control of the function $g(t,\alpha)$. For example, if $g(t,\alpha) = t^\alpha$, we have sub-diffusion kinetics when $0 < \alpha < 1$ and super-diffusion kinetics when $\alpha > 1$. This is in agreement with the CTRW-method.

In the above-mentioned way, by choosing a function $g(t,\alpha)$, we can obtain other suitable solutions of our problem to control a very strong super-diffusion process. For example, if we take $g(t,\alpha) = \exp(t^\alpha)$ with $\alpha > 1$, perhaps (4.14) would yield a new model for the dynamics of the properties of super-conductivity of some materials under certain boundary conditions.

Now we can put the corresponding derivative model with the solution (4.14). It is clear that we have the problem to find some derivative operator $\mathbf{D}^\alpha$ (fractional, singular with a memory, or others) with the following property:

$$\mathbf{D}^\alpha T(t,\alpha) = (\mathbf{a}\rho)^2 T(t,\alpha), \tag{4.15}$$

where

$$T(t,\alpha) \equiv T_n(t,\alpha) = \exp\left[-(\mathbf{a}\rho)^{2/\alpha} g(t,\alpha)\right] \quad (n \in \mathbb{N}).$$

Such a problem is solved by taking $\mathbf{D}^\alpha = D^\alpha_{-,g(t,\alpha)}$, where $D^\alpha_{-;g(t,\alpha)} f$, as a function of t, is the fractional derivative of a function f with respect to $g(t,\alpha)$ defined in (3.6). Indeed, if we suppose that $g(t,\alpha)$ (for any fixed α) is an increasing and monotonic function of $t \in \mathbb{R}_+ = (0,\infty)$ having the continuous derivative $\frac{\partial}{\partial t} g(t,\alpha) \neq 0$, then, in accordance with (3.9),

$$\left(D^\alpha_{-;g(t,\alpha)} \exp\left[-(\mathbf{a}\rho)^{2/\alpha} g(\tau,\alpha)\right]\right)(t) = (\mathbf{a}\rho)^2 \exp\left[-(\mathbf{a}\rho)^{2/\alpha} g(t,\alpha)\right],$$

and hence (4.15) holds true for $\mathbf{D}^\alpha = D^\alpha_{-;g(t,\alpha)}$. Therefore, the new fractional diffusion model can be presented by the fractional differential equation:

$$D^\alpha_{-;g(t,\alpha)} U(x,t) = \rho^2 \frac{\partial^2}{\partial x^2} U(x,t) \quad (\alpha > 0;\ t > 0;\ x \in \mathbb{R}). \tag{4.16}$$

If we additionally suppose that $\lim_{t\to 0+} g(t,\alpha) = k(\alpha) \in \mathbb{R}$ for any $\alpha > 0$, then the explicit solution of the problem modelled by (4.16), (4.4) and (4.5) has the form (4.14):

$$U(x,t) = \frac{1}{k(\alpha)} \sum_{n=1}^{\infty} c_n \sin(\mathbf{a}x) \exp\left[-(\mathbf{a}\rho)^{2/\alpha} g(t,\alpha)\right], \tag{4.17}$$

provided that $g(t, \alpha)$ is so chosen that $U(x, t)$, $D^{\alpha}_{-;g(t,\alpha)}U(x, t)$, and $U_{xx}(x, t)$ are convergent double series on $[0, l] \times [0, \infty)$, c_n being given in (4.12).

For example, when $g(t, \alpha) = t^{\alpha}$ $(\alpha > 0)$, we can prove the above convergence just as in the ordinary case when $\alpha = 1$. In this case, $D^{\alpha}_{-;g(t,\alpha)} = D^{\alpha}_{-;t^{\alpha}}$ is the fractional derivative (3.11), and the equation (4.16) takes the form:

$$D^{\alpha}_{-;t^{\alpha}}U(x, t) = \rho^2 \frac{\partial^2}{\partial x^2} U(x, t) \quad (\alpha > 0;\ t > 0;\ x \in \mathbb{R}), \tag{4.18}$$

and the explicit solution of the problem modelled by (4.18), (4.4), and (4.5) is represented by

$$U(x, t) = \sum_{n=1}^{\infty} c_n \sin(\mathbf{a}x) \exp\left[-(\mathbf{a}\rho)^{2/\alpha} t^{\alpha}\right]. \tag{4.19}$$

In a similar manner, we can consider the case when $g(t, \alpha) = \exp(t^{\alpha})$.

The model (4.16) can be extended to the space variable $\bar{x} \in \mathbb{R}^n$ if we replace the partial derivative $\partial^2/\partial x^2$ by the Laplace operator (3.14) with respect to $\bar{x}$:

$$D^{\alpha}_{-;g(t,\alpha)}U(\bar{x}, t) = \rho^2 \Delta_{\bar{x}} U(\bar{x}, t) \quad (\alpha > 0;\ t > 0;\ \bar{x} \in \mathbb{R}^n). \tag{4.20}$$

In particular, the model (4.18) is extended to the form:

$$D^{\alpha}_{-;t^{\alpha}}U(\bar{x}, t) = \rho^2 \Delta_{\bar{x}} U(\bar{x}, t) \quad (\alpha > 0;\ t > 0;\ \bar{x} \in \mathbb{R}^n). \tag{4.21}$$

Analogous to (4.17) and (4.19), the explicit solutions of these equations under the corresponding boundary and initial conditions can be also deduced.

It seems that the above approach, based on the use of the generalized fractional derivative (3.6), can also be applied in order to construct the corresponding generalized fractional calculus models for the Fokker-Planck and advection-diffusion equations. We think that these models will be compatible with those for the corresponding generalizations including, for example, the Riesz operator D^{α} acting on the space variable.

We note that there are many more possibilities in order to control the exponential-type decay of the fundamental solutions of the generalized fractional diffusion equations, and that (in this way) the same mathematical technique is possible to be used to study new fractional calculus models for the control of anomalous wave processes.

Acknowledgements

This work was supported, in part, by Belarusian Fundamental Research Fund, by MCYT (BFM2000-0797), by DGUI of G.A.CC, by ULL Research Funds, and by NSERC-Canada Research Grants. The authors are appreciative of these generous supports to their present investigation.

References

[1] Addison, P. S., Qu, B., Nisbet, A., and Pender, G., A non-Fictian, particle-tracking diffusion model based on fractional Brownian motion, *Internat. J. Numer. Meth. Fluids*, **25** (1997), 1373–1384.

[2] Agrawal, O. P., A new Lagrangian and a new Lagrange equation of motion for fractionally damped systems, *J. Appl. Mech.*, **68** (2001), 339–341.

[3] Allegrini, P., Buiatti, M., Grigolini, P., and West, B. J., Fractional Brownian motion as a nonstationary process: An alternative paradigm for DNA sequences, *Phys. Rev. E*, **57** (1998), no. 4, 4558-4567.

[4] Bagley, R. L., *Applications of Generalised Derivatives to Viscoelasticity*, Ph.D. Thesis in Materials, Air Force Institute of Technology and Material Laboratory, AFML-TR-79-4103, 1979.

[5] Bagley, R. L. and Torvik, P. L., A theoretical basis for the application of fractional calculus to viscoelasticity, *J. Rheol.*, **27** (1983), no. 3, 201–210.

[6] Bagley, R. L. and Torvik, P. L., On appearance of the fractional derivative in the behaviour of real materials, *J. Appl. Mech.*, **51** (1984), 294–298.

[7] Bagley, R. L. and Torvik, P. L., On the fractional calculus models of viscoelastic behaviour, *J. Rheol.*, **30** (1986), no. 1, 133–155.

[8] Baillie, R. T. and King, M. L., Fractional differencing and long memory processes, *J. Econometr.*, **73** (1996), no. 1, 1–3.

[9] Balescu, R., Anomalous transport in turbulent plasmas and continuous time random walks, *Phys. Rev. E*, **51** (1995), no. 5, 4807–4822.

[10] Barkai, E., Fractional Fokker-Planck equation, solution and application, *Phys. Rev. E*, **63** (2001), no. 4, article no. 046118, Part 2.

[11] Barkai, E., Metzler, R., and Klafter, J., From continuous time random walk to the fractional Fokker-Planck equation, *Phys. Rev. E*, **61** (2000), no. 1, 132–138.

[12] Barkai, E. and Silbey, R. J., Fractional Kramers equation, *J. Phys. Chem. B*, **104** (2000), no. 16, 3866–3874.

[13] Barlow, M. T., Diffusions on fractals. In: *Lectures on probability theory and statistics* (Saint-Flour, 1995), Lecture Notes in Math., **1690**, 1–121. Springer-Verlag, Berlin, 1998.

[14] Ben-Avraham, D. and Havlin, S., *Diffusion and Reactions in Fractals and Disordered Systems.* Cambridge University Press, Cambridge, 2000.

[15] Benson, D. A., *The Fractional Advection-Dispersion Equation: Development and Applications*, Ph.D. Thesis in Hydrology, University of Nevada, U.S.A., 1998.

[16] Berkowits, B. and Scher, H., Theory of anomalous chemical transport in random fracture networks, *Phys. Rev. E*, **57** (1998), no. 5, 5858–5869.

[17] Blumen, A., Gurtovenko, A. A., and Jespersen, S., Anomalous dynamics of model polymer systems, *J. Lumin.*, **94** (2001), 437–440.

[18] Bouchaud, J. P. and Georges, A., Anomalous diffusion in disordered media: Statistical mechanics, models and physical applications, *Phys. Rep.*, **195** (1990), no. 4–5, 127–293.

[19] Bouchaud, J. P. and Potters, M., *Theory of Financial Risks.* Cambridge University Express, Cambridge, 1999.

[20] Carpinteri, A. and Mainardi, F. (editors), *Fractals and Fractional Calculus in Continuum Mechanics* (Udine, 1996), CIAM Courses and Lectures, **378**. Springer-Verlag, New York, 1997.

[21] Chechkin, A. V., Gonchar, V. Y., and Szydlowski, M., Fractional kinetics for relaxation and superdiffusion in a magnetic field, *Phys. Plas.*, **9** (2002), no. 1, 78–88.

[22] Compte, A., Stochastic foundations of fractional dynamics, *Phys. Rev. E*, **53** (1996), no. 4, 4191–4193.

[23] Compte, A. and Jou, D., Non-equilibrium thermodynamics and anomalous diffusion, *J. Phys. A: Math. Gen.*, **29** (1996), no. 15, 4321–4329.

[24] del-Castillo-Negrete, D., Chaotic transport in zonal flows in anomalous geophysical and plasma systems, *Phys. Plas.*, **7** (2000), no. 5, 1702–1711.

[25] Enelund, M. and Olsson, P., Damping described by fading memory: analysis and application to fractional derivative models, *Internat. J. Sol. Struct.*, **36** (1999), no. 7, 939–970.

[26] Engheta, N., On the role of fractional calculus in electromagnetic theory, *IEEE Antenn. Propag.*, **39** (1997), no. 4, 35–46.

[27] Engheta, N., On fractional paradigm and intermediate zones in electromagnetism. I: Planar observation, *Microw. Optim. Tech. Lett.*, **22** (1999), no. 4, 236–241.

[28] Engheta, N., On fractional paradigm and intermediate zones in electromagnetism. II: Cylindrical and spherical observations, *Microw. Optim. Tech. Lett.*, **23** (1999), no. 2, 100–103.

[29] Ertas, D. and Kardar, M., Dynamic roughening of directed lines, *Phys. Rev. Lett.*, **69** (1992), no. 6, 929–932.

[30] Feller, W., *An Introduction to Probability Theory and Its Applications.* John Wiley and Sons, New York, 1971.

[31] Floriani, E., Trefan, G., Grigolini, P., and West, B. J., A dynamical approach to anomalous conductivity, *J. Statist. Phys.*, **84** (1996), 1043–1066.

[32] Fogedby, C., Langevin-equations for continuous time Levy flights, *Phys. Rev. E*, **50** (1994), no. 2, 1657–1660.

[33] Fujita, Y., Integrodifferential equation which interpolates the heat equation and the wave equation, *Osaka J. Math.*, **27** (1990), 309–321.

[34] Fujita, Y., Energy inequalities for integro-partial differential equations with Riemann-Liouville integral, *SIAM J. Math. Anal.*, **23** (1992), no. 5, 1182–1188.

[35] Glockle, W. G. and Nonnenmacher, T. F., Fractional integral-operators and Fox functions in the theory of viscoelasticity, *Macromolecules J.*, **24** (1991), no. 24, 6426–6434.

[36] Glockle, W. G. and Nonnenmacher, T. F., Fractional relaxation and the time temperature superposition principle, *Rheol. Acta*, **33** (1994), no. 4, 337–343.

[37] Glockle, W. G. and Nonnenmacher, T. F., A fractional calculus approach to seff-similar protein dynamics, *Biophys. J.*, **68** (1995), no. 1, 46–53.

[38] Gnedenko, B. V. and Kolmogorov, A. N., *Limit Distributions for Sums of Independent Random Variables.* Addison-Wesley Publishing Company, Inc., Cambridge, Massachusetts, 1954.

[39] Gorenflo, R. and Mainardi, F., Fractional calculus and stable probability distributions, *Arch. Mech.*, **50** (1998), 377–388.

[40] Gorenflo, R., Mainardi, F., and Srivastava, H. M., Special functions in fractional relaxation-oscillation and fractional diffusion-wave phenomena. In: *Proceedings of the Eighth International Colloquium on Differential Equations* (Plovdiv, 1997), 195–202. VSP, Utrecht, 1998.

[41] Hanyga, A., Wave propagation in media with singular memory, *Math. Comput. Modelling*, **34** (2001), no. 12-13, 1399–1421.

[42] Hanyga, A., Multidimensional solutions of space-time-fractional equations, *Proc. Roy. Soc. London Ser. A Math. Phys. Sci.*, **458** (2002), 429–450.

[43] Hanyga, A., Multidimensional solutions of time-fractional diffusion-wave equations, *Proc. Roy. Soc. London Ser. A Math. Phys. Sci.*, **458** (2002), 933–957.

[44] Hanyga, A., Wave propagation in poroelasticity: equations and solutions. In: Shang, E. C., Gao, T. F., and Li, Q. (Eds.), *Theoretical and Computational Acoustics 2001*. World Sci. Publ., London, 2002.

[45] Haus, J. W. and Kehr, K. W., Diffusion in regular and disordered lattices, *Phys. Lett.*, **150** (1987), no. 5-6, 263–406.

[46] Havlin, S. and Ben-Avraham, D., Diffusion in disordered media, *Adv. Phys.*, **36** (1987), no. 6, 695–798.

[47] Havlin, S., Buldyrev, S. V., Goldberger, A. L., Maniegna, R. N., Ossadnik, S. M., Peng, C. K., Simon, M., and Stanley, H. E., Fractals in biology and medicine, *Chaos Sol. Fract.*, **6** (1995), 171–201.

[48] Henry, B. I. and Wearne, S. L., Fractional reaction-diffusion, *Phys. A*, **276** (2000), no. 3-4, 448-455.

[49] Henry, B. I. and Wearne, S. L., Existence of turing instabilities in two-species fractional reaction-diffusion system, *SIAM J. Appl. Math.*, **62** (2002), no. 3, 870–887.

[50] Herrchen, M. P., *Stochastic Modelling of Dispersive Diffusion by Non-Gaussian Noise*, Ph.D. Thesis in Natural Sciences, Swiss Federal Institute of Technology, 2001.

[51] Hilfer, R. (editor), *Applications of Fractional Calculus in Physics*. World Scientific Publishing Company, Singapore, London, 2000.

[52] Hughes, B. D., *Random Walks and Random Environments*, Vol. 1. Oxford University Press, Oxford, 1995.

[53] Hughes, B. D. *Random Walks and Random Environments*, Vol. 2. Oxford University Press, Oxford, 1996.

[54] Hwang, C., Leu, J. F., and Tsay, S. Y., A note on time-domain simulation of feedback fractional-order systems, *IEEE Trans. Automat. Control*, **47** (2002), no. 4, 625–631.

[55] Isichenko, M. B., Percolation, statistical topography and transport in random media, *Rev. Mod. Phys.*, **64** (1992), no. 4, 961–1043.

[56] Jumarie, G., A new approach to complex-valued fractional Brownian motion via rotating white noise, *Chaos Sol. Fract.*, **9** (1998), no. 6, 881–893.

[57] Kilbas, A. A. and Trujillo, J. J., Differential equations of fractional order: methods, results and problems. I, *Appl. Anal.*, **78** (2001), no. 1-2, 157–192.

[58] Kilbas, A. A. and Trujillo, J. J., Differential equations of fractional order: methods, results and problems. II, *Appl. Anal.*, (2002), **81** (2002), no. 2, 435–493.

[59] Klafter, J., Blumen, A., and Shlesinger, M. F., Stochastic pathway to anomalous diffusion, *Phys. Rev. A*, **35** (1987), no. 7, 3081–3085.

[60] Klafter, J., Shlesinger, M. F., and Zumofen, G., Beyond Brownian motion, *Phys. Today*, **49** (1996), no. 2, 33–39.

[61] Klafter, J. and Silbey, R., Derivation of the continuous time random walk equation, *Phys. Rev. Lett.*, **44** (1980), no. 2, 55–58.

[62] Koeller, R. C., Application of fractional calculus to theory of viscoelasticity, *ASME J. Appl. Mech.*, **51** (1984), 299–307.

[63] Kochubei, A. N., Parabolic pseudo-differential equations, hypersingular integrals and Markov processes, *Math. USSR-Izv.*, **33** (1989), no. 2, 233–259.

[64] Kochubei, A. N., The Cauchy problem for evolution equations of fractional order, *Diff. Equat.*, **25** (1989), no. 8, 967–974.

[65] Kochubei, A. N., Diffusion of fractional order, *Diff. Equat.*, **26** (1990), no. 4, 485–492.

[66] Kochubei, A. N., Fractional differentiation operartor over an infinite extension of local field. In: *Proceedings of International Conference of p-Adic Functional Analysis* (Poznań, 1998), Lecture Notes in Pure and Appl. Math., **207**, 167–178. Dekker, New York, 1999.

[67] Kochubei, A. N., *Pseudo-Differential Equations and Stochastics over non-Archimedean Fields*, Monographs and Textbooks in Pure and Applied Mathematics, **244**. Marcel Dekker, Inc., New York, 2001.

[68] Kotomin, E. and Kuzovkov, V., *Modern Aspects of Diffusion-Controlled Reactions: Cooperative Phenomena in Bimolecular Processes*, Comprehensive Chemical Kinetics, **34**. Elsevier, Amsterdam, 1996.

[69] Krall, N. and Trivelpiece, A. W., *Principles of Plasma Physics*. Academic Press, New York, London, 1973.

[70] Lagutin, A. A., Nikulin, Y. A., and Uchaikin, W., The "knee" in the primary cosmic ray spectrum as consequence of the anomalous diffusion of the particles in the fractal interstellar medium, *Nuclear Phys. B Proc. Suppl.*, **97** (2001), 267–270.

[71] Latora, V., Rapisarda, A., and Ruffo, S., Super-diffusion and out-of-equilibrium chaotic dynamics with many degrees of freedoms, *Phys. Rev. Lett.*, **83** (1999), no. 11, 2104–2107.

[72] Latora, V., Rapisarda, A., and Ruffo, S., Chaotic dynamics and super-diffusion in a Hamiltonian system with many degrees of freedom, *Phys. A*, **280** (2000), no. 1-2, 81–86.

[73] Lee, C., Hoopes, M. F., Diehl, J., Gilliland, W., Huxel, G., Leaver, E. V., McCann, K., Umbanhowar, J., and Mogilner, A., Non-local concepts and models in biology, *J. Theoret. Biol.*, **210** (2001), 201–219.

[74] Leoncini, X., Kuznetsov, L., and Zaslavsky, G. M., Chaotic advection near a three-vortex collapse, *Phys. Rev. E*, **63** (2001), no. 3, article no. 036224, part 2.

[75] Lévi, P., *Théorie de l'Addition des Variables Aleatories*. Gauthier-Villars, Paris, 1937.

[76] Lévi, P., *Processus Stochastiques et Mouvement Brownien*. Gauthier-Villars, Paris, 1948.

[77] Lutz, E., Fractional Langevin equation, *Phys. Rev. E*, **64** (2001), no. 5, article no. 051106.

[78] Mainardi, F., Raberto, M., Gorenflo, R., and Scala, E., Fractional calculus and continuous time finance. II: The waking-time distribution, *Phys. A*, **287** (2000), no. 3-4, 468–481.

[79] Mainardi, F. and Tomirotti, M., Seismic pulse propagation with constant Q and stable probability distributions, *Ann. Geofis.*, **40** (1997), 1311–1328.

[80] Makris, N. and Constantinou, M. C., Fractional derivative Maxwell model for viscous dampers, *J. Struct. Engrg.*, **117** (1991), no. 9, 2708–2724.

[81] Makris, N. and Constantinou, M. C., Spring-viscous damper systems for combined seismic and vibration isolation, *Earthquake Engrg. Struct. Dyn.*, **21** (1992), 649–664.

[82] Makris, N., Dargush, C. F., and Constantinou, M. C., Dynamic analysis of generalized viscoelastic fluids, *J. Engrg. Mech.*, **119** (1993), 1663–1679.

[83] Mallamace, F. and Stanley, H. E. (editors), *The Physics of Complex Systems*, International School of Physics Enrico Fermi, Vol. 134. IOS Press, Amsterdam, 1997.

[84] Mandelbrot, B. B., *The Fractal Geometry of Nature.* Freeman, New York, 1982.

[85] Mandelbrot, B. B., *Fractals and Scaling in Finance.* Springer-Verlag, New York, 1997.

[86] Mandelbrot, B. B. and Van Ness, J., Fractional Brownian motions, fractional noises and applications, *SIAM Rev.*, **10** (1968), no. 4, 422–437.

[87] Margulies, T. S., Acoustic wave processes in viscoelastic porous media. In: *Proceedings of the 130th ASA Meeting* (St. Louis, Missouri), 1995.

[88] Metzler, R., Barkai, E., and Klafter, J., Anomalous diffusion and relaxation close to thermal equillibrium: a fractional approach, *Phys. Rev. Lett.*, **82** (1999), no. 19, 3563–3567.

[89] Metzler, R., Glockle, W. G., and Nonnenmacher, T. F., Fractional model equation for anomalous diffusion, *Phys. A*, **211** (1994), 13–24.

[90] Metzler, R. and Klafter, J., The random walk's guide to anomalous diffusion: a fractional dynamic approach, *Phys. Rep.*, **339** (2000), no. 1, 1–77.

[91] Metzler, R. and Klafter, J., Accelerating Brownian motion: a fractional dynamics approach to fast diffusion, *Europhys. Lett.*, **51** (2000), no. 5, 492–498.

[92] Metzler, R. and Nonnenmacher, T.-F., Space- and time-fractional diffusion and wave equations, fractional Fokker-Planck equations, and physical motivation, *Chem. Phys.* (2002), **284** (2002), no. 1–2, 67–90.

[93] Miller, K. S. and Ross, B., *An Introduction to the Fractional Calculus and Fractional Differential Equations.* John Wiley and Sons, New York, 1993.

[94] Montroll, E. W. and Weiss, G. H., Random walk on lattices. II, *J. Math. Phys.*, **6** (1965), 167–181.

[95] Montseny, G., Audounet, J., and Magtinon, D., Diffusive representation for pseudo-differentially damped nonlinear systems. In: *Nonlinear control in the year 2000*, Vol. 2 (Paris), Lecture Notes in Control and Inform. Sci., **259**, 163–182. Springer-Verlag, London, 2001.

[96] Naumkin, P. I. and Shishmarev, I. A., *Nonlinear Nonlocal Equations in the Theory of Waves*, Translations of Mathematical Monographs, **133**. AMS, Providence, R.I., 1994.

[97] Nigmatullin, R. R., On the theory of relaxation with "remnant" memory, *Phys. Statist. Sol. B*, **124** (1984), 389-393.

[98] Nikias, C. L. and Shao, M., *Signal Processing with Alpha-Stable Distributions and Applications.* John Wiley and Sons, New York, 1995.

[99] Novak, M. M., *Paradigms of Complexity.* World Scientific Publishing Company, Singapore, London, 2000.

[100] Oldham, K. B. and Spanier, J., The replacement of Fick's law by a formulation involving semidifferentiation, *J. Electroanal. Chem. Interf. Electr.*, **26** (1970), 331–341.

[101] Oldham, K. B. and Spanier, J., *The Fractional Calculus.* Academic Press, New York, London, 1974.

[102] Osada, H., Self-similar diffusions on a class of infinitely ramified fractals, *J. Math. Soc. Japan*, **47** (1995), no. 4, 591–616.

[103] Osada, H., A family of diffusion processes on Sierpinski carpets, *Probab. Theory Related Fields*, **119** (2001), no. 2, 275–310.

[104] Ozaktas, H. M., Zalevsky, Z., and Kutai, M. A., *The Fractional Fourier Transform with Applications in Optics and Signal Processing.* John Wiley and Sons, New York, 2000.

[105] Pachepsky, Y., Benson, D., and Rawls, W., Simulating scale-dependent solute transport in soils with the fractional advective-dispersive equation, *Soil. Sci. Soc. Amer. J.*, **64** (2000), no. 4, 1234–1243.

[106] Paradisi, P, Cesari, R., Mainardi, F., and Tampieri, F., The fractional Fick's law for non-local transport processes, *Phys. A*, **293** (2001), no. 1-2, 130–142.

[107] Pekalski, A. and Sznajd-Weron, K. (editors), *Anomalous Diffusion from Basics to Applications*. Springer-Verlag, Berlin, 1999.

[108] Peliti, L. and Vulpiani, A. (editors), *Measures of Complexity*. Springer-Verlag, Berlin, 1988.

[109] Pfister, G. and Scher, H., Time dependent electrical transport in amorphous solids-AS2SE, *Phys. Rev. B*, **15** (1977), no. 4, 2062–2083.

[110] Pfister, G. and Scher, H., Dispersive (non-Gaussian) transient transport in disordered solids, *Adv. Phys.*, **27** (1978), no. 5, 747–798.

[111] Plonka, A., Recent developments in dispersive kinetics, *Progr. React. Kinet. Mech.*, **25** (2000), no. 2, 109–217.

[112] Podlubny, I., *Fractional Differential Equations*, Mathematics in Sciences and Engineering, **198**. Academic Press, San-Diego, CA, 1999.

[113] Richardson, L. F., Atmosphere diffusion shown on a distance neighbour graph, *Proc. Roy. Soc. London Ser. A Math. Phys. Sci.*, **110** (1926), 709–737.

[114] Riesz, M., L'intégrale de Riemann-Liouville et le problem de Cauchy, *Acta Math.*, **81** (1949), 1–223.

[115] Riewe, F., Mechanics with fractional derivative, *Phys. Rev. E*, **55** (1997), no. 3, 3581–3592.

[116] Rocco, A. and West, B. J., Fractional calculus and the evolution of fractal phenomena, *Phys. A*, **265** (1999), no. 3–4, 535–546.

[117] Ross, B., The development of fractional calculus: 1695–1900, *Hist. Math.*, **4** (1977), 75–89.

[118] Rossikhin, Y. A. and Shitikova, M. V., Applications of fractional calculus to dynamic problems of linear and nonlinear hereditary mechanics of solids, *Appl. Mech. Rev.*, **50** (1997), no. 1, 15–67.

[119] Saichev, A. I. and Zaslavsky, G. M., Fractional kinetic equations: solutions and applications, *Chaos*, **7** (1997), no. 4, 753–764.

[120] Samko, S. G., Spaces of Riesz potentials, *Math. URSS-Izv.*, **10** (1976), no. 5, 1089–1117.

[121] Samko, S. G., *Hypersingular Integrals and their Applications*. Taylor and Francis Inc., London, 2002.

[122] Samko, S. G., Kilbas, A. A., and Marichev, O. I., *Fractional Integrals and Derivatives. Theory and Applications*. Gordon and Breach Science Publishers, Yverdon, 1993.

[123] Sánchez, A. and Vázquez, L., Nonlinear wave propagation in disordered media, *Internat. J. Mod. Phys. B*, **5** (1991), no. 18, 2825–2882.

[124] Scher, H., and Montroll, E. W., Anomalous transit-time dispersion in amorphous solids, *Phys. Rev. B*, **12** (1975), no. 6, 2455–2477.

[125] Scher, H., Shlesinger, M. F., and Bendler, J. T., Time-scale invariance in transport and relaxation, *Phys. Today*, **44** (1991), no. 1, 26–34.

[126] Schneider, W. R. and Wyss, W., Fractional diffusion and wave equations, *J. Math. Phys.*, **30** (1989), 134–144.

[127] Schumer, R., Benson, D. A., Meerscaert, M. M., and Wheatcraft, S. W., Eulerian derivation of fractional advection-dispertion equation, *J. Cont. Hydr.*, **48** (2001), 69–88.

[128] Schwartz, L., *Théorie des Distributions*, Vols. I and II. Hermann, Paris, 1957.

[129] Scott, A., *Nonlinear Science: Emergence and Dynamics of Coherent Structures*. Oxford University Press, Oxford, 1999.

[130] Shimizu, N. and Zhang, W., Fractional calculus approach to dynamic problems of viscoelastic materials, *JSME Internat. J. Ser. C*, **42** (1999), no. 4, 825–837.

[131] Shlesinger, M. F., Asymptotic solutions of continuous time random walks, *J. Statist. Phys.*, **10** (1974), no. 5, 421–433.

[132] Shlesinger, M. F., West, B. J., and Klafter, J., Lévi dynamics of enhanced diffusion: application to turbulence, *Phys. Rev. Lett.*, **58** (1987), no. 11, 1100–1103.

[133] Shlesinger, M. F., Zaslavsky, G. M., and Frisch, U. (editors), *Lévi Flight and Related Topics in Physics*, Proceedings of the International Workshop held in Nice, June 27–30, 1994, Lecture Notes in Physics, **450**. Springer-Verlag, Berlin, 1995.

[134] Shlesinger, M. F., Zaslavsky, G. M., and Klafter, J., Asymptotic solutions of continuous time random walks, *J. Statist. Phys.*, **10** (1974), no. 5, 421–434.

[135] Shlesinger, M. F., Zaslavsky, G. M., and Klafter, J., Strange kinetics, *Nature*, **363**(6424) (1993), 31–37.

[136] Sjoberg, M. and Kari, L., Non-linear behaviour of a rubber isolator system using fractional derivatives, *Vehicle Syst. Dyn.*, **37** (2002), no. 3, 217–236.

[137] Smirnov, I. P., Virovlyansky, A. L., and Zaslavsky, G. M., Theory and applications of ray chaos to underwater acoustics, *Phys. Rev. E*, **64** (2002), no. 3, article no. 036221, Part 2.

[138] Soddemann, T., Schiessel, H., and Blumen, A., Molecular-dynamics simulations of polyampholytes: instabilities due to excess charges and external fields, *Phys. Rev. E*, **57** (1998), no. 2, 2081–2090.

[139] Sokolov, I. M., Thermodynamics and fractional Fokker-Planck equations, *Phys. Rev. E*, **63** (2001), no. 5, article no. 056111, 2001.

[140] Sokolov, I. M., Blumen, A., and Klafter, J., Linear response in Complex Systems: CTRW and the fractional Fokker-Planck equations, *Phys. A*, **302** (2001), no. 1-4, 268–278.

[141] Stiassnie, M., On the applications of fractional calculus for the formulation of viscoelastic models, *Appl. Math. Modelling*, **3** (1979), 300–302.

[142] Sung, Y. M., Barkai, E., Silbey, R. J., and Lee, S., Fractional dynamics approach to diffusion-assisted reactions in disordered media, *J. Chem. Phys.*, **116** (2002), no. 6, 2338–2341.

[143] Telban, R. J., Cardullo, F. M., and Guo, L., Investigation of mathematical models of otolith organs for human centered motion cueing algorithms, *J. Amer. Inst. Aeronaut. Astronaut.* (2000), article no. AIAA-2000-4291.

[144] Tsallis, C., General statistical mechanics connections with thermodinamics, *J. Phys. A: Math. Gen.*, **24** (1991), no. 2, 69–72.

[145] Tsallis, C., Generalized entropy-based criterion for consistent testing, *Phys. Rev. E*, **58** (1998), no. 2, 1442–1445.

[146] Tsallis, C., Entropic nonextensivity: a possible measure of complexity chaos, *Sol. Fract.*, **13** (2002), no. 3, 371–391.

[147] Turchetti, G., Usero, D., and Vazquez, L., Hamiltonian systems with fractional time derivative, *Tamsui Oxford J. Math. Sci.*, **18** (2002), no. 1, 31–44.

[148] Turing, A. M., The chemical basis of morphogenesis, *Philos. Trans. Roy. Soc. London Ser. B*, **237** (1952), 37–72.

[149] Upadhyaya, A., Rieu, J. P., Glazier, J. A., and Sawada, Y., Anomalous diffusion and non-Gaussian velocity distribution of Hydra cells in cellular aggregates, *Phys. A*, **293** (2001), no. 3-4, 549–558.

[150] Verkman, A. S., Solute and macromolecule diffusion in cellular aqueous compartments, *Trends in Biochem. Sci.*, **27** (2002), no. 1, 27–33.

[151] Weiss, G. H., *Aspects and Applications of the Random Walk.* North-Holland, Amsterdam, 1994.

[152] Weiss, G. H. and Rubin, R. J., Ramdom walks theory and selected applications, *Adv. Chem. Phys.*, **52** (1983), 363–505.

[153] Werner, D. H. and Mittra, R. (editors), *Frontiers in Electromagnetics.* IEEE Press and John Wiley and Sons, New York, 1999.

[154] West, B. J., *Fractal Physiology and Chaos in Medicine.* World Scientific Publishing Company, Singapore, London, 1990.

[155] West, B. J., *Physiology, Promiscuity and Prophecy at the Millennium: A Tale of Tails.* World Scientific Publishing Company, Singapore, London, 1999.

[156] West, B. J., Grigolini, P., Metzler, R., and Nonnemacher, R., Fractional diffusion and Lévi stable processes, *Phys. Rev. E*, **55** (1997), no. 1, 99–106.

[157] West, B. J., and Picozzi, S., Fractional Langevin model of memory in financial time series, *Phys. Rev. E*, **65** (2002), no. 3, article no. 037106, Part 2B.

[158] Yakushevich, L. V., *Nonlinear Physics of DNA.* John Wiley and Sons, New York, 1990.

[159] Zaslavsky, G. M., Fractional kinetic equation for Hamiltonian chaos, *Phys. D*, **76** (1994), no. 1-3, 110–122.

[160] Zaslavsky, G. M. and Benkadda, S. (editors), *Chaos, Kinetics and Nonlinear Dynamics in Fluids and Plasmas*, Proceedings of the workshop held in Carry-Le Rouet, June 16–21, 1997. Lectures Notes in Physics, **511**. Springer-Verlag, Berlin, 1998.

[161] Zaslavsky, G. M., Edelman, M., and Niyazov, B. A., Self-similarity, renormalization and phase space nonuniformity of Hamiltonian chaotic dynamics, *Chaos*, **7** (1997), no. 1, 159–181.

BOUNDARY VALUE PROBLEMS FOR ANALYTIC AND HARMONIC FUNCTIONS OF SMIRNOV CLASSES IN DOMAINS WITH NON-SMOOTH BOUNDARIES

Vakhtang Kokilashvili
A. Razmadze Mathematical Institute of the Georgian Academy of Sciences, Georgia
kokil@rmi.acnet.ge

Zurab Meshveliani
Sukhumi branch of I. Djavakhishvili Tbilisi State University, Georgia
paata@imath.acnet.ge

Vakhtang Paatashvili
A. Razmadze Mathematical Institute of the Georgian Academy of Sciences, Georgia
paata@imath.acnet.ge

Abstract The Riemann boundary value problem for analytic functions with boundary values from various function spaces is solved. The results are applied to the Dirichlet problem for harmonic functions of weighted Smirnov classes in domains with non-smooth boundaries.

Keywords: Analytic functions, Riemann boundary value problem, Cauchy type integral, piecewise smooth boundary, Smirnov class, harmonic functions, Dirichlet problems, Zygmund class.

Introduction

In this paper we present a complete picture of the solvability of linear boundary value problems for analytic functions and the Dirichlet problem for harmonic functions from Smirnov type classes in "bad" plane domains. We give the solution of the Riemann boundary value problem for an analytic function when the boundary function belongs to the weighted Zygmund class. As for the Dirichlet problem in a simply connected finite domain with an arbitrary rectifiable Jordan boundary, a criterion of the unique solvability for any boundary function from the L^p $(1 < p < \infty)$ class is established. In the case of an arbitrary piecewise smooth boundary, the non-Fredholm cases are exposed; the influence of geometric properties of the boundary on the solvability is revealed. Similar problems are solved in weighted Smirnov classes of harmonic functions. The weight is an arbitrary power function. A special section is devoted to the solution of the Dirichlet problem with the boundary function from the weighted Zygmund class. The solution is sought in a class of harmonic functions representable by the real part of Cauchy type integrals. In all the cases of solvability of the above-mentioned problems, formulas for the solution in terms of Cauchy type integrals as well as for conformal mapping functions are given explicitly.

At present, boundary value problems for analytic functions in the case where the unknown functions are continuous up to the boundary or their continuity is violated only at a finite number of points are well studied and described in the fundamental monographs [35] and [10]. The problems in which the unknown functions may fail to possess boundary values in an infinite set of boundary points have been investigated mainly in classes of analytic functions which are subclasses of the functions representable by Cauchy-Lebesgue type integrals. As an example of such sets, we have considered the $\mathcal{K}^p(\Gamma, \rho)$ classes, i.e., the set of analytic functions which are representable by a Cauchy type integral with density from the weighted Lebesgue space $L^p(\Gamma, \rho)$, where $1 < p < \infty$ and ρ is a power function of the type

$$\rho(t) = \prod_{n}^{k=1} |t - c_k|^{\alpha_k}, \quad -\frac{1}{p} < \alpha_k < \frac{1}{p'}, \quad p' = \frac{p}{p-1};$$

here c_k are the boundary points of the domain ([1], [3], [12], [19], [21], [22], [25], [39]). The problem mentioned above is usually called a discontinuous boundary value problem ([22]). The problem is stated as follows. Let Γ be a simple closed rectifiable curve bounding a finite domain D^+ and an infinite domain D^-. Suppose $G(t)$ and $g(t)$ are measurable functions given on Γ, $g \in L^p(\Gamma, \rho)$, $1 \leq p < \infty$. We have to find a function ϕ from the class $K^p(\Gamma, \rho)$ whose angular boundary values $\phi^+(t)$ (from D^+) and $\phi^-(t)$ (from D^-) satisfy the condition

$$\phi^+(t) = G(t)\phi^-(t) + g(t) \tag{1}$$

a.e. on Γ.

When investigating the above-stated problem (1), it is necessary to restrict both the set of curves Γ and the set of coefficients G.

In the earliest works in this direction, Γ was assumed to be a Lyapunov curve and G to be piecewise continuous ([12], [21], [22], [39]). Considerable extensions of the classes of boundary curves and coefficients as well as of the weight functions have been achieved later on (see [1], [3], [12], [14], [19], [21]–[25], [39], etc.).

It should be noted that when g is only integrable, problem (1) has no solution in the classes $K^1(\Gamma) = K^1(\Gamma; 1)$ even in the simplest cases (e.g., when $G \equiv -1$).

Therefore, one is naturally led to the problem of restricting the set of unknown functions so as to get a sufficiently complete picture of the solvability of problem (1), including some classes of discontinuous coefficients G.

There are a great number of applications of problem (1) to boundary value problems of mathematical physics in continuous and piecewise continuous cases (see [35] and [10]), but not many in discontinuous cases ([3], [21], [39]); the latter mainly concern domains with smooth boundaries. The interest in boundary value problems with partial differential equations in domains with non-smooth boundaries has appreciably increased in the recent years (see, e.g., [2], [4], [6], [8], [9], [31], [41]).

When investigating boundary value problems for harmonic functions of two variables it becomes possible to use the methods and results of the theory of analytic functions. The fruitfulness of such an approach in the case of continuous problems has been shown by N. Muskhelishvili (see [34], [35]), who obtained solutions of a number of boundary value problems for harmonic functions by reducing them to problem (1) with an additional requirement for the function ϕ.

In this paper we present a further development of N. Muskhelishvili's method with the aim to investigate boundary value problems for those harmonic functions which are real parts of the functions from the $\mathcal{K}^p(\Gamma,\rho)$ class. Moreover, the case $p=1$ is not excluded. At this stage we have solved a number of problems connected with properties of conformal mapping of a disk onto a simply connected domain with a non-smooth boundary, investigated problem (1) with oscillating coefficients G, as well as weighted problems for singular Cauchy type integrals. Problem (1) is studied when the function g belongs to the Zygmund class, and so on.

1. The Riemann Problem with Boundary Values from the Zygmund Class

Let Γ be a rectifiable oriented simple closed curve bounding a bounded domain D^+ and an unbounded domain D^-, and let ρ be a given function on Γ which is measurable and different from zero a.e. By $L^p(\Gamma;\rho)$ we denote the set of functions f for which $f\rho$ belongs to the Lebesgue class $L^p(\Gamma)$. Let

$$\mathcal{K}^p(\Gamma,\rho)=\left\{\Phi:\Phi(z)=\frac{1}{2\pi i}\int_\Gamma\frac{f(t)dt}{t-z},\quad z\notin\Gamma,\ f\in L^p(\Gamma,\rho)\cap L(\Gamma)\right\}\tag{1.1}$$

and $K^p(\Gamma)=K^p(\Gamma,1)$. A function $\phi(z)$ from $K^p(\Gamma,\rho)$ has at almost every point of Γ angular boundary values $\phi^\pm(t)$ (see, e.g., [19, p. 22]).

If Γ is a curve such that the Cauchy singular operator

$$S=S_\Gamma:f\to S_\Gamma f,\quad (S_\Gamma f)(t)=\frac{1}{\pi i}\int_\Gamma\frac{\varphi(\tau)d\tau}{\tau-t},\quad t\in\Gamma,\tag{1.2}$$

is continuous in the space $L^p(\Gamma)$, $p>1$, then $\mathcal{K}^p(\Gamma)$ coincides with the set of functions which are analytic on the plane, cut along Γ, and whose restrictions on

D^+ and D^- belong, respectively, to the Smirnov classes $E^p(D^+)$ and $E^p(D^-)$ (see [15], [36], and also [19], p. 30). Here $E^p(D)$ is the class of functions which are analytic in D and for which

$$\sup_{0<r<1} \int_{\Gamma_r} |\Phi(z)|^p |dz| < \infty,$$

where Γ_r are the images of the circumferences of radius r under the conformal mapping of the unit disk $U = \{w : |w| < 1\}$ onto D (see, e.g., [38, p. 205] or [13, p. 422]). It is known that $\phi \in E^p(D)$, if there is a sequence of curves $\Gamma_n \subset D$ converging to the boundary D such that the p-th integral means of φ on Γ_n are uniformly bounded (see [17] and also [13], p. 422).

It is proved in [5] that the operator S_Γ is continuous in the spaces $L^p(\Gamma)$, $p > 1$, if and only if Γ is a regular curve, i.e.,

$$\sup_{\rho>0,\zeta\in\Gamma} l_\zeta(\rho)\rho^{-1} < \infty, \tag{1.3}$$

where $l_\zeta(\rho)$ is the arc-length measure of that part of Γ which is contained in the disk with center at ζ and radius ρ. The set of regular curves will be denoted by R.

Denote by $E^p(\Gamma)$ the set of functions which are analytic in the plane cut along Γ and whose restrictions to D^+ and D^- belong to $E^p(D^+)$ and $E^p(D^-)$, respectively. From the above-mentioned result it follows that if $\Gamma \in R$, then

$$K^p(\Gamma) = E^p(\Gamma), \quad p > 1. \tag{1.4}$$

However, if $p = 1$, then it can be easily shown that

$$E^1(\Gamma) \subset K^1(\Gamma), \quad E^1(\Gamma) \neq K^1(\Gamma). \tag{1.5}$$

Let $\Gamma \in R$ and let ν be the arclength measure on Γ. For $z \in \Gamma$ and $r > 0$ we denote by $B(z,r)$ the disk of radius r centered at the point z. Let $D(z,r) = B(z,r) \cap \Gamma$.

Definition 1.1 A measurable a.e. positive function w is said to be in $A_p(\Gamma)$, $1 < p < \infty$, if

$$\sup_{z\in\Gamma} \frac{1}{\nu D(z,r)} \int_{D(z,r)} w(t) d\nu \left(\frac{1}{\nu D(z,r)} \int_{D(z,r)} w^{1-p'}(t) d\nu \right)^{p-1} < \infty$$

and in $A_1(\Gamma)$, if

$$\frac{1}{\nu(D(z,r))} \int_{D(z,r)} w(\tau) d\nu \leq c \operatorname{ess\,inf}_{t\in D(z,r)} w(t),$$

where c does not depend on z and r.

Functions $w_0(t) = |t-t_0|^\alpha$, $-1 < \alpha \leq 0$ are important examples of functions in $A_1(\Gamma)$.

Let $\alpha > 0$, and let $w \geq 0$ be a weight function. We say that a function f given on Γ belongs to the Zygmund class $Z_{\alpha,w}(\Gamma)$ if

$$\int_\Gamma |f(t)| \log^\alpha \left(2 + \frac{|f(t)|}{w}\right) d\nu < \infty.$$

If $w(t) \equiv 1$, then we write $Z_{\alpha,1}(\Gamma) = Z_\alpha(\Gamma)$, $Z_1(\Gamma) = Z(\Gamma)$. Set $Z_0(\Gamma) = L(\Gamma)$.

Theorem 1.2 *If* $\Gamma \in R$, $w \in A_1(\Gamma)$, $\alpha \geq 1$, *then the operator*

$$T : f \to Tf, \quad (Tf)(t) = \frac{w(\tau)}{\pi i} \int_\Gamma \frac{f(\tau)}{w(\tau)} \frac{d\tau}{\tau - t} \tag{1.6}$$

maps the class $Z_{\alpha,w}(\Gamma)$ *into the class* $Z_{\alpha-1,w}(\Gamma)$. *Moreover*

$$\begin{aligned} &\int_\Gamma w(t) \left| S\left(\frac{f}{w}\right)(t)\right| \cdot \log^{\alpha-1}\left(2 + \left|S\left(\frac{f}{w}\right)(t)\right|\right) d\nu \\ \leq\ & c_1 \int_\Gamma |f(t)| \log^\alpha \left(2 + \frac{|f(t)|}{w(t)}\right) d\nu + c_2 \end{aligned} \tag{1.7}$$

where the constants c_1 *and* c_2 *do not depend on the function* f.

In our proof of this theorem given in [29], we also derived the inequality

$$\begin{aligned} &\int_\Gamma |Sf(t)| \log^{\alpha-1}(2 + |Sf(t)|) w(t) d\nu \\ \leq\ & c_1 \int_\Gamma |f(t)|\ \log^\alpha (2 + |f(t)|) w(t) d\nu + c_2, \end{aligned} \tag{1.8}$$

Let $E^{1,\alpha}(\Gamma)$ be the subset of functions $\phi \in E^1(\Gamma)$ for which $\phi^\pm \in Z_\alpha(\Gamma)$, $\alpha \geq 0$.

We consider the problem of finding a solution of problem (1) in the class $E^{1,\alpha}(\Gamma)$. We assume that

$$\Gamma \in R, \quad G(t) \in H(\Gamma), \quad G(t) \neq 0, \quad g \in Z_{\alpha+1}(\Gamma). \tag{1.9}$$

and denote

$$æ = \frac{1}{2\pi}[\arg G]_\Gamma, \tag{1.10}$$

where $[\,]_\Gamma$ is the increment of $\arg G$ along Γ (see [35]), and $G_æ(t) = G(t)(t-a)^{-æ}$, $a \in D^+$. We also denote

$$\begin{aligned} \gamma(z) &= \frac{1}{2\pi i} \int_\Gamma \frac{\ln G_æ(t) dt}{t - z}, \\ X(z) &= \begin{cases} \exp\gamma(z), & z \in D^+, \\ (z-a)^{-æ}\exp\gamma(z), & z \in D^-. \end{cases} \end{aligned} \tag{1.11}$$

It is not difficult to check that all the possible solutions of the above-posed problem are in the set of functions given by the equality

$$\phi(z) = \frac{X(z)}{2\pi i}\int_\Gamma \frac{g(\tau)}{X^+(\tau)}\frac{d\tau}{\tau - z} + P_{æ-1}(z)X(z), \tag{1.12}$$

where $P_{æ-1}(z)$ is an arbitrary polynomial of order $æ - 1$ for $æ - 1 \geq 0$ and $P_{æ-1}(z) = 0$ for $æ - 1 < 0$.

By Theorem 1.2, we find that this function belongs to $E^{1,\alpha}(\Gamma)$. Thus the following theorem is valid.

Theorem 1.3 *Let conditions* (1.9) *be fulfilled and let the integer æ be defined by* (1.10). *Then*

(i) *for* $æ \geq 0$, *the boundary value problem* (1) *is unconditionally solvable in the class* $E^{1,\alpha}(\Gamma)$, $\alpha \geq 0$, *and all the solutions are given by* (1.12), *where* $P_{æ-1}(z)$ *is an arbitrary polynomial of order* $æ - 1$, $(P_{-1}(z) \equiv 0)$, *and* $X(z)$ *is defined by* (1.11);

(ii) *for* $æ < 0$, *the problem is solvable in* $E^{1,\alpha}(\Gamma)$, $\alpha \geq 0$ *if and only if the conditions*

$$\int_\Gamma g(t)X^+(t)t^k dt = 0, \quad k = 0, 1, \ldots, -æ - 1$$

are fulfilled. If this is the case, then the problem has a unique solution given by (1.12) *with* $P_{æ-1} \equiv 0$.

Along with the classes $E^{1,\alpha}(\Gamma)$ let us consider also the following class of functions

$$K^{1,\alpha}(\Gamma) = \left\{\phi : \phi(z) = \frac{1}{2\pi i}\int_\Gamma \frac{\varphi(t)dt}{t - z}, \quad \varphi \in Z_\alpha(\Gamma)\right\}. \tag{1.13}$$

This class is wider than $E^{1,\alpha}(\Gamma)$.

It is not difficult to show that any solution of problem (1) in $K^{1,\alpha}(\Gamma)$ must belong to the set of functions given by equality (1.12). Therefore, by Theorem 1.3, the following statement is also valid.

Theorem 1.4 *If conditions* (1.9) *are fulfilled, then any solution of problem* (1) *in the class* $K^{1,\alpha}(\Gamma)$ *belongs also to the class* $E^{1,\alpha}(\Gamma)$, *and thus for problem* (1) *in the class* $K^{1,\alpha}(\Gamma)$ *all the statements of Theorem* 1.3 *are valid.*

As we can see from the above, if $G \in H(\Gamma)$, then in problem (1) we can take an arbitrary regular curve as boundary curve. However, when G is a piecewise Hölder function, then we have to restrict the set of admissible curves.

Let $A_1, A_2, \ldots, A_n$ be points on Γ. If a function G belongs to a Hölder class on closed arcs A_kA_{k+1}, $k = 1, 2, \ldots, n$ $(A_{n+1} = A_1)$, then we write $G \in H(\Gamma, A_1, A_2, \ldots, A_n)$. To formulate the precise results on solutions of problem (1), we have to define first some weight function. To this end, we single out a branch of the function $\arg G(t)$. We assume that the points A_k,

$k = 1, 2, \ldots, n$, are enumerated according to the arc coordinate increment. Arbitrarily fixing the value $\arg G(A_1+)$ and moving along Γ in the positive direction, we have continuously varying values of $\arg G(t)$ on the arc A_1A_2 and as a result define the value of $\arg G(A_2-)$. We define $\arg(A_2+)$ so as to have

$$-1 < \beta_1 = \frac{\arg G(A_2-) - \arg G(A_2+)}{2\pi} \leq 0. \tag{1.14}$$

We continue to define $\arg G(t)$ on the arcs A_kA_{k+1} so as to have the relations

$$-1 < \beta_k = \frac{\arg G(A_k-) - \arg G(A_k+)}{2\pi} \leq 0, \quad k = 2, \ldots, n, \tag{1.15}$$

at the points A_k, and arrive at the arc A_nA_1 with the naturally defined value of $\arg G(A_1-)$. Let

$$h = \frac{\arg G(A_1-) - \arg G(A_1+)}{2\pi} \tag{1.16}$$

and

$$æ = \begin{cases} h & \text{if } h \in \mathbf{Z}, \\ [h] - 1 & \text{if } h \overline{\in} \mathbf{Z}, \end{cases} \tag{1.17}$$

where $[h]$ is the integral part of h and Z is the set of integers.

By $R(A_1, A_2, \ldots, A_n)$ we denote the set of curves from R which pass through the points $A_1, A_2, \ldots, A_n$ and are smooth in some one-sided neighborhoods of the points A_k.

We consider the following problem: find a function $\phi \in E^1(\Gamma)$ whose angular boundary values $\phi^{\pm}$ belong to $Z_{\alpha,w}(\Gamma)$. We denote the class of such functions by $E_w^{1,\alpha}(\Gamma)$. Assume that

$$\begin{gathered} \Gamma \in R(A_1, A_2, \ldots, A_n),\ G \in H(\Gamma, A_1, \ldots, A_n),\ G \neq 0, \\ w(t) = \textstyle\prod_{k=1}^n |t - A_k|^{\beta_k},\ g \in Z_{\alpha+1,w}(\Gamma),\ \alpha \geq 0, \end{gathered} \tag{1.18}$$

where the numbers β_k are defined by (1.14)–(1.15).

Let $X(z)$ be the function defined by (1.11). We have

$$\ln G_æ(t) = -æ \ln(t - a) + \ln |G(t)| + i \arg G(t),$$

where $\arg G(t)$ is the above-defined branch. Taking into account that for $\Gamma \in R$ and $\delta < 1$ the function

$$F_\delta(t) = \int_\Gamma \frac{d\nu(\tau)}{|\tau - t|^\delta} \tag{1.19}$$

is bounded, we can conclude that for z in the vicinity of Γ

$$X(z) = \begin{cases} \displaystyle\prod_{k=1}^n (z - A_k)^{\beta_k} X_0(z), & z \in D^+, \\ \displaystyle\prod_{k=1}^n (z - A_k)^{\beta_k} (z - a)^{-æ} X_0(z), & z \in D^-,\quad a \in D^+, \end{cases} \tag{1.20}$$

where

$$0 < \inf_{z \in \overline{D}} |X_0(z)| \le \sup_{z \in \overline{D}} |X_0(z)| < \infty. \tag{1.21}$$

It is known that for $\Gamma \in R$ and a bounded function φ, the function $\exp\{\frac{1}{2\pi i}\int_\Gamma \frac{\varphi(t)dt}{t-z}\}$ belongs to $E^\delta(D^\pm)$ for some $\delta > 0$, see [24]. Therefore, from the above and Theorem 1.2 it follows that the solution of problem (1) is again given by equality, i.e. the following theorem is valid.

Theorem 1.5 *Let $\Gamma \in R(A_1, A_2, \dots, A_n)$, $G \in H(\Gamma, A_1, A_2, \dots, A_n)$, $G \neq 0$, and let æ and w be defined by equality* (1.17) *and* (1.18), *respectively.*

Then the assertions of Theorem 1.3 *are valid for the problem of linear conjugation* (1) *in the class $E_w^{1,\alpha}(\Gamma)$.*

Corollary 1.6 *If $g \in Z(\Gamma)$, $G \in H(\Gamma, A_1, A_2, \dots, A_n), \Gamma \in R(A_1, A_2, \dots, A_n)$, then the assertions of Theorem* 1.3 *are valid for problem* (1) *in the class $E^1(\Gamma)$.*

Remark 1.7 Analogously to the above, one can investigate the linear conjugation problem in the class

$$\widetilde{E}_w^{1,\alpha}(\Gamma) = \{\Phi : \Phi = \Phi_1 + \text{const}, \ \ \Phi_1 \in E_w^{1,\alpha}(\Gamma)\}.$$

In this case the number æ is replaced by the number $\widetilde{æ} = æ + 1$.

Remark 1.8 It is not difficult to verify that Theorem 1.5 remains in force if Γ is a curve of the class R such that

$$0 < m_\pm = \inf_{Z \in D^\pm} \prod_{k=1}^n \frac{|(z-A_k)^{\beta_k}|}{|z-A_k|^\beta} \le \sup_{Z \in D^\pm} \prod \frac{|(z-A_k)^{\beta_k}|}{|z-A_k|^{\beta_k}} = M_\pm < \infty,$$

where β_k are defined by (1.15).

Remark 1.9 It is easy to show that the index æ of a function G in the class $E_w^{1,\alpha}(\Gamma)$ is not smaller than the index $æ_p(G)$ of the same function in the classes $E^p(\Gamma)$, $p > 1$ (for the definition of $æ_p(G)$ see [22] or [3]).

For details of the proofs of the results from this section we refer to [29].

2. The Dirichlet Problem for Harmonic Functions from the Smirnov Classes $e^p(D)$ and $e^p(D, \rho)$

Let D be a bounded domain limited by a simple closed piecewise smooth curve Γ.

Let

$$e^p(D) = \{u : \ u = \text{Re}\Phi, \ \ \Phi \in E^p(D)\}. \tag{2.1}$$

We consider the Dirichlet problem: find a function u satisfying the conditions

$$\left.\begin{array}{c}\Delta u = 0, \quad u \in e^p(D), \quad p > 1, \\ u^+(t) = f(t), \quad t \in \Gamma, \quad f \in L^p(\Gamma).\end{array}\right\} \tag{2.2}$$

Let $u = \mathrm{Re}\Phi$ and

$$\Psi(w) = \sqrt[p]{z'(w)}\Phi(z(w)), \quad |w| < 1, \tag{2.3}$$

where $z = z(w)$ is the function mapping the disk $U = \{w : |w| < 1\}$ conformally onto D. Following [35, §§ 40–41], we assume that

$$\begin{gathered}\Omega(w) = \begin{cases}\Psi(w), & |w| < 1, \\ \overline{\Psi(\frac{1}{\overline{w}})}, & |w| > 1,\end{cases} \\ \Omega_*(w) = \overline{\Omega}\Big(\frac{1}{\overline{w}}\Big), \quad |w| \neq 1.\end{gathered} \tag{2.4}$$

Let γ be the unit circle. Then $\Omega \in \widetilde{\mathcal{K}}^p(\gamma)$, where

$$\widetilde{\mathcal{K}}^p(\gamma) = \{\phi : \phi = \phi_1 + \text{const}, \quad \phi_1 \in K^p(\gamma)\}. \tag{2.5}$$

It is easy to establish that problem (2.2) is equivalent to the problem

$$\left.\begin{array}{r}\Omega^+(\tau) = -\dfrac{\sqrt[p]{z'(\tau)}}{\overline{\sqrt[p]{z'(\tau)}}}\Omega^-(\tau) + g(\tau), \; g(\tau) = 2f(z(\tau))\sqrt[p]{z'(\tau)}, \\ \Omega \in \widetilde{\mathcal{K}}^p(\gamma), \quad \Omega(w) = \Omega_*(w)\end{array}\right\} \tag{2.6}$$

in the sense that every solution $u = \mathrm{Re}\Phi$ generates, according to (2.3) and (2.4), a function Ω satisfying conditions (2.6), and vice versa, any solution Ω of problem (2.6) generates a solution of problem (2.2),

$$u = \mathrm{Re}[\Psi(w(z))\sqrt[p]{w'(z)}],$$

where ψ is the restriction of Ω onto U and $w = w(z)$ is the function inverse to $z = z(w)$ (see, e.g., [19], [37]).

Now we present a scheme for investigating problem (2.6). We restrict ourselves to the case where $\Gamma \in C_A^\nu$, i.e. Γ is a piecewise smooth curve with one angular point A at which the size of the interior angle is equal to $\nu\pi$, $0 \leq \nu \leq 2$. Let $w(A) = c$, $c \in \gamma$. We consider all the possible cases below.

(i) Let $0 < \nu < p$. Suppose that

$$X(w) = \begin{cases}-\sqrt[p]{z'(w)}, & |w| < 1, \\ \overline{\sqrt[p]{z'(\frac{1}{\overline{w}})}}, & |w| > 1.\end{cases} \tag{2.7}$$

Then it is obvious that X belongs to the Hardy class H^p, since $z' \in H^1$ (see, e.g., [13, p. 405]). Moreover,

$$X^+(\tau) = (\tau - c)^{\frac{\nu-1}{p}} z_0^{\frac{1}{p}}(\tau), \quad c = w(A), \tag{2.8}$$

$$z_0^{\pm 1} \in \cap_{\delta>1} H^\delta, \quad (z_0^+)^{\pm 1} \in \cap_{\delta>1} A_\delta(\gamma). \tag{2.9}$$

On the basis of the above facts we find that in the case under consideration the problem has a unique solution, which is given by the equality

$$u(z) = \frac{1}{2}\operatorname{Re}\left(\frac{1}{2\pi i}\int_\gamma \frac{f(z(\zeta))}{\zeta}\frac{\zeta + w(z)}{\zeta - w(z)}d\zeta\right). \tag{2.10}$$

(ii) Let $p < \nu$. Assume that

$$X_1(w) = X(w)(w-c)^{-1}, \tag{2.11}$$

where X is the function defined by (2.7). This time we have

$$X_1(w) \sim (w-c)^{\frac{\nu-p-1}{p}} z_1(w), \quad \text{where} \quad z_1(w) = z_0^{\frac{1}{p}}(w).$$

Therefore, all the solutions of problem (2.6) belong to the set of functions of the form

$$\Omega(w) = \frac{X_1(w)}{2\pi i}\int_\gamma \frac{g(\tau)}{X_1^+(\tau)}\frac{d\tau}{\tau - w} + (\alpha w + \beta)X_1(w), \tag{2.12}$$

where α and β are arbitrary complex constants.

For the homogeneous problem we have $\Omega_0(w) = (\alpha w + \beta)X_1$.

The condition $\Omega_0(w) = [\Omega_0(w)]_*$ implies that $\overline{\beta}c - \alpha = 0$, $\overline{\alpha}c - \beta = 0$ and hence $\beta = \overline{\alpha}c$. Therefore, for the solution of the homogeneous problem we have

$$u_0(z(w)) = \operatorname{Re}\left[\frac{w\overline{\beta}c + \beta}{w - c}\right].$$

This implies

$$u_0(z) = M\operatorname{Re}\frac{c + w(z)}{c - w(z)}, \tag{2.13}$$

where M is an arbitrary real number.

A particular solution of the inhomogeneous problem has the form

$$u_f(z) = \operatorname{Re}\left[\frac{1}{c - w(z)}\left(\frac{1}{2\pi i}\int_\gamma \frac{f(z(\zeta))(\zeta - c)}{\zeta - w(z)}d\zeta - \frac{cw^2(z)}{2\pi i}\int_\gamma \frac{f(z)(\overline{\zeta} - \overline{c})}{\zeta(\zeta - w(z))}d\zeta\right)\right]. \tag{2.14}$$

Consequently, the general solution of problem (2.2) is given by the equality

$$u(z) = u_0(z) + u_f(z), \tag{2.15}$$

where u_0 and u_f are given by equalities (2.13)–(2.14).

(iii) Let $\nu = p$. In this case a solution of the inhomogeneous problem (2.6) is the function $\Omega_0(w) = \alpha X(w) + \beta X_1(w)$, where α and β are arbitrary complex constants.

Further, for constructing a solution of the homogeneous problem (2.2) we proceed from the integral properties of the function X_1. Namely, if $X_1 \in H^p$, then the solution is again given by formula (2.14), and for $X_1 \overline{\in} H^p$ it is trivial.

It should be noted that in the case of piecewise Lyapunov curves we always have $X_1 \overline{\in} H^p$.

As for the inhomogeneous problem, it is, generally speaking, unsolvable, i.e. there exists an $f_0 \in L^p(\Gamma)$ such that for $f = f_0$ problem (2.2) has no solutions, see [26].

For problem (2.2) to be solvable for the given boundary function $f \in L^p(\Gamma)$, it is necessary and sufficient that

$$\frac{X^+(t)}{\pi i} \int_\gamma \frac{g(\tau)}{X^+(\tau)} \frac{d\tau}{\tau - t} \in L^p(\gamma). \tag{2.16}$$

If this condition is fulfilled, we have

$$u(z) = \frac{1}{2} \mathrm{Re} \left[\frac{1}{2\pi i} \int_\gamma \frac{f(z(\zeta))}{\zeta} \frac{\zeta + w(z)}{\zeta - w(z)} d\zeta \right] + u_0(z), \tag{2.17}$$

where

$$u_0(z) = \begin{cases} 0 & \text{for } X_1 \overline{\in} H^p, \\ M \mathrm{Re} \frac{w(z)+w(A)}{w(z)-w(A)} & \text{for } X_1 \in H^p. \end{cases} \tag{2.18}$$

Let us consider in more detail some classes of functions for which condition (2.16) is fulfilled. In the case of arbitrary piecewise smooth curves we can, for example, prove that if $f(t) \ln |w(t) - A| \in L^p(\Gamma)$, then condition (2.16) is fulfilled.

Moreover, if the boundary Γ is piecewise Lyapunov, we can give a necessary and sufficient condition on the boundary function for problem (2.2) to be solvable. For a measurable function φ given on the circumference γ we introduce the function

$$^y\varphi(e^{i\theta}) = \begin{cases} \varphi(e^{i\theta}) & \text{for } |\varphi(e^{i\theta})| > y, \\ 0 & \text{for } |\varphi(e^{i\theta})| \le y. \end{cases} \tag{2.19}$$

Let now the function g be defined by (2.6) and

$$h(\theta) = \frac{g(e^{i\theta})}{|e^{i\theta} - c|^{1/p'}}.$$

It turns out that in the case of piecewise Lyapunov boundaries Γ problem (2.2) is solvable for $\nu = p$ if and only if

$$\int_0^\infty \left| \int_{-\pi}^{\pi} {}^y h(\theta) d\theta \right|^p \frac{dy}{y} < \infty. \tag{2.20}$$

(iv) Let now $\nu = 0$. In this case

$$X(w) = O\Big((w-c)^{-\frac{1}{p}} z_0^{\frac{1}{p}}(w)\Big)$$

and therefore problem (2.2) is solvable only under condition (2.16).

Thus the following theorem is valid.

Theorem 2.1 *Let Γ be a simple, closed, piecewise smooth curve containing one angular point A with interior angle of size $\nu\pi$, $0 \le \nu \le 2$ and let $X_1(w) = (w-c)X(w)$, where X is given by (2.7). Then the Dirichlet problem (2.2)*

- is uniquely solvable for $0 < \nu < p$;

- has a set of solutions depending on one parameter for $p < \nu$;

- is, in general, unsolvable for $p = \nu$ and becomes solvable if and only if condition (2.16) is fulfilled. Moreover, it has a unique solution if $X_1 \overline{\in} H^p$ and a set of solutions depending on only one parameter for $X_1 \in H^p$;

- is, in general, unsolvable for $\nu = 0$. For the solvability it is necessary and sufficient that conditions (2.16) be fulfilled. If condition (2.16) is fulfilled, it has a unique solution.

When the solution exists, it is given by (2.10) for $0 \le \nu < p$, by (2.13)–(2.15) for $p < \nu$ and by (2.17) and (2.18) for $p = \nu$.

On the basis of the above result, we can easily describe the picture of solvability of the Dirichlet problem (2.2) for domains bounded by arbitrary piecewise smooth curves.

An analysis of the above arguments leads to the following observation: in the case of piecewise smooth curves, problem (2.2) is uniquely solvable if $|z'(t)| \in A_p(\gamma)$, however, if there exists a point $c \in \gamma$ such that $|z'(t)||t-c|^{-\frac{1}{p}} \in A_p(\gamma)$, then it has a set of solutions depending on only one parameter. The same phenomenon turns out to be typical in the general case as well.

In [33] the following theorem was proved.

Theorem 2.2 *Let D be a simply connected domain bounded by a rectifiable Jordan curve Γ and let $z = z(w)$ be a function mapping the unit disk conformally onto D. For the Dirichlet problem (2.2) to be solvable for any function $f \in L^p(\Gamma)$ and the corresponding homogeneous problem to have n $(n \ge 0)$ linearly independent solutions forming its general solution, it is necessary and sufficient that there exists a polynomial $P(w)$ of order n with zeros on γ such that $\rho(\zeta) = \left|\frac{z'(\zeta)}{P^p(\zeta)}\right| \in A_p(\gamma)$. Particularly, for problem (2.2) to be uniquely* *solvable for any* $f \in L^p(\Gamma)$, *it is necessary and sufficient that the function* $\rho(\zeta) = |z'(\zeta)|$ *belong to the class* $A_p(\gamma)$.

We remark that we make use of the following result when proving the above theorem [33, pp.72-73].

Theorem 2.3 *Let*

$$\Omega(w) = \frac{1}{2\pi i}\int_\gamma \frac{f(\xi)d\xi}{\xi - w} + a, \quad |w| \neq 1, \quad f \in L^p(\gamma), \quad a = \text{const}.$$

Then for the equality $\Omega(w) = \Omega_*(w)$, $|w| \neq 1$ *to take place, it is necessary and sufficient that*

$$\text{Re} f(\xi) = 0 \quad \textit{and} \quad \int_\gamma f(e^{i\theta})d\theta = -i4\pi \text{Im} a. \tag{2.21}$$

We mention an application of Theorem 2.2. Let $\mu(s)$ be an angle formed by an oriented tangent at the point $t(s)$ and the x-axis. We say that a Jordan rectifiable oriented curve Γ belongs to the class $T(\alpha)$, $0 < \alpha \leq \pi$, if for every its point there exists an arc neighborhood in which the values of the function $e^{i\alpha(s)}$ lie in the angle of the size less than α with the vertex at the origin.

Let $p > 1$ and $\overline{p} = \max(p, p')$. If $\Gamma \in T(\frac{\pi}{\overline{p}})$, then $z' \in A_p$ (see [20]), and therefore Theorem 2.2 yields the following corollary.

Corollary 2.4 *If* Γ *is a simple closed curve of the class* $T(\frac{\pi}{\overline{p}})$, *then the Dirichlet problem* (2.2) *is uniquely solvable. If, however,* Γ *consists both of a family of a finite number of curves of the class* $T(\frac{\pi}{\overline{p}})$ *and of piecewise smooth curves having no angles with values* $\nu = 0$ *and* $\nu = p$, *then problem* (2.2) *is solvable and its general solution contains as many arbitrary real constants as there are angles with vertices greater than* $p\pi$.

Let $\Gamma \in C_A^\nu$, let β be an arbitrary real number, Γ_A a simple curve containing points $z = A$, $z = \infty$ for which $\overline{D} \cap \Gamma_A = \{A\}$ and let the function $\rho(z) = (z - A)^\beta$ be analytic in the plane with a cut on Γ_A. Let, moreover, $z = z(w)$ be a conformal mapping of the disk U onto the domain D, $w(z)$ be its inverse and $c = w(A)$.

We introduce the following classes of functions:

$$E^p(D;\rho) = \{\phi(z) : \phi(z) = \Psi(z)\rho^{-1}(z), \quad z \in D, \quad \Psi \in E^p(D)\},$$
$$e^p(D,\rho) = \{u : u = \text{Re}\phi, \quad \phi \in E^p(D,\rho)\}.$$

Consider now the problem

$$\left.\begin{array}{r} \Delta u = 0, \quad u \in e^p(D,\rho), \quad p > 1, \\ u^+(t) = f(t), \quad t \in \Gamma, \quad f \in L^p(\Gamma,\rho). \end{array}\right\} \tag{2.22}$$

This problem is reduced to

$$\left.\begin{aligned}\Omega^+(\zeta) &= -\frac{\sqrt[p]{z'(\zeta)}(z(\zeta)-z(c))^\beta}{\sqrt[p]{z'(\zeta)}(z(\zeta)-z(c))^\beta}\Omega^-(\zeta)+g(\zeta),\\ \Omega \in \widetilde{\mathcal{K}}^p(\gamma),\ \ |w|\neq 1,&\ \ G\in L^p(\gamma),\ \ \Omega(w)=\Omega_*(w).\end{aligned}\right\} \tag{2.23}$$

If we take into account (2.8) and (2.9) and also the equality

$$z(w)-z(c)=(w-c)^\nu z_1(w), \quad z_1\neq 0,$$

proved in [20], [42], we can investigate problems (2.22) and (2.23) using the scheme proposed above, and get a complete picture of the solvability of problem (2.22).

Let $[x]$ denote the integral part of the number x. The following theorem is valid [28].

Theorem 2.5 *Let*

$$k^*=\frac{\nu}{p}+\nu\beta, \quad k=[k^*]. \tag{2.24}$$

Then

I. *For $k^*\overline{\in}Z$, $k\geq 0$ the Dirichlet problem* (2.22) *is solvable, and its general solution has the form*

$$u(z)=u_f(z)+u_0(z), \tag{2.25}$$

where

$$u_f(z)=\mathrm{Re}\left[\frac{1}{2\pi i[w(z)-w(A)]^k}\int_\Gamma\frac{f(t)(w(t)-w(A))^k}{w(t)-w(z)}\left(1+\frac{w^{k+1}(z)}{w^{k+1}(t)}\right)w'(t)dt\right], \tag{2.26}$$

$$u_0(z)=\mathrm{Re}\left[Q_k(w(z))(w(z)-w(A))^{-k}\right]$$

and $Q_k(w)=\sum_{j=0}^k a_jw^j$ is an arbitrary polynomial whose coefficients satisfy the conditions

$$c_k\overline{a}_j=a_{k-j}, \quad c_k=(-1)^{k+1}c^k. \tag{2.27}$$

II. *For $k^*\overline{\in}Z$, $k\leq -1$:*

(a) *if $k=-1$, then the problem is unconditionally and uniquely solvable. The solution has the form $u(z)=u_f(z)$, where $u_f(z)$ is defined by equality* (2.26)*;*

(b) *if $k\leq -2$, then for the problem to be solvable, it is necessary and sufficient that the conditions*

$$\int_\Gamma f(t)(w(t)-w(A))^kw^j(t)w'(t)dt=0, \quad j=0,\dots,-k-2, \tag{2.28}$$

be fulfilled, and if these conditions are fulfilled, then its unique solution is the function given by equality (2.26).

When $k^* \in Z$, *problem* (2.22) *is unsolvable in the general case.*

III. *For* $k^* \in Z$, $k \geq 0$ *for the problem to be solvable, it is necessary and sufficient that*

$$\frac{1}{(w(t)-w(A))^k}\int_\Gamma \frac{f(\tau)(w(\tau)-w(A))^k}{w(\tau)-w(t)} w'(\tau)d\tau \in L^p(\Gamma;\rho). \tag{2.29}$$

If this condition is fulfilled, then:

(a) *if* $\rho(t)[w(t)-w(A)]^{-k} \in L^p(\Gamma)$, *the solutions are given by equalities* (2.25)–(2.27);

(b) *if* $\rho(t)[w(t)-w(A)]^{-k} \overline{\in} L^p(\Gamma)$, *the solution is again given by equalities* (2.25)–(2.27), *if in addition to conditions* (2.28) *we have* $Q_k(a) = 0$.

IV. *For* $k^* \in Z$, $k \leq -1$:

(a) *if* $k = -1$, *then for the problem to be solvable, it is necessary and sufficient that condition* (2.29) *be fulfilled, and if this condition is fulfilled, problem* (2.22) *is uniquely solvable and its solution is the function* u_f;

(b) *if* $k \leq -2$, *then for the problem to be solvable, it is necessary and sufficient that conditions* (2.28)–(2.29) *be fulfilled. If they are satisfied, the solution is the function* u_f.

Remark 2.6 After the works [26]–[28], [32] and [19], in [7] some results analogous to those presented above (regarding the character of solvability even without construction of solutions) were in a somewhat different way established for the particular case of piecewise Lyapunov curves and weight functions $\rho = |t-c|^\beta$, $-\frac{1}{p} < \beta < \frac{1}{p'}$.

3. The Dirichlet Problem in the Class $e^1(D)$ when the Boundary Function Belongs to the Zygmund Class

It is not difficult to see that if one looks for solutions of problem (2.2) in the space $e^1(D)$, instead of $e^p(D)$, $p > 1$, and, respectively, takes arbitrary functions $f \in L(\Gamma)$ as boundary values, the problem is far from being solvable for all f even if D is a disk. Therefore, to get a more or less visible picture of the solvability as well as to cover the cases with non-smooth boundary curves, we have to restrict the set of admissible functions f. The results presented in Section 1 allow us to realize this purpose.

We consider the Dirichlet problem which is formulated as follows: find a function u satisfying the conditions

$$\begin{cases} \Delta u = 0, & u \in e^1(D), \\ u^+(t) = f(t), & t \in \Gamma, \quad f \in Z(\Gamma). \end{cases} \tag{3.1}$$

Since $u \in e^1(D)$, there exists a Φ such that $u(z) = \mathrm{Re}\phi(z)$, $\phi \in E^1(D)$. If we define the functions Ψ and Ω by equalities (2.3) and (2.4), then it can be easily seen that $\Omega \in \widetilde{E}^1(\gamma)$, and following the arguments of Section 2, we obtain that to solve problem (3.1) is equivalent to solving problem

$$\begin{cases} \Omega^+(\tau) = -\frac{\overline{z'(\tau)}}{z'(\tau)}\Omega^-(\tau) + g(\tau), & \Omega \in \widetilde{E}^1(\gamma) \\ \Omega_*(w) = \Omega(w), & |w| \neq 1, \end{cases} \tag{3.2}$$

where $g(\tau) = 2f(z(\tau))z'(\tau)$, and Ω_* is defined by equality (2.4).

In order to apply the results of Section 1 to that problem, we have to assume that the coefficient $G(t) = -\frac{\overline{z'(t)}}{z'(t)}$ in the problem (3.2) of linear conjugation is piecewise Hölder continuous. Therefore, in what follows the boundary Γ of the domain D will be supposed to belong to $C^{1,\nu_1,\nu_2,\dots,\nu_n}_{A_1,A_2,\dots,A_n}$, i.e., Γ is a Jordan piecewise Lyapunov curve with angular points $A_1, A_2, \dots A_n$ and angle sizes $\pi\nu_k$ at those points.

First of all we assume that $n = 1$, $A_1 = A$, $\nu_1 = \nu$, $0 < \nu \leq 2$. Let $w(A) = c$. We consider separately the cases

$$\text{(i) } \nu > 1, \qquad \text{(ii) } 0 < \nu < 1 \qquad \text{and} \qquad \text{(iii) } \nu = 1.$$

(i) $\nu > 1$. Assume that

$$X(w) = \begin{cases} -z'(w)(w-c)^{-1}, & |w| < 1, \\ \overline{z'(\frac{1}{\overline{w}})}(w-c)^{-1}, & |w| > 1. \end{cases} \tag{3.3}$$

Then the boundary condition from (3.2) takes the form

$$\Omega^+(\tau)[X^+(\tau)]^{-1} = \Omega^-(\tau)[X^-(\tau)]^{-1} + g(\tau)[X^+(\tau)]^{-1}, \quad \tau \in \gamma.$$

In the case under consideration

$$z'(w) = (w-c)^{\nu-1} z_0(w), \tag{3.4}$$

where $z_0 \in H(\gamma)$, $z_0(w) \neq 0$ (see [20], [42] and also [19], p. 146). This makes it possible to show that $g \in Z_{1,w}(\gamma)$, $w = |z'(\tau)|$ and $g \in Z(\gamma)$.

It follows from (3.3) and (3.4) that for points w close to γ,

$$X(w) = (w-c)^{\nu-2} z_0(w), \quad z_0(w) \neq 0, \quad z_0 \in H(\gamma), \tag{3.5}$$

where under the assumption (i) we have $-1 < \nu - 2 \leq 0$. As usual (see, e.g., [22, Ch.IV]), this implies that all the solutions of the problem of linear conjugation from (3.2) are contained in the set of functions

$$\Omega(w) = \frac{X(w)}{2\pi i}\int_\gamma \frac{g(\zeta)}{X^+(\zeta)}\frac{d\zeta}{\zeta - w} + (Bw + C)X(w) = \Omega_g(w) + \Omega_0(w), \tag{3.6}$$

where $\Omega_0(w) = X(w)(Bw + C)$, and B and C are arbitrary complex constants. The condition $\Omega_0(w) = [\Omega_0(w)]_*$ results in

$$u_0(z) = M\mathrm{Re}\left[\frac{c + w(z)}{c - w(z)}\right], \quad c = w(A), \tag{3.7}$$

where M is an arbitrary real constant.

Since $g \in Z(\gamma)$ and $|X^+(\tau)| \sim |\tau - c|^{\nu-2}$ for $\nu - 2 \in (-1, 0]$, the function $\Omega_g^+(t)$ belongs to $L(\Gamma)$ by Theorem 1.2, and it is not difficult to show that in the disk U the function Ω belongs to the Hardy class H^1. Hence, we obtain that $\Omega \in \widetilde{E}^1(\gamma)$. Thus Ω is a particular solution of the problem of conjugation from (3.2). Therefore, the function $\frac{1}{2}[\Omega_g + (\Omega_g)_*]$ will satisfy all the conditions from (3.2). Finally, for $\Gamma \in C_A^{1,\nu}$, $\nu > 1$ all the solutions of problem (3.1) are given by the equality

$$u(z) = u_f(z) + u_0(z), \tag{3.8}$$

where the functions $u_f(z)$ and $u_0(z)$ are defined by equalities (2.14) and (2.13).

(ii) $-1 < \nu - 1 < 0$. Let

$$Y(w) = \begin{cases} \overline{-z'(w)}, & |w| < 1, \\ z'(\frac{1}{\overline{w}}), & |w| > 1. \end{cases} \tag{3.9}$$

Then all the solutions of problem (3.2) are contained in the set of functions

$$\Omega(w) = \frac{Y(w)}{2\pi i} \int_\gamma \frac{g(\tau)}{Y^+(\tau)} \frac{d\tau}{\tau - w} + CY(w), \quad C \in \mathbf{C}. \tag{3.10}$$

Just in the same way as in [19, pp. 157–158] we establish that $C = 0$ and hence $u_0 = 0$. As for the inhomogeneous problem, this time we are unable to state that $g \in Z(\gamma)$, since in this case z' is an unbounded function. Therefore, the function f is required to meet harder conditions (than belonging to $Z(\Gamma)$), which guarantee the inclusion $g \in Z(\gamma)$. Assume that the condition

$$f(t) \log \left(2 + \left|\frac{f(t)}{w(t) - w(A)}\right|\right) \in L(\Gamma), \tag{3.11}$$

is fulfilled. One can prove that this condition already guarantees the inclusion $g \in Z(\gamma)$. Therefore, $\Omega_g \in \widetilde{E}^1(\gamma)$, and by virtue of this fact the solution of problem (3.1) is given by equality (2.10).

(iii) $\nu = 1$. In this case Γ is a Lyapunov curve. Then $z'(w)$ is a non vanishing Hölder function, and from the condition $f \in Z(\Gamma)$ it follows that $g \in Z(\gamma)$. Therefore, (2.10) provides the unique solution of problem (3.1). But in the case under consideration the condition $g \in Z(\gamma)$ is not necessary. The matter is that the function $f \in Z(\Gamma)$ given by equality (2.10) is the solution even in the case where $S_\gamma(\frac{g(\tau)}{\tau}) \in L(\gamma)$. For this to be valid, it suffices in its turn that $S_\Gamma f \in L(\Gamma)$.

Simple calculations show that condition (3.11) for $\nu \in (0, 1)$ is equivalent to the pair of conditions

$$f \in Z(\Gamma), \quad |f(t)| \log \left(2 + \left|\frac{f(t)}{w(t) - w(A)}\right|\right) \in L(\Gamma), \tag{3.11$_1$}$$

as well as to the condition

$$f(t)\log\left(2+\frac{|f(t)|}{|t-A|}\right)\in L(\Gamma), \qquad (3.11_2),$$

or to the conditions

$$f\in Z(\Gamma),\quad |f(t)|\log\left(2+\frac{1}{|t-A|}\right)\in L(\Gamma). \qquad (3.11_3).$$

As a result, we have the following statement.

Theorem 3.1 *Let Γ be a simple, piecewise Lyapunov curve with only one angular point A of angle size $\pi\nu$, $0<\nu\leq 2$, confining a bounded domain D. Moreover, let $z=z(w)$ be a conformal mapping of the unit disk onto D, let $w=w(z)$ be the inverse function and $w(A)=c$. Then:*
(i) if $1<\nu\leq 2$ and $f\in Z(\Gamma)$, the Dirichlet problem in the class $e^1(D)$ is solvable and has a solution depending on one real parameter; the solution is given by equalities (2.13)–(2.15);
(ii) if $0<\nu<1$ and one of conditions (3.11), (3.11_1)–(3.11_3) (equivalent to each other) is fulfilled, then problem (3.1) is uniquely solvable, and its solution is given by equality (2.10);
(iii) if $\nu=1$, $f\in L(\Gamma)$ and $(S_\Gamma f)\in L(\Gamma)$, the problem is uniquely solvable, and its solution is again given by equality (2.10).

Basing on Theorem 3.1, we can easily investigate the Dirichlet problem in the class $e^1(D)$ for a common piecewise Lyapunov curve, as well. Namely, the following theorem is valid.

Theorem 3.2 *Let $\Gamma\in C^{1,\nu_1,\nu_2,\ldots,\nu_n}_{A_1,A_2,\ldots,A_n}$, $0<\nu_k\leq 2$, $k=\overline{1,n}$, where the points A_k are enumerated so that*

$$1<\nu_1\leq 2,\ 1<\nu_2\leq 2,\ldots,1<\nu_j\leq 2,\ \ 0<\nu_{j+1}<1,\ldots,\ \ 0<\nu_n<1.$$

Assume that f satisfies the conditions

$$f\in Z(\Gamma)\quad \textit{and}\quad f(t)\log\left(2+\frac{|f(t)|}{\prod_{i=j+1}^{n}|t-A_j|}\right)\in L^1(\Gamma). \qquad (3.12)$$

Then problem (3.1) is solvable, and its general solution is given by the equality

$$u(z)=u_f(z)+\sum_{i=1}^{j}M_i\mathrm{Re}\left[\frac{c_i+w(z)}{c_i-w(z)}\right],\quad c_i=w(A_i), \qquad (3.13)$$

where

$$\begin{aligned}u_f(z) &= \mathrm{Re}\left[\frac{1}{\rho(w(z))}\left(\frac{1}{2\pi i}\int_\gamma\frac{f(z(\tau))\rho(\tau)}{\tau-w(z)}\,d\tau\right.\right.\\ &\left.\left.+\frac{(-1)^j w^{j+1}(z)}{2\pi i}\prod_{k=1}^{j}A_i\int_\gamma\frac{f(z(\tau))\overline{\rho}(\tau)}{\tau(\tau-w(z))}\,d\tau\right)\right].\end{aligned} \qquad (3.14)$$

and $\rho(w(z)) = \prod_{k=1}^{j}(w(z) - c_k)$ *if* $j \neq 0$ *and* $\rho(w(z)) = 1$ *for* $j = 0$.

Consider the Dirichlet problem in the weight class $e^1(D, \rho)$, $\rho(t) = |w'(t)|$, where $w = w(z)$ is the conformal mapping of the domain D onto the unit disk. If the boundary Γ of D is a piecewise Lyapunov curve with one angular point A of angle size $\nu\pi$, $0 < \nu \leq 2$, then $|w'(t)| \sim |t - A|^{\frac{1}{\nu}-1}$.

Thus, the problem is to find a function u in the domain D, bounded by a simple piecewise Lyapunov curve Γ with one angular point A of angle size $\nu\pi$, $1 < \nu \leq 2$, by the conditions

$$\left.\begin{aligned} \Delta u = 0, \quad u \in e^1(D, \rho), \quad \rho = |w'(t)| \sim |t - A|^{\frac{1}{\nu}-1}, \\ u\big|_\Gamma = f, \quad f \in Z_{1,\rho}(\Gamma). \end{aligned}\right\} \tag{3.15}$$

By the definition of Smirnov and Zygmund classes, it immediately follows that
(i) if $u = \text{Re}\phi$, then the analytic function $\Psi = \phi(z)z'(w)$ belongs to the Hardy class H^1, and the function Ψ^+ to the class $Z(\gamma)$;
(ii) $g(\tau) = 2f(z(\tau))z'(\tau) \in Z(\gamma)$.

Since for any $f \in Z_{1,\rho}(\Gamma)$ and an arbitrary value ν the function g belongs to $Z(\gamma)$, it is not necessary to introduce additional restrictions on f, and it is not difficult to conclude that the following result holds.

Theorem 3.3 *For problem* (3.15) *the following conclusions are valid:*
(i) *if* $\nu \in (1, 2]$, *then the problem is solvable and the set of its solutions given by equalities* (2.13)–(2.15) *depends on one parameter only;*
(ii) *if* $\nu \in (0, 1)$, *then the problem is uniquely solvable and its solution is given by equality* (2.10).

In the case of an arbitrary piecewise Lyapunov curve an analogue of Theorem 3.2, in which condition (3.12) is replaced by the condition $f \in Z_{1,\rho}(\Gamma)$, is valid.

The proofs of the results of Section 3 can be found in [30].

Acknowledgement

The authors are grateful to the referee for many encouraging comments.

References

[1] Böttcher, A. and Karlovich, Yu. I., *Carleson Curves, Muckenhoupt Weights and Toeplitz Operators*, Progress in Mathematics, **154**. Birkhäuser Verlag, Basel, Boston, Berlin, 1997.

[2] Chkadua, O., The Dirichlet, Neumann and mixed boundary value problems of the theory of elasticity in n-dimensional domains with boundaries containing closed cuspidal edges, *Math. Nachr.*, **189** (1998), 61–105.

[3] Danilyuk, I., *Nonregular Boundary Value Problems in the Plane.* Nauka, Moscow, 1975 (in Russian).

[4] Dauge, M., *Elliptic Boundary Value Problems in Corner Domains*, Lecture Notes in Math., **1341**. Springer-Verlag, Berlin, Heidelberg, New York, 1997.

[5] David, G., Opérateurs intégraux singuliers sur certaines courbes du plan complexe, *Ann. Sci. École Norm. Sup.*, **17** (1984), 157–189.

[6] Duduchava, R., Latsabidze, T., and Saginashvili, A., Singular integral operators with the complex conjugation on curves with cusps, *Integr. Equat. Oper. Theor.*, **22** (1995), 1–36.

[7] Duduchava, R. and Silbermann, B., Boundary value problems in domains with peaks, *Mem. Differential Equations Math. Phys.*, **21** (2000), 1–122.

[8] Fabes, E. B., Jodeit, Jr. M., and Lewis, J. E., Double layer potentials for domains with corners and edges, *Indiana Univ. Math. J.*, **26** (1977), 95–114.

[9] Fabes, E. B., Jodeit, Jr. M., and Rivière, N. M., Potential techniques for boundary value problems on C^1 domains, *Acta Math.*, **141** (1978), 165–186.

[10] Gakhov, F. D., *Boundary Value Problems.* Pergamon Press, Oxford, New York, Paris, 1966. Extended Russian edition: Nauka, Moscow, 1977.

[11] Genebashvili, I., Gogatishvili, A., Kokilashvili, V. and Krbec, M., *Weight Theory for Integral Transforms on Spaces of Homogeneous Type*, Pitman Monographs and Surveys in Pure and Applied Mathematics, **92**. Addison Wesley Longman, Harlow, 1998.

[12] Gokhberg, I. and Krupnik, N. Ya., *Introduction to the Theory of One-Dimensional Singular Integral Operators.* Shtiintsa, Kishinev, 1973 (in Russian).

[13] Goluzin, G. M., *Geometric Theory of Functions of a Complex Variable.* Nauka, Moscow, 1966 (in Russian).

[14] Gordadze, E., On boundary value problem of linear conjugation in the case of non-smooth lines and some measurable coefficients, *Proc. A. Razmadze Math. Inst. Georgian Acad. Sci.*, **123** (2000), 149–151.

[15] Havin, V. P., Boundary properties of Cauchy type integrals and harmonically conjugated functions in domains with rectifiable boundary, *Mat. Sb.*, **68** (1965), no. 4, 499–517 (in Russian).

[16] Hofmann, S., Weighted norm inequalities and vector valued inequalities for certain rough operators, *Indiana Univ. Math. J*, **42** (1993), 1–14.

[17] Keldysh, M. V. and Lavrent'yev, M. A., Sur la représentation conforme des domines limites par des courbes rectifiables, *Ann. Ecole Norm.*, **54** (1937), no. 3, 1–38.

[18] Khuskivadze, G. A., On conjugate functions and Cauchy type integrals, *Trudy Tbiliss. Mat. Inst. Razmadze Akad. Nauk Gruzin. SSR*, **31** (1966), 5–54 (in Russian).

[19] Khuskivadze, G., Kokilashvili, V., and Paatashvili, V., Boundary value problems for analytic and harmonic functions in domains with nonsmooth boundaries. Applications to conformal mappings. *Mem. Differential Equations Math. Phys.*, **14** (1998), 1–195.

[20] Khuskivadze, G. and Paatashvili, V., On conformal mapping of a circle onto the domain with piecewise smooth boundary, *Proc. A. Razmadze Math. Inst. Georgian Acad. Sci.*, **116** (1998), 123–132.

[21] Khvedelidze, B. V., Linear discontinuous boundary value problems of function theory, singular integral equations, and some of their applications, *Trudy Tbiliss. Mat. Inst. Razmadze Akad. Nauk Gruzin. SSR*, **23** (1956), 3–158 (in Russian).

[22] Khvedelidze, B. V., The method of Cauchy type integrals for discontinuous boundary value problems of the theory of holomorphic functions of one complex variable, *Itogi Nauki i Tekhniki, Sovrem. Probl. Mat.*, **7** (1975), 5–162 (in Russian). English translation: *J. Sov. Math.*, **7** (1977), 309–414.

[23] Kokilashvili, V., A survey of recent results of Georgian mathematicians on boundary value problems for holomorphic functions, *Mem. Differential Equations Math. Phys.*, **23** (2001), 85–138.

[24] Kokilashvili, V. M. and Paatashvili, V. A., On boundary value problem of linear conjugation with measurable coefficients, *Trudy Tbiliss. Mat. Inst. Razmadze Akad. Nauk Gruzin. SSR*, **55** (1977), 59–92 (in Russian).

[25] Kokilashvili, V. M. and Paatashvili, V. A., Discontinuous problem of linear conjugation and singular integral equations, *Differentsial'nye Uravneniya*, **16** (1980), no. 9, 1650–1659 (in Russian). English translation: *Differential Equations*, **16** (1980), no. 9, 1067–1075 (1981).

[26] Kokilashvili, V. and Paatashvili, V., On the Riemann–Hilbert problem in the domain with a non-smooth boundary, *Georgian Math. J.*, **4** (1997), no. 3, 279–302.

[27] Kokilashvili, V. and Paatashvili, V., Neumann problem in a class of harmonic functions in domains with a piecewise Lyapunov boundary, *Mem. Differential Equations Math. Phys.*, **12** (1997), 114–121.

[28] Kokilashvili, V. and Paatashvili., V., On the Dirichlet and Neumann problems in the domains with piecewise smooth boundaries, *Bull. Georg. Acad. Sci.*, **159** (1999), no. 2, 181–184.

[29] Kokilashvili, V. and Paatashvili, V., A problem of linear conjugation for analytic functions with boundary values from the Zygmund class, *Georgian Math. J.*, **9** (2002), no. 2, 309–324.

[30] Kokilashvili, V. and Paatashvili, V., The Dirichlet problem for functions of Smirnov class with boudary function from the Zygmund class, *Georgian Math. J.*, to appear.

[31] Maz'ya, V. G. and Soloviev, A., On the integral equation of the Dirichlet problem in a plane domain with cusps on the boundary, *Mat. Sb.*, **180** (1989), no. 9, 1211–1233 (in Russian).

[32] Meshveliani, Z., Dirichlet problem for functions of Smirnov weighted class analytic in that domain with piecewise smooth boundaries, *Rep. Enlarged Ses. Sem. I. Vekua Inst. Appl. Math.*, **13** (1998), 30–32.

[33] Meshveliani, Z. and Paatashvili, V., On Smirnov classes of harmonic functions, and Dirichlet problem, *Proc. A. Razmadze Math. Inst. Georgian Acad. Sci.*, **123** (2000), 61–91.

[34] Muskhelishvili, N. I., *Applications des Intégrales Analogues à Celles de Cauchy à Quelques Problèmes de la Physique Mathématique.* Tbilisi University Press, Tbilisi, 1922.

[35] Muskhelishvili, N. I., *Singular Integral Equations, Boundary Value Problems and Some of Their Applications in Mathematical Physics.* Nauka, Moscow, 1968 (in Russian).

[36] Paatashvili, V. A., The membership in the classes E^p of analytic functions representable by a Cauchy type integral, *Trudy Tbiliss. Mat. Inst. Razmadze Akad. Nauk Gruzin. SSR*, **42** (1972), 87–94 (in Russian).

[37] Paatashvili, V. A. and Khuskivadze, G. A., The Riemann-Hilbert problem in domains with nonsmooth boundaries, *Proc. A. Razmadze Math. Inst.*, **106** (1993), 105–122 (in Russian).

[38] Privalov, I. I., *Boundary Properties of Analytic Functions.* Gostekhizdat, Moscow, Leningrad, 1950 (in Russian).

[39] Simonenko, I. B., The Riemann boundary-value problem for n pairs of functions with measurable coefficients and its application to the study of singular integrals in L_p spaces with weights, *Izv. Akad. Nauk SSSR Ser. Mat.*, **28** (1964), 277–306 (in Russian).

[40] Simonenko, I. B., Some general questions of the theory of the Riemann boundary value problem, *Math. USSR Izv.*, **2** (1968), 1091–1099.

[41] Soloviev, A. A., Estimate in L^p of integral operators connected with the spaces of analytic and harmonic functions, *Sibirsk. Mat. Zh.*, **26** (1985), no. 3, 168–191 (in Russian).

[42] Warschawskiĭ, S. E., Über das Randverhatlen der Abletung der Abbildungsfunktion der konformer Abbildung, *Math. Zeitschr.*, **35** (1932), no. 3-4, 321–456.

[43] Zygmund, A., *Trigonometric Series.* Cambridge Univ. Press, London, New York, 1968.

AN ESTIMATE FOR THE DIMENSION OF THE KERNEL OF A SINGULAR OPERATOR WITH A NON-CARLEMAN SHIFT

Viktor G. Kravchenko
Área Departamental de Matemática, F.C.T., Universidade do Algarve, Portugal
vkravch@ualg.pt

Rui C. Marreiros
Área Departamental de Matemática, F.C.T., Universidade do Algarve, Portugal
rmarrei@ualg.pt

Dedicated to Professor Georgii Litvinchuk on the occasion of his 70th birthday

Abstract An estimate for the dimension of the kernel of the singular integral operator $I - cUP_+$, $L_2^n(\mathbb{T}) \to L_2^n(\mathbb{T})$, with a non-Carleman shift is obtained, where P_+ is the Cauchy projector, U is the isometric shift operator and $c(t)$ is a continuous matrix function. It is supposed that the shift has a finite set of fixed points and all the eigenvalues of the matrix $c(t)$ at the fixed points, simultaneously belong either to the interior of the unit circle $\mathbb{T}$ or to its exterior.

Keywords: singular integral operators, shift operators.

1. Introduction

We consider the singular integral operator $T : L_2^n(\mathbb{T}) \to L_2^n(\mathbb{T})$,

$$T = I - cUP_+ \tag{1.1}$$

on the unit circle $\mathbb{T}$, with a non-Carleman shift (diffeomorphism) $\alpha : \mathbb{T} \to \mathbb{T}$, which has a finite set of fixed points $\{\tau_1, ..., \tau_s\}, s \geq 1$, I is the identity operator, $c \in C^{n \times n}(\mathbb{T})$ is a continuous matrix function,

$$(U\varphi)(t) = |\alpha'(t)|^{\frac{1}{2}} \varphi(\alpha(t)) \tag{1.2}$$

is the isometric shift operator and

$$P_{\pm} = \frac{1}{2}(I \pm S)$$

are the mutually complementary projection operators, where

$$(S\varphi)(t) = (\pi i)^{-1} \int_{\mathbb{T}} \varphi(\tau)(\tau - t)^{-1} d\tau$$

is the operator of singular integration.

We note that for the singular integral operator with shift

$$T(A_1, A_2) = A_1 P_+ + A_2 P_-,$$

where

$$A_1 = a_1 I + b_1 U, \quad A_2 = a_2 I + b_2 U$$

and $a_1, a_2, b_1, b_2 \in C^{n \times n}(\mathbb{T})$, only its Fredholm theory is known [4]. It can be formulated as follows: the operator $T(A_1, A_2)$ is Fredholm in $L_2^n(\mathbb{T})$ if and only if the functional operators A_1 and A_2 are continuously invertible in $L_2^n(\mathbb{T})$. The spectral properties of the operator $T(A_1, A_2)$ are less studied, even for the case of a Carleman shift. For the case of a non-Carleman shift, there exists only the work [1], up to authors' knowledge. In that paper an estimate for the defect number $\dim \ker T$ (with $n = 1$) was obtained in the case when $|c(\tau_j)| < 1$ (one of the two alternative sets of Fredholm conditions for the operator T).

We can also write T in the form

$$T = (I - cU)P_+ + P_-.$$

So the question of Fredholmeless of the operator T is reduced to the question of continuous invertibility of the operator $I - cU$. Keeping in mind the necessary and sufficient invertibility conditions [4] for this operator, we consider the following two extreme forms of these conditions:

1) First we assume c to be a continuous matrix function whose spectrum at the fixed points is contained in the open unit disk. This part of our paper is a natural generalization, from the scalar case to the matrix one, of the results of [1].

2) Afterwards, considering a linear fractional shift, we suppose c such that its spectrum at the fixed points is contained in the exterior of the unit disk and we obtain an estimate for $\dim \ker T$. This result is new even for the scalar case.

2. An estimate for $\dim \ker T$

Wherever r arises, it will denote a polynomial matrix $n \times n$ satisfying the conditions

$$P_+ r^{\pm 1} P_+ = r^{\pm 1} P_+. \tag{2.1}$$

As usual, $\mathbb{T}_+(\mathbb{T}_-)$ denotes the interior (exterior, respectively) of the unit disk, $\sigma(g)$ is the spectrum of a matrix $g \in C^{n\times n}$ and $\rho(g)$ is the spectral radius of g.

Proposition 2.1 *For any continuous matrix function $d \in C^{n\times n}(\mathbb{T})$ such that*

$$\sigma[d(\tau_j)] \subset \mathbb{T}_+, \quad j = \overline{1,s}, \tag{2.2}$$

there exist an induced matrix norm $\|.\|_0$ and a polynomial matrix r satisfying the condition

$$\max_{t\in\mathbb{T}} \left\|r(t)d(t)r^{-1}(\alpha(t))\right\|_0 < 1. \tag{2.3}$$

Proof. Let

$$\rho_j \equiv \rho[d(\tau_j)], \quad j = \overline{1,s}.$$

Under condition (2.2) there exists some $q > 0$ such that

$$\rho_j < q < 1, \quad j = \overline{1,s}.$$

For each $d(\tau_j)$ there exists a matrix norm $\|.\|_j$ such that (see, for instance, Lemma 5.6.10, in [3])

$$\rho_j \le \|d(\tau_j)\|_j \le q < 1, \quad j = \overline{1,s}.$$

It is easy to show that the function

$$\|g\|_* = \max_{j=\overline{1,s}}\{\|g\|_j\}$$

defines a matrix norm. From here there exists an induced matrix norm $\|\cdot\|_0$ such that (see, for instance, Theorem 5.6.26, in [3])

$$\|g\|_0 \le \|g\|_* .$$

Thus we have

$$\|d(\tau_j)\|_0 \le \|d(\tau_j)\|_* \le q < 1, \quad j = \overline{1,s}.$$

Now we consider

$$r = r_0 E_n,$$

where

$$r_0(t) = \prod_{k=1}^{m}(t-\lambda_k), |\lambda_k| > 1, k = \overline{1,m}$$

and E_n is the identity matrix. Then the proof proceeds analogously to the proof of Lemma 2.1 in [1]. □

Proposition 2.2 *Let*

$$\begin{aligned} N &= I - aUP_+, r \\ R &= rP_- + (I - aU)P_+, \\ M &= I - rP_+r^{-1}P_-N^{-1}, \end{aligned} \tag{2.4}$$

where $a \in C^{n\times n}(\mathbb{T})$, N is an invertible operator, r is a polynomial matrix $n \times n$ satisfying the conditions (2.1) *and*

$$l = \sum_{i=1}^{m} \max_{j=\overline{1,m}} l_{i,j}, \tag{2.5}$$

where $l_{i,j}$ is the degree of the element $r_{i,j}$ of the polynomial matrix r. Then the relations

$$\dim \ker R = \dim \ker M \leq l \tag{2.6}$$

hold.

Proof. It is easy to verify that

$$RK = M,$$

where $K = (P_+ + P_-r^{-1}P_-)N^{-1}$ is an invertible operator. Thus

$$\dim \ker R = \dim \ker M$$

Further we have

$$\begin{aligned} p \in \ker M &\Leftrightarrow p = rP_+r^{-1}P_-N^{-1}p \\ &\Leftrightarrow p = -P_+rP_-r^{-1}P_-N^{-1}p, \end{aligned}$$

which means that p belongs to the image of the finite dimension operator P_+rP_-. It is easy to see that

$$\dim \mathrm{im} P_+rP_- = l$$

So $\dim \ker M \leq l$ and, therefore, (2.6) holds. □

Proposition 2.3 *If $c \in C^{n\times n}(\mathbb{T})$ in* (1.1) *satisfies the condition* (2.2), *then the estimate*

$$\dim \ker T \leq l, \tag{2.7}$$

where l is defined by (2.5) *and r is a polynomial matrix $n \times n$ satisfying the conditions* (2.1) *and* (2.3), *holds.*

Proof. According to Proposition 2.1, there exist an induced matrix norm $\|.\|_0$ and a polynomial matrix r such that $a(t) = r(t)c(t)r^{-1}[\alpha(t)]$ satisfies the condition (2.3). From here we can show that the operator (2.4) is invertible.

Now we write T as a product of the operators

$$T = (I - cUP_+) = r^{-1}[rP_- + (I - aU)P_+](P_- + rP_+).$$

Both operators $r^{-1}I$ and $(P_- + rP_+)$ are continuously invertible so that their kernels are trivial. Then

$$\dim \ker T = \dim \ker[rP_- + (I - aU)P_+] = \dim \ker R.$$

Now we apply Proposition 2.2. This implies (2.7). □

Now we consider a linear fractional non-Carleman shift preserving the orientation on $\mathbb{T}$

$$\alpha(t) = \frac{at + b}{\bar{b}t + \bar{a}} : \mathbb{T} \to \mathbb{T}$$

which has fixed points τ_1 and τ_2 on $\mathbb{T}$ and it is assumed that $|a|^2 - |b|^2 = 1$, $a, b \in \mathbb{C}$.

We can write $\alpha(t)$ in the form

$$\alpha(t) = \alpha_+(t)t\alpha_-(t),$$

where we put

$$\alpha_+(t) = \frac{1}{\bar{b}t + \bar{a}}, \quad \alpha_-(t) = \frac{at + b}{t}.$$

We see that $\alpha_\pm, \alpha_\pm^{-1}$ are analytic in $\mathbb{T}_\pm$ and continuous in the closure of $\mathbb{T}_\pm$, respectively.

For the linear fractional shift $\alpha(t)$, it is convenient to consider the isometric shift operator

$$(V\varphi)(t) = \alpha_+(t)\varphi(\alpha(t)),$$

instead of the operator U defined by (1.2), because V satisfies the additional property $VS = SV$. Since both operators, U and V, are isometric, all the anterior results hold under the substitution of U by V.

Then we consider the operator

$$T = I - cVP_+,$$

where we suppose now that $c \in C^{n\times n}(\mathbb{T})$ has the properties

$$\sigma[c(\tau_j)] \subset \mathbb{T}_-, \quad j = 1, 2, \tag{2.8}$$

$$\det c(t) \neq 0, \quad \forall t \in \mathbb{T}. \tag{2.9}$$

Under condition (2.9) the continuous matrix function c admits the factorization in $L_2^{n\times n}(\mathbb{T})$ (see, for instance, Theorem 1.1, Ch. VIII, in [2]; see also [5])

$$c = c_- \Lambda c_+, \tag{2.10}$$

where

$$c_-^{\pm 1} \in \left[L_2^-(\mathbb{T})\right]^{n\times n}, \quad c_+^{\pm 1} \in \left[L_2^+(\mathbb{T})\right]^{n\times n}, \quad \Lambda = \mathrm{diag}\{t^{\varkappa_j}\},$$

$\varkappa_j$ are the partial indices of the factorization with $\varkappa_1 \geq \varkappa_2 \geq \ldots \geq \varkappa_n$, and, as usual, $L_2^+ = P_+ L_2, L_2^- = P_- L_2 \oplus \mathbb{C}$. Moreover, we assume that

$$c_\pm^{\pm 1} \in L_\infty^{n\times n}(\mathbb{T}).$$

Proposition 2.4 *If conditions* (2.9) *and* (2.10) *are fulfilled, then the estimate*

$$\dim\ker T \leq \dim\ker(I - c^{-1}V^{-1}P_+) + \sum_{\varkappa_j<0} |\varkappa_j| \tag{2.11}$$

holds.

Proof. We consider the product of T and a certain continuously invertible operator, whose kernel is trivial:

$$\begin{aligned} &(I - cVP_+)(c_-P_- - c_+^{-1}(\alpha_{-1})V^{-1}P_+) \\ = \;& c_-\Lambda(\Lambda^{-1}P_- + P_+ - \Lambda^{-1}c_-^{-1}c_+^{-1}(\alpha_{-1})V^{-1}P_+). \end{aligned}$$

As $c_-\Lambda$ is a continuously invertible operator, we get

$$\dim\ker T = \dim\ker B,$$

where

$$B = \Lambda^{-1}P_- + P_+ - \Lambda^{-1}c_-^{-1}c_+^{-1}(\alpha_{-1})V^{-1}P_+. \tag{2.12}$$

To treat the operator (2.12) we consider first the scalar operators (k-components of the operators B and $I - c^{-1}V^{-1}P_+$)

$$\begin{aligned} B_k &= t^k P_- + P_+ - bV^{-1}P_+, \\ A &= I - bV^{-1}P_+, \end{aligned}$$

where $b \in L_\infty(\mathbb{T})$.

In the case $k > 0$, the inclusion

$$\varphi = \varphi_- + \varphi_+ \in \ker B_k,$$

with $\varphi_\pm \in L_2^\pm$, yields

$$f = t^k\varphi_- + \varphi_+ \in \ker A$$

under the condition

$$P_-(t^k\varphi_-) = t^k\varphi_-. \tag{2.13}$$

The number of linear independent solutions which do not satisfy condition (2.13) can not be greater than k. So

$$\dim\ker B_k \leq \dim\ker A + k.$$

Returning to (2.12), we observe that Λ^{-1} is a diagonal matrix and $k = -\varkappa_j$. This implies (2.11). □

Now we analyze the operator $I - c^{-1}V^{-1}P_+$ in (2.11).

Taking into account (2.8), we can see that c^{-1} satisfies the conditions of Proposition 2.1, i.e., for c^{-1} there exists a polynomial matrix r satisfying the conditions (2.1) and (2.3). For this matrix r, let l be defined by (2.5). Since V^{-1} and V verify similar conditions, the operator $I - c^{-1}V^{-1}P_+$ satisfies all conditions of Proposition 2.3. Thus

$$\dim\ker(I - c^{-1}V^{-1}P_+) \leq l.$$

With (2.11) we get the following estimate.

Proposition 2.5 *If the conditions of Proposition* 2.4 *and* (2.8) *are fulfilled, then the estimate*

$$\dim\ker T \leq l + \sum_{\varkappa_j<0} |\varkappa_j|,$$

holds.

Acknowledgements

This research was partially supported by Centro de Matemática Aplicada at Instituto Superior Técnico (Portugal) and Fundação para a Ciência e Tecnologia (Portugal), project PRAXIS/2/2.1/FIS/286/94.

References

[1] Baturev, A. A., Kravchenko, V. G., and Litvinchuk, G. S., Approximate methods for singular integral equations with a non-Carleman shift, *J. Integral Equat. Appl.*, **8** (1996), no. 1, 1–17.

[2] Clancey, K. and Gohberg, I., *Factorization of Matrix Functions and Singular Integral Operators*, Operator Theory: Advances and Applications, **3**. Birkhäuser Verlag, Basel, 1981.

[3] Horn, R. A. and Johnson, C. R., *Matrix Analysis*. Cambridge University Press, Cambridge, 1996.

[4] Kravchenko, V. G. and Litvinchuk, G. S., *Introduction to the Theory of Singular Integral Operators with Shift*, Mathematics and its Applications, **289**. Kluwer Academic Publishers, Dordrecht, 1994.

[5] Litvinchuk, G. S. and Spitkovskii, I. M., *Factorization of Measurable Matrix Functions*, Operator Theory: Advances and Applications, **25**. Birkhäuser Verlag, Basel, 1987.

ON THE SOLUTION OF INTEGRAL EQUATIONS ON THE CIRCULAR DISK BY USE OF ORTHOGONAL POLYNOMIALS

Frank Penzel*
Darmstadt, Germany
Frank.Penzel@t-online.de

To Professor G. S. Litvinchuk on the occasion of his 70th birthday

Abstract In this paper results on the approximation of special pseudodifferential operator equations on the circular disk by the use of orthogonal polynomials and a fully discretized Galerkin scheme are presented.

Keywords: boundary integral equations, fully discretized Galerkin method, error estimates.

1. Introduction

This paper deals with a fully discretised spectral approximation of pseudodifferential operators which appear in a boundary integral equation approach when applied to the scattering problem of time-harmonic waves at a circular screen in $\mathbb{R}^3$. The eigenfunctions of the underlying pseudodifferential operators in the case of the Laplacian are studied in Sobolev spaces. These eigenfunctions are used for the definition of a Galerkin scheme for approximation of the Helmholtz equation with real wave number ω. Error estimates are obtained.

In this paper we present some results which have been obtained in [9], where complete proofs are presented.

*This work was completed during the author's visit at the Instituto Superior Técnico. The author likes to thank the members of the Centro de Matemática Aplicada for the invitation and for the opportunity to do common research.

We study the convergence of a fully discretised Galerkin method for the approximation of the pseudodifferential operators

$$D_\omega = \frac{1}{2} r_\Omega F^{-1} \sqrt{|\xi|^2 - \omega^2} F : \tilde{H}^s(\Omega) \to H^{s-1}(\Omega), 0 < s < 1, \tag{1.1}$$

$$V_\omega = \frac{1}{2} r_\Omega F^{-1} \frac{1}{\sqrt{|\xi|^2 - \omega^2}} F : \tilde{H}^{-s}(\Omega) \to H^{1-s}(\Omega), 0 < s < 1, \tag{1.2}$$

where $\omega > 0$ is assumed, $\sqrt{|\xi|^2 - \omega^2} > 0$, if $|\xi| > \omega$, $i\sqrt{|\xi|^2 - \omega^2} > 0$, if $|\xi| < \omega$, F denotes the Fourier-transformation

$$(Ff)(\xi) = \int_{\mathbb{R}^2} e^{ix\xi} f(x)\, dx \quad (\xi \in \mathbb{R}^2), x\xi = \sum_{j=1}^{2} x_j \xi_j, \tag{1.3}$$

defined by continuous extension for all $f \in S'(\mathbb{R}^2)$ [12], Ω denotes the unit disk, i.e.

$$\Omega = \{(x_1, x_2, 0) : x_1^2 + x_2^2 < 1\}, \tag{1.4}$$

and r_Ω denotes the restriction of distributions from $S'(\mathbb{R}^n)$ to Ω [3]. It was proved in [14] that the operators $D_\omega : \tilde{H}^s(\Omega) \to H^{s-1}(\Omega)$ and $V_\omega : \tilde{H}^{s-1}(\Omega) \to H^s(\Omega)$ are continuously invertible operators for $s \in (0,1)$, see Section 2 for the definition of the underlying Sobolev spaces $\tilde{H}^s(\Omega), H^s(\Omega)$. The analysis of Galerkin's method for the solution of pseudodifferential equations (1.1) and (1.2) is well established [14].

In this paper we make use of the knowledge about the analytical formulas for the explicit inverses of the operators D_0 and V_0, which have been studied in a Sobolev space setting in [7, 8]. Let $w(x) = 1/\sqrt{1 - |x|^2}$, for $x \in \mathbb{R}^2, |x| \neq 1$. We shall prove here that the operators $D_0 w^{-1}$ and $V_0 w$ have a complete set of eigenfunctions spanned by a countable set of polynomials. These eigenfunctions were obtained earlier in [15] and [4], but Sobolev spaces have not been used in these papers. Here we obtain expansions for solutions in the spaces $\tilde{H}^s(\Omega)$ for $s = \pm\frac{1}{2}$ which are the natural spaces for the boundary values of the weak solution of the Helmholtz equation.

The possibility to use orthogonal polynomials for approximation of (1.1), (1.2) was mentioned in the introduction of [5]. In that paper it was pointed out that this approach will be a generalisation of results from [2] and [13] in which collocation methods for the one-dimensional analogues of equations (1.1) and (1.2) in case of $\Omega = (-1, 1)$ were analysed. The use of global polynomials for the approximation of the solution of singular integral equations on the interval have been studied by many authors, see [6] and the references cited therein. In the recent paper [6] solvability results in weighted Sobolev spaces for the generalized airfoil equation on the interval are obtained and stability and error estimates for a numerical procedure using polynomials are proved. The generalized airfoil equation is defined by a combination of the finite Hilbert transform, the integral operator of logarithmic type, a multiplication operator and a class of compact operators.

Our paper is organised as follows. In Section 2 we give some basic definitions and results concerning mainly the explicit inverses of the pseudodifferential operators D_0 and V_0 and we prove formulas for the eigenfunctions of $D_0\, w^{-1}$ and $V_0\, w$. We introduce in Section 3 a fully discretised Galerkin method for the numerical approximation of the operator V_ω and we analyse its convergence in Section 4.

2. Basic results

We use the scale of Sobolev spaces [3] for $s \in \mathbb{R}$

$$H^s(\mathbb{R}^2) = \{f \in S'(\mathbb{R}^2) : Ff \in L^2_{loc}(\mathbb{R}^2), \int_{\mathbb{R}^2} |Ff(\xi)|^2(1+|\xi|^2)^s d\xi < \infty\},$$

$$H^s(\Omega) = \{r_\Omega f : f \in H^s(\mathbb{R}^2)\},$$

$$\tilde{H}^s(\Omega) = \{f \in H^s(\mathbb{R}^2) : \operatorname{supp} f \subset \overline{\Omega}\},$$

the norms in these spaces are given by

$$||f||^2_{H^s(\mathbb{R}^2)} = \int_{\mathbb{R}^2} |Ff(\xi)|^2(1+|\xi|^2)^s d\xi,$$

$$||f||_{H^s(\Omega)} = \inf\{||g||_{H^s(\mathbb{R}^2)} : g \in H^s(\mathbb{R}^2), r_\Omega g = f\},$$

$$||f||_{\tilde{H}^s(\Omega)} = ||f||_{H^s(\mathbb{R}^2)}.$$

For $f \in C_0^\infty(\Omega) \subset \tilde{H}^s(\Omega)$ and $g \in C^\infty(\overline{\Omega}) \subset H^{-s}(\Omega)$ we have that

$$|(f,g)| = |\int_\Omega f(x)\overline{g(x)}dx| \leq ||f||_{\tilde{H}^s(\Omega)}||g||_{H^s(\Omega)}.$$

By $(.,.)$ we shall denote the continuous extension of this sesquilinear form to $\tilde{H}^s(\Omega) \times H^{-s}(\Omega)$. The operators D_0 and V_0 are convolution operators with homogeneous symbols $|\xi|^{\pm 1}$, [3], this motivates the importance of the following result for our approach.

Lemma 2.1 *Given $s \in (-1,1)$ there are numbers $C_s > 0$ and $D_s > 0$ such that for all $f \in \tilde{H}^s(\Omega)$ the following inequalities hold.*

$$C_s||f||^2_{\tilde{H}^s(\Omega)} \leq \int_{\mathbb{R}^2} |Ff(\xi)|^2|\xi|^{2s}d\xi \leq D_s||f||^2_{\tilde{H}^s(\Omega)}.$$

The following one-dimensional fractional integral operators are introduced for $\nu > 0, \alpha > 0, \eta \in \mathbb{R}$, by

$$(I_{\nu,\alpha,\eta}f)(x) = \frac{\nu x^{-\nu(\eta+\alpha-1)}}{\Gamma(\alpha)} \int_0^x (x^\nu - t^\nu)^{\alpha-1}t^{\nu\eta-1}f(t)\,dt,$$

$$(J_{\nu,\alpha,\eta}f)(x) = \frac{\nu x^{\nu\eta}}{\Gamma(\alpha)} \int_x^\infty (t^\nu - x^\nu)^{\alpha-1}t^{-\nu(\alpha+\eta-1)-1}f(t)\,dt.$$

For a given function $f \in L^2(\Omega)$ we define the function $f_R : [0,1]\times[0,2\pi) \to \mathbb{C}$ by

$$f_R(r,\phi) = f(r\cos\phi, r\sin\phi) \ \ (r \in [0,1], \phi \in [0,2\pi]).$$

We define the discrete Fourier-transform $\mathcal{F}f : \mathbb{Z} \times [0,1] \to \mathbb{C}$ of a function f defined on Ω by

$$(\mathcal{F}f)(k,r) = \frac{1}{2\pi}\int_0^{2\pi} f_R(r,\phi)e^{ik\phi}d\phi \ \ (k \in \mathbb{Z}, 0 < r < 1).$$

It is known that

$$f_R(r,\phi) = \sum_{k\in\mathbb{Z}} (\mathcal{F}f)(k,r)e^{-ik\phi} \ a.e.$$

We define two-dimensional integral operators $\mathcal{I}_{\nu,\alpha,\eta}$, $\mathcal{J}_{\nu,\alpha,\eta}$ by

$$(\mathcal{I}_{\nu,\alpha,\eta}f)(x) = \sum_{k\in\mathbb{Z}} (I_{\nu,\alpha,\frac{|k|}{2}+\eta}\mathcal{F}f)(k,r)e^{-ik\varphi},$$
$$(\mathcal{J}_{\nu,\alpha,\eta}f)(x) = \sum_{k\in\mathbb{Z}} (J_{\nu,\alpha,\frac{|k|}{2}+\eta}\mathcal{F}f)(k,r)e^{-ik\varphi}.$$

The Mellin transformation $\mathcal{MF}f$ of $\mathcal{F}f$ is given for $z \in \mathbb{C}$ by

$$(\mathcal{MF}f)(k,z) = \int_0^\infty \rho^{z-1}\mathcal{F}f(k,\rho)\,d\rho \ \ (k \in \mathbb{Z}).$$

Given $\mu \in \mathbb{R}$ we define the Mellin transformation $M_\mu\mathcal{F}f$ of $\mathcal{F}f$ at $t \in \mathbb{R}$ by

$$(M_\mu\mathcal{F}f)(k,t) = (\mathcal{MF}f)(k,\frac{\mu}{2}+it) \ \ (k \in \mathbb{Z}).$$

We use weighted Hilbert-spaces

$$L^2_\mu(\mathbb{R}^+) = \left\{f : \int_0^\infty |f(r)|^2 r^{\mu-1}dr < \infty\right\}$$

with the sesquilinearform

$$(f,g)_\mu = \int_0^\infty f(r)\,\overline{g(r)}\,r^{\mu-1}dr.$$

Given $\eta \in \mathbb{R}$, $f : \mathbb{R}^n \to \mathbb{C}$ let m_η be the following operator of multiplication

$$(m_\eta f)(x) = |x|^\eta f(x) \ \ (x \in \mathbb{R}^n).$$

The following results were obtained in [10, Corollary 4.1]

$$(\mathcal{M}I_{\nu,\alpha,\eta}f)(z) = \frac{\Gamma(\eta - \frac{z}{\nu})}{\Gamma(\eta + \alpha - \frac{z}{\nu})}(\mathcal{M}f)(z), \operatorname{Re} z = \frac{\mu}{2},$$
$$(\mathcal{M}J_{\nu,\alpha,\eta}f)(z) = \frac{\Gamma(\eta + \frac{z}{\nu})}{\Gamma(\eta + \alpha + \frac{z}{\nu})}(\mathcal{M}f)(z), \operatorname{Re} z = \frac{\mu}{2}.$$

Given $|s| < 1$ we introduce the subset $R^s(\mathbb{R}^2)$ of all tempered distributions f having a finite norm

$$||f||_{R^s(\mathbb{R}^2)} = \left(\int_{\mathbb{R}^2} |Ff(\xi)|^2 |\xi|^{2s} d\xi \right)^{\frac{1}{2}}. \tag{2.1}$$

$\tilde{R}^s(\Omega)$ is the set of all distributions from $R^s(\mathbb{R}^2) \cap \tilde{H}^s(\Omega)$. The norm in this set is given by (2.1). $R^s(\Omega)$ is the set of restrictions of distributions from $R^s(\mathbb{R}^2)$ to Ω. The norm for given f in $R^s(\Omega)$ is

$$||f||_{R^s(\Omega)} = \inf\{||g||_{R^s(\mathbb{R}^2)} : g \in R^s(\mathbb{R}^2), r_\Omega g = f\}.$$

By Lemma 2.1 we can identify the spaces $\tilde{R}^s(\Omega)$ and $\tilde{H}^s(\Omega)$ in case of $|s| < 1$ and therefore the same holds for $R^s(\Omega)$ and $H^s(\Omega)$. The following characterization of the norms in $\tilde{R}^s(\Omega)$ was obtained in [8], namely, for all $f \in C_0^\infty(\Omega)$ we have that

$$||f||^2_{\tilde{R}^s(\Omega)} = 4\pi^2 \sum_{k\in\mathbb{Z}} \int_{\mathbb{R}} 2^{2s} \left| \frac{\Gamma(\frac{|k|}{2} + \frac{s+1}{2} + \frac{it}{2})}{\Gamma(\frac{|k|}{2} - \frac{s-1}{2} - \frac{it}{2})} (M_{2-2s}\mathcal{F}f)(k,t) \right|^2 dt.$$

Obviously, the operator $\mathcal{J}_{\nu,\alpha,\eta}$ maps functions supported in $\overline{\Omega}$ onto functions having this property, too. We construct operators which are one-to-one from $\tilde{R}^s(\Omega)$ onto $\tilde{R}^t(\Omega)$ by the aid of these operators in the following lemma.

Lemma 2.2 *The operator*

$$m_\gamma \mathcal{J}_{2,\gamma,\eta} : \tilde{R}^{s-\gamma}(\Omega) \to \tilde{R}^s(\Omega)$$

is one-to-one and onto, if $0 < \eta + \frac{1-s+\gamma}{2}, 0 < \eta + \frac{1-s+3\gamma}{2}, |s| < 1$. *In case of* $s = 0, \gamma = \frac{1}{2}, \eta = -\frac{1}{2}$ *we obtain the equation*

$$||m_{\frac{1}{2}} \mathcal{J}_{2,\frac{1}{2},\frac{-1}{2}} f||_{\tilde{\mathbb{R}}^0(\Omega)} = 2||f||_{\tilde{\mathbb{R}}^{-\frac{1}{2}}(\Omega)}.$$

Corollary 2.3 *The operator*

$$m_\gamma \mathcal{I}_{2,\gamma,\frac{\gamma}{2}+1+\eta} : R^{-s}(\Omega) \to R^{\gamma-s}(\Omega)$$

is one-to-one and onto under the restrictions of Lemma 2.2 upon γ, η and s.

The monomials $x_1^l x_2^m$ are expressed in polar coordinates by

$$x_1^l x_2^m = r^{l+m} \cos^l \varphi \sin^m \varphi = \sum_{k \in \mathbb{Z}} \alpha_{l,m,k} r^{l+m} e^{-ik\varphi}$$

and it is easily seen that

$$\alpha_{l,m,k} = \int_0^{2\pi} \cos^l \varphi \sin^m \varphi e^{ik\varphi} d\varphi = 0$$

if and only if $l + m - |k|$ is an odd integer. From these observations and the density of the polynomials in $C^0(\overline{\Omega})$ it is easily seen that the following set of polynomials span a dense subset in the space $C^0(\overline{\Omega})$

$$r^{2n+|k|} e^{-ik\varphi}, k \in \mathbb{Z}, n \in \mathbb{N}_0.$$

Let $P_l^{(\alpha,\beta)}$ be the Gauss-Jacobi-polynomials [1], which satisfy the orthogonality relations for all $l, l' \in \mathbb{N}_0, \alpha, \beta > -1$

$$\int_{-1}^{1} (1-\rho)^\alpha (1+\rho)^\beta P_l^{(\alpha,\beta)}(\rho) P_{l'}^{(\alpha,\beta)}(\rho) d\rho$$
$$= \delta_{l,l'} \frac{2^{\alpha+\beta+1}\Gamma(l+\alpha+1)\Gamma(l+\beta+1)}{(2l+\alpha+\beta+1)\Gamma(l+1)\Gamma(l+\alpha+\beta+1)}.$$

Lemma 2.4 *The following sets of functions are two complete orthonormal systems in $L^2(\Omega)$*

$$q_{l,k}(r,\varphi) = a_{l,k} r^{|k|+\frac{1}{2}} P_l^{(0,|k|+\frac{1}{2})}(2r^2-1) e^{-ik\varphi}, l \in \mathbb{N}_0, k \in \mathbb{Z},$$
$$t_{l,k}(r,\varphi) = b_{l,k} r^{|k|-\frac{1}{2}} P_l^{(0,|k|-\frac{1}{2})}(2r^2-1) e^{-ik\varphi}, l \in \mathbb{N}_0, k \in \mathbb{Z},$$

with the constants $a_{l,k} = \sqrt{\frac{2l+|k|+\frac{3}{2}}{\pi}}$ and $b_{l,k} = \sqrt{\frac{2l+|k|+\frac{1}{2}}{\pi}}$.

We use the following abbreviations

$$I_- f(r,\varphi) = r^{\frac{1}{2}} \sum_{k \in \mathbb{Z}} I_{2,\frac{1}{2},\frac{|k|}{2}+1} \mathcal{F} f(r,\varphi) e^{-ik\varphi},$$

$$\tilde{I}_- f(r,\varphi) = r^{\frac{1}{2}} \sum_{k \in \mathbb{Z}} I_{2,\frac{1}{2},\frac{|k|}{2}+\frac{3}{4}} \mathcal{F} f(r,\varphi) e^{-ik\varphi},$$

$$K_+ f(r,\varphi) = r^{\frac{1}{2}} \sum_{k \in \mathbb{Z}} J_{2,\frac{1}{2},\frac{|k|}{2}-\frac{1}{4}} \mathcal{F} f(r,\varphi) e^{-ik\varphi},$$

$$\tilde{K}_+ f(r,\varphi) = r^{\frac{1}{2}} \sum_{k \in \mathbb{Z}} J_{2,\frac{1}{2},\frac{|k|}{2}-\frac{1}{2}} \mathcal{F} f(r,\varphi) e^{-ik\varphi}.$$

As in [8] we obtain the following theorem.

Theorem 2.5 *The following identities hold true*

$$D_0^{-1} = K_+ I_- : R^{-\frac{1}{2}} \to \tilde{R}^{\frac{1}{2}}, V_0 = \frac{1}{4}\tilde{I}_- \tilde{K}_+ : \tilde{R}^{-\frac{1}{2}} \to R^{\frac{1}{2}}.$$

Lemma 2.6 *The following equations hold true for $l \geq 0, k \in \mathbb{Z}, r \in (0,1), \varphi \in (0, 2\pi), r_1 = 2r^2 - 1, \rho_1 = 2\rho^2 - 1$,*

$$m_{\frac{1}{2}} I_{2,\frac{1}{2},\frac{|k|}{2}+\frac{3}{4}} r^{|k|-\frac{1}{2}} P_l^{(0,|k|-\frac{1}{2})}(r_1) = \frac{\Gamma(|k|+l+\frac{1}{2})}{\Gamma(|k|+l+1)} \rho^{|k|} P_l^{(-\frac{1}{2},|k|)}(\rho_1),$$

$$m_{\frac{1}{2}} J_{2,\frac{1}{2},\frac{|k|}{2}-\frac{1}{2}} r^{|k|} \frac{P_l^{(-\frac{1}{2},|k|)}(r_1)}{\sqrt{1-r^2}} = \frac{\Gamma(l+\frac{1}{2})}{\Gamma(l+1)} \rho^{|k|-\frac{1}{2}} P_l^{(0,|k|-\frac{1}{2})}(\rho_1),$$

$$m_{\frac{1}{2}} I_{2,\frac{1}{2},\frac{|k|}{2}+1} r^{|k|} P_l^{(\frac{1}{2},|k|)}(r_1) = \frac{\Gamma(|k|+l+1)}{\Gamma(|k|+l+\frac{3}{2})} \rho^{|k|+\frac{1}{2}} P_l^{(0,|k|+\frac{1}{2})}(\rho_1),$$

$$m_{\frac{1}{2}} J_{2,\frac{1}{2},\frac{|k|}{2}-\frac{1}{4}} r^{|k|+\frac{1}{2}} P_l^{(0,|k|+\frac{1}{2})}(r_1) = \frac{\Gamma(l+1)}{\Gamma(l+\frac{3}{2})} \rho^{|k|} \sqrt{1-\rho^2} P_l^{(\frac{1}{2},|k|)}(\rho_1),$$

$$V_0 r^{|k|} \frac{P_l^{(-\frac{1}{2},|k|)}(r_1)}{\sqrt{1-r^2}} e^{-ik\varphi} = \frac{1}{4} \frac{\Gamma(l+\frac{1}{2})\Gamma(|k|+l+\frac{1}{2})}{\Gamma(l+1)\Gamma(|k|+l+1)} \rho^{|k|} P_l^{(-\frac{1}{2},|k|)}(\rho_1) e^{-ik\varphi},$$

$$D_0 r^{|k|} P_l^{(\frac{1}{2},|k|)}(r_1) \sqrt{1-r^2} e^{-ik\varphi} = \frac{\Gamma(l+\frac{3}{2})\Gamma(|k|+l+\frac{3}{2})}{\Gamma(l+1)\Gamma(|k|+l+1)} \rho^{|k|} P_l^{(\frac{1}{2},|k|)}(\rho_1) e^{-ik\varphi}.$$

As a consequence we obtain a characterisation of distributions in $\tilde{R}^{-1/2}(\Omega)$ by their coefficients with respect to a basis of orthogonal polynomials.

Corollary 2.7 *The restriction $r_\Omega f$ of a distribution $f \in S'(\mathbb{R}^2)$ is in $\tilde{R}^{-1/2}(\Omega)$, if and only if there is a sequence $\{c_{l,k}\}_{l\in\mathbb{N}_0, k\in\mathbb{Z}}$ of complex numbers such that for all $g \in R^{1/2}(\Omega)$*

$$(f, g) = \sum_{0 \leq l, k \in \mathbb{Z}} c_{l,k} (w p_{l,k}, g)$$

and

$$\sum_{0 \leq l, k \in \mathbb{Z}} \frac{|c_{l,k}|^2}{(2l + |k| + \frac{1}{2})(l+1)} < \infty.$$

Then the numbers $c_{l,k}$ are given by

$$c_{l,k} = \frac{(f, p_{l,k})}{(w p_{l,k}, p_{l,k})}.$$

For each function $h \in L^2_w(\Omega) = \{g| \int_\Omega |g(x)|^2 w(x)dx < \infty\}$ *the function* wh *represents a distribution from* $\tilde{R}^{-1/2}(\Omega)$ *and the mapping*

$$h \in L^2_w(\Omega) \to wh \in \tilde{R}^{-1/2}(\Omega)$$

is continuous.

Corollary 2.8 *For all* $n \in \mathbb{N}$ *and all polynomials* f *which satisfy* $deg(f) \leq n$ *(*$deg(f)$ *denotes the degree of* f*) the estimate*

$$||f||_{L^2_w(\Omega)} \leq C_1\sqrt{n}||wf||_{\tilde{R}^{-1/2}(\Omega)}$$

holds, where

$$C_1 = \min_{n\in\mathbb{N}, 0\leq 2l+|k|\leq n} n\frac{\Gamma(l+\frac{1}{2})\Gamma(l+|k|+\frac{1}{2})}{\Gamma(l+1)\Gamma(l+|k|+1)}. \tag{2.2}$$

3. The fully discretised Galerkin method

In this section we analyse the convergence of a fully discretised Galerkin method which is defined by the use of the singular functions $wp_{l,k}$,

$$w^{-1}(r) = \sqrt{1-r^2}, 0 < r < 1,$$

$$p_{l,k}(r,\varphi) = r^{|k|}P_l^{(-\frac{1}{2},|k|)}(2r^2-1)e^{-ik\varphi},\ 0 \leq \varphi < 2\pi,$$

and which is applied to the boundary integral equation

$$V_\omega u = f. \tag{3.1}$$

This equation is of importance in the solution of acoustic scattering problems.

We define a Galerkin method for the integral equation (3.1) by the following linear system with unknowns $\alpha_{l,k}$

$$\sum_{(l,k)\in M_n} \alpha_{l,k}(V_\omega wp_{l,k}, wp_{l',k'}) = (f, wp_{l',k'}),\ (l',k') \in M_n \tag{3.2}$$

using the definition

$$M_n = \{(l,k) : 0 \leq l \leq n, -n \leq k \leq n, 0 \leq 2l+|k| \leq n, l,k \in \mathbb{Z}\}.$$

We proved in Section 2 that the system of polynomials times w form a complete basis of $\tilde{R}^{-1/2}(\Omega)$ and therefore standard estimates guarantee the convergence of the solution $\sum_{(l,k)\in M_n} \alpha_{l,k}(wp_{l,k})$ to the exact solution of equation (3.1), which is the Galerkin discretization for the coercive operator V_ω. The first problem in the numerical realisation of equation (3.2) is the computation of the scalar products, specifically those on the left-hand side of (3.2). We use the following splitting

$$(V_\omega wp_{l,k}, wp_{l',k'}) = (R_\omega wp_{l,k}, wp_{l',k'}) + i(I_\omega wp_{l,k}, wp_{l',k'}),$$

where the convolution operators R_ω and I_ω are defined by

$$R_\omega u(x) = \int_\Omega \frac{\cos(\omega|x-y|)u(y)}{4\pi|x-y|}\,dy, \quad I_\omega u(x) = \int_\Omega \frac{\sin(\omega|x-y|)u(y)}{4\pi|x-y|}\,dy.$$

For approximation of the scalar products in equation (3.2) we introduce quadrature schemes. Let τ_n be the Gauss-Jacobi points and w_n are the Gauss-Jacobi weights in the quadrature formula

$$\int_{-1}^{1} \frac{f(\tau)}{\sqrt{1-\tau}}\,d\tau \approx \sum_{n=1}^{N} w_n f(\tau_n),$$

which is an exact approximation for $deg(f) \le 2N-1$. For functions $f \in C^0(\overline{\Omega})$ we introduce the linear functional $Q_{M,N}(f)$ by

$$\frac{2\pi}{M+1}\frac{\sqrt{2}}{4}\sum_{n=1}^{N}\sum_{m=0}^{M} w_n f\left(\sqrt{\frac{(1+\tau_n)}{2}}\cos\frac{2\pi m}{M+1}, \sqrt{\frac{(1+\tau_n)}{2}}\sin\frac{2\pi m}{M+1}\right).$$

Lemma 3.1 *If $d \le \min(M, 4N-2)$ then the formula*

$$\int_\Omega \frac{f(x)}{\sqrt{1-|x|^2}}\,dx = Q_{M,N}(f) \tag{3.3}$$

holds for all polynomials f of degree less or equal d.

Remark 3.2 In case of $d = M = 4N-2$ a quadrature formula of degree $4N-2$ with $(4N-1)N$ points is defined in Lemma 3.1. A lower bound for the minimal number of points in a quadrature rule of even degree $4N-2$ is known to be $N(2N+1)$, see [11]. Nevertheless, in a simple formula we would use as many points as the dimension of the space of polynomials of degree $\le 4N-2$ is, namely $2N(4N-1)$ points.

If functions $f, g \in C^0(\overline{\Omega})$ are given and $h \in C^0(\overline{\Omega} \times \overline{\Omega})$ is defined by $h(x,y) = f(x)g(y), x, y \in \Omega$ we introduce the operator $Q^{(2)}_{M,N}$ by $Q^{(2)}_{M,N}h = (Q_{M,N}f)(Q_{M,N}g)$ where $Q_{M,N}$ is given by (3.3). We use the notation $Q^{(2)}_{M,N}$ for the unique continuous extension of this operator to $C^0(\overline{\Omega} \times \overline{\Omega})$ and we shall use the notation $Q^{(2)}_{M,N}[h_1, f, g]$ for the application of $Q^{(2)}_{M,N}$ to the function $h_1(x,y)f(x)g(y)$, where $h_1 \in C^0(\overline{\Omega} \times \overline{\Omega})$ and $f, g \in C^0(\overline{\Omega})$. For $x, y \in \Omega, x \neq y$, we define the function $W_\omega(x,y) = \frac{\sin\omega|x-y|}{4\pi|x-y|}$. We introduce an approximation of the imaginary part of the scalar products in equation (3.2) by the formula

$$(I_\omega wp_{l,k}, wp_{l',k'}) = \int_\Omega\int_\Omega \frac{\sin\omega|x-y|}{4\pi|x-y|}\, wp_{l,k}(x)\,\overline{wp_{l',k'}}(y)\,dxdy$$
$$\approx Q^{(2)}_{M,N}[W_\omega, wp_{l,k}, \overline{wp_{l',k'}}].$$

The approximation of the operator R_ω is more difficult. Let $Z_\omega(x,y) = \cos(\omega|x-y|)$. In a first step we approximate the functions

$$\cos(\omega|x-y|)wp_{l,k}(x)\overline{wp_{l',k'}(y)}$$

by an approximation of its projection to a space of polynomials, namely we use with $2n \le 2n_1 \le \min(M, 4N-2)$

$$\cos(\omega|x-y|)w(|x|)p_{l,k}(x)\overline{w(|y|)p_{l',k'}(y)}$$
$$\approx \sum_{(\mu,\nu),(\mu',\nu')\in M_{n_1}} (\frac{\delta_{k+\nu,k'+\nu'}}{b_{\mu,\nu}b_{\mu',\nu'}} Q^{(2)}_{M,N}[Z_\omega, wp_{l,k}\overline{p_{\mu,\nu}}, \overline{wp_{l',k'}}p_{\mu',\nu'}],$$

$$wp_{\mu,\nu}(x)\,\overline{wp_{\mu',\nu'}}(y)), b^2_{\mu,\nu} = \int_\Omega w(x)|p_{\mu,\nu}(x)|^2dx.$$

We obtain the following approximation for the scalar products defined by the operator R_ω

$$(R_\omega wp_{l,k}, wp_{l',k'}) = \int_\Omega\int_\Omega \frac{\cos(\omega|x-y|)}{4\pi|x-y|}\, wp_{l,k}(x)\,\overline{wp_{l',k'}}(y)\,dxdy$$
$$\approx \sum_{(\mu,\nu),(\mu',\nu')\in M_{n_1}} \frac{\delta_{k+\nu,k'+\nu'}}{b^2_{\mu,\nu}b^2_{\mu',\nu'}} Q^{(2)}_{M,N}[Z_\omega, wp_{l,k}\overline{p_{\mu,\nu}}, \overline{wp_{l',k'}}p_{\mu',\nu'}],$$

$$(V_0 wp_{\mu,\nu}, wp_{\mu',\nu'}) = \sum_{(\mu,\nu),(\mu',\nu')\in M_{n_1}} \frac{\delta_{k+\nu,k'+\nu'}}{b_{\mu,\nu}b_{\mu',\nu'}} Q^{(2)}_{M,N}[Z_\omega, wp_{l,k}\overline{p_{\mu,\nu}}, \overline{wp_{l',k'}}p_{\mu',\nu'}],$$

$$\delta_{\nu,\nu'}\delta_{\mu,\mu'}\lambda_{\mu,\nu} = \sum_{(\mu,\nu)\in M_{n_1}} \frac{\delta_{k,k'}}{b^2_{\mu,\nu}} Q^{(2)}_{M,N}\left[Z_\omega, wp_{l,k}\overline{p_{\mu,\nu}}, \overline{wp_{l',k'}}p_{\mu',\nu'}\right]\lambda_{\mu,\nu}$$
$$= Q^{(2)}_{M,N} Z_{l,k,l',\omega},$$

defining numbers $\lambda_{\mu,\nu}$ and the function $Z_{l,k,l',\omega}$ by

$$\lambda_{\mu,\nu} = \frac{\Gamma(\mu+\frac{1}{2})\Gamma(|\nu|+\mu+\frac{1}{2})}{\Gamma(\mu+1)\Gamma(|\nu|+\mu+1)},$$

$$Z_{l,k,l',\omega}(x,y) = \sum_{(\mu,\nu)\in M_{n_1}} \frac{\lambda_{\mu,\nu}}{b^2_{\mu,\nu}} (Z_\omega(x,y)(wp_{l,k}\overline{p_{\mu,\nu}})(x)(\overline{wp_{l',k}}p_{\mu,\nu})(y)).$$

We introduce a discretisation of the exact Galerkin method (3.2), namely, we look for a function $v_{\omega,n} = \sum_{(l,k)\in M_n} \beta_{l,k} w p_{l,k}$ to be the solution of the fully discretised Galerkin equations

$$\sum_{(l,k)\in M_n} \beta_{l,k} Q^{(2)}_{M,N}[W_\omega, wp_{l,k}, \overline{wp_{l',k'}}]$$
$$+i \sum_{(l,k)\in M_n} \beta_{l,k}(Q^{(2)}_{M,N} Z_{l,k,l',\omega}) = Q_{M,N}(f\overline{p_{l',k'}}),\ (l',k') \in M_n. \qquad (3.4)$$

4. Error estimates

In this section we present several auxiliary results that are useful to obtain Theorem 4.4, in which the convergence of the discretised Galerkin method (3.4) is proved. Let us denote the space of polynomials of degree $\leq n$ restricted to Ω by P^n.

Lemma 4.1 *Let $f \in C^0(\overline{\Omega \times \Omega})$ be arbitrary and let q be a polynomial of degree $\leq s$. Then we obtain that for $u, v \in P^n$*

$$\left| \int_{\Omega\times\Omega} f(x,y)\,(wu)(x)\,\overline{wv}(y)dxdy - Q^{(2)}_{M,N}[f, wu, \overline{wv}] \right|$$
$$\leq 4\pi\, ||f-q||_{L^\infty(\Omega\times\Omega)}\, ||u||_{L^2_w(\Omega)}\, ||v||_{L^2_w(\Omega)},$$

in the case that $\max(2n, s+n) \leq \min(M, 4N-2)$.

Corollary 4.2 *If $\sum_{l,k} \alpha_{l,k} p_{l,k} \in P^n$ and $s \in \mathbb{N}$ is given and if* $\max(2n, s+n)) \leq \min(M, 4N-2)$, *then the inequality*

$$\left| \sum_{(l,k),(l',k')\in M_n} \alpha_{l,k} \left((I_\omega wp_{l,k}, wp_{l',k'}) - Q^{(2)}_{M,N}[W_\omega, wp_{l,k}, \overline{wp_{l',k'}}] \right) \overline{\alpha_{l',k'}} \right|$$
$$\leq C_1^2 n \frac{2^{s-1}\omega^s}{s!} \left\| \sum_{(l,k)\in M_n} \alpha_{l,k} wp_{l,k} \right\|^2_{\tilde{R}^{-\frac{1}{2}}(\Omega)}$$

holds with the constant C_1 defined in (2.2) being independent of n and ω.

Lemma 4.3 *Let $u, v \in P^n$, $s+n+n_1 \leq 2n_1 \leq \min(M, 4N-2)$, then there is a constant $C > 0$ such that*

$$\left| \int_{\Omega\times\Omega} \frac{Z_\omega(x,y)\, wu(x)\, w\overline{v}(y)}{|x-y|} dxdy - \sum_{(\mu,\nu)\in M_{n_1}} Q^{(2)}_{M,N}[Z_\omega, \overline{p_{\mu,\nu}}u, p_{\mu,\nu}\overline{v}] \frac{\lambda_{\mu,\nu}}{b^2_{\mu,\nu}} \right|$$
$$\leq C_1^2(n||V_0||_{L^2\to L^2} + C_2 n n_1) \frac{(2\omega)^{s+2}}{(s+2)!} ||wu||_{\tilde{R}^{-\frac{1}{2}}(\Omega)} ||wv||_{\tilde{R}^{-\frac{1}{2}}(\Omega)},$$

where C_1 is defined by (2.2) and

$$C_2 = \sup_{n_1 \in \mathbb{N}} \sum_{(\mu,\nu)\in M_{n_1}} \frac{\Gamma(\mu+\frac{1}{2})\Gamma(|\nu|+\mu+\frac{1}{2})}{n_1\Gamma(\mu+1)\Gamma(|\nu|+\mu+1)}.$$

Theorem 4.4 *There is a constant $1 \le C_0 \le \frac{2}{1-e^{-2}}$ such that*

$$||u||^2_{\tilde{R}^{-\frac{1}{2}}(\Omega)} \le C_0 \max(1,\omega)\,|(V_\omega u, u)|$$

for all $u \in \tilde{H}^{-1/2}(\Omega)$ and for all $\omega \in [0,\infty)$. If

$$C_0 \max(1,\omega)\,(C_1^2 n \frac{2^{s-1}\omega^s}{s!} + (C_1^2 n ||V_0||_{L^2\to L^2} + C_2 n n_1)\frac{(2\omega)^{s+2}}{(s+2)!}) < 1, \quad (4.1)$$

then the equations (3.4) are uniquely solvable.

Let $s+n \le n_1$, $\max(2n, 2n_1, s+n+n_1) \le \min(M, 4N-2)$, $f \in R^{\frac{1}{2}}(\Omega) \cap C^0(\overline{\Omega})$ and let $u_{\omega,n} = \sum_{l,k\in M_n} \alpha_{l,k} w p_{l,k}$ be the solution of the exact Galerkin equations (3.2) and let $v_{\omega,n} = \sum_{(l,k)\in M_n} \beta_{l,k} w p_{l,k}$ be the solution of the fully discretised Galerkin equations (3.4). Then there exists a constant $C > 0$ independent of n and ω such that

$$\begin{aligned} ||u_{\omega,n} - v_{\omega,n}||_{\tilde{R}^{-\frac{1}{2}}(\Omega)} \le\; & C(n\frac{2^{s-1}\omega^s}{s!} + n(1+n_1)\frac{(2\omega)^{s+2}}{(s+2)!})||u_{\omega,n}||_{\tilde{R}^{-\frac{1}{2}}(\Omega)} \\ & + C\max(1,\omega) \inf_{p_n \in P^n} ||f - p_n||_{R^{\frac{1}{2}}(\Omega)}. \end{aligned} \quad (4.2)$$

Remark 4.5 1. It is well-known that under the assumptions of Theorem 4.4 the norm of $u_{\omega,n}$ does converge to the norm of $V_\omega^{-1} f$ with the same order than the second summand in the right-hand side of equation (4.2), specifically the sequence $||u_{\omega,n}||^2_{\tilde{R}^{-\frac{1}{2}}(\Omega)}$ is bounded with respect to $n \in \mathbb{N}$. Therefore we obtain convergence of the fully discretised Galerkin scheme, if we choose $t > 1$ and $\omega > 0$ fixed and we choose $n_1 = [tn]$, $s = n_1 - n$, for all $n \in \mathbb{N}$. Then we obtain a positive constant $C > 0$ such that $\frac{(2\omega)^s}{s!} \le C(n_1-n)^{-\frac{n_1-n}{2}}$, hence (4.1) is fulfilled for sufficiently large $n \in \mathbb{N}$. By the estimate (4.2) we have in that case that the perturbation term caused by the discretisation of the Galerkin scheme by a quadrature rule satisfying the assertions of Theorem 4.4 is exponentially small in the asymptotic range.

2. The results presented in Section 3 and Section 4 for the single layer potential can be generalised to cover error estimates for discretised Galerkin methods defined by use of orthogonal polynomials applied to the hypersingular integral operator D_ω. Furthermore, the error estimates obtained here for the single layer potential operator on the circular disk can be obtained in the same way for the single layer potential defined on those open manifolds $\mathcal{M}$ which

can be represented by $\overline{\mathcal{M}} = \kappa(\overline{\Omega})$ with a real function $\kappa \in C^\infty(\overline{\Omega})$. In that case, the kernel function

$$\frac{\cos i\omega|\kappa(x) - \kappa(y)|}{|\kappa(x) - \kappa(y)|}$$

can be approximated by a discretisation of the orthogonal projection to $P_1^n \otimes P_1^n$ using $Q_{M,N} \otimes Q_{M,N}$. We conjecture that the use of orthogonal polynomials on the circular disk will in general not lead to high order of convergence even for given smooth data functions on a open smooth manifold $\mathcal{M}$.

References

[1] Abramowitz, M. and Stegun, I. A., *Handbook of Mathematical Functions.* National Bureau of Standards, Washington, 1965.

[2] Ervin, V. J. and Stephan, E. P., Collocation with Chebyshev polynomials for a hypersingular integral equation on an interval, *J. Comput. Appl. Math.*, **43** (1992), 221–229.

[3] Èskin, G. I., *Boundary Value Problems for Elliptic Pseudodifferential Operators*, Translations of Mathematical Monographs, **52**. AMS, Providence, R.I., 1981.

[4] Krenk, S., A circular crack under asymmetric loads and some related integral equations, *J. Appl. Mech.*, **46** (1979), 821–826.

[5] Martin, P. A., Mapping flat cracks onto penny-shaped cracks, with application to somewhat circular tensile cracks, *Q. Appl. Math.*, **54** (1996), 663–675.

[6] Okada, S. and Prössdorf, S., On the solution of the generalized airfoil equation, *J. Integral Equat. Appl.*, **9** (1997), 71–98.

[7] Penzel, F., Sobolev space methods for the Laplace equation in the exterior of the disk, *Integr. Equat. Oper. Theor.*, **17** (1993), 123–136.

[8] Penzel, F., Gemischte Fourier-Randelementmethoden zur Berechnung zeitharmonischer Streufelder an Schirmen, *Math. Nachr.*, **175** (1995), 269–324.

[9] Penzel, F., On the solution of integral equations on the circular disk by use of orthogonal polynomials, Departamento de Matemática, Instituto Superior Técnico, Preprint 1/98.

[10] Rooney, P. G., On the ranges of certain fractional integrals, *Can. J. Math.*, **24** (1972), 1198–1216.

[11] Schmid, H. J., On cubature formulae with a minimal number of knots, *Numer. Math.*, **31** (1978), 281–297.

[12] Schwartz, L., *Theorié des Distributions.* Hermann, Paris, 1966.

[13] Sloan, I. H. and Stephan, E. P., Collocation with Chebyshev polynomials for Symm's integral equation on an interval, *J. Austral. Math. Soc. Ser. B*, **34** (1992), no. 2, 199–211.

[14] Stephan, E. P., Boundary integral equations method for screen problems in $\mathbb{R}^3$, *Integr. Equat. Oper. Theor.*, **10** (1987), 236–257.

[15] Wolfe, P. Eigenfunctions of the integral equation for the potential of the charged disk, *J. Math. Phys.*, **12** (1971), 1215–1218.

SINGULAR AND FREDHOLM INTEGRAL EQUATIONS FOR DIRICHLET BOUNDARY PROBLEMS FOR AXIAL-SYMMETRIC POTENTIAL FIELDS*

Sergiy Plaksa
Institute of Mathematics of National Academy of Sciences of Ukraine, Ukraine
plaksa@imath.kiev.ua

Dedicated to G. S. Litvinchuk on the occasion of his 70th birthday

Abstract We develop new methods for solving boundary problems for spatial axial-symmetric potential solenoid fields depending on the nature and specific features of axial-symmetric problems. The Dirichlet problems for the axial-symmetric potential and the Stokes flow function in a simply connected domain of the meridian plane are reduced to the Cauchy singular integral equations. If the boundary of a domain is a smooth curve satisfying certain additional requirements, then these singular integral equations are reduced to the Fredholm integral equations.

Keywords: axial-symmetric potential, Stokes flow function, Dirichlet boundary problem.

1. Introduction

A spatial potential solenoid field symmetric with respect to the axis Ox is described in its meridian plane xOy in terms of the axial-symmetric potential φ and the Stokes flow function ψ satisfying the following system of equations:

$$y\frac{\partial\varphi(x,y)}{\partial x} = \frac{\partial\psi(x,y)}{\partial y}, \qquad y\frac{\partial\varphi(x,y)}{\partial y} = -\frac{\partial\psi(x,y)}{\partial x}. \tag{1.1}$$

*This research was supported in part by INTAS-99-00089 project.

From the system (1.1) it follows that the function φ satisfies the equation

$$y\Delta\varphi(x,y) + \frac{\partial\varphi(x,y)}{\partial y} = 0 \tag{1.2}$$

and the function ψ satisfies the equation

$$y\Delta\psi(x,y) - \frac{\partial\psi(x,y)}{\partial y} = 0, \tag{1.3}$$

where $\Delta := \frac{\partial^2}{\partial x^2} + \frac{\partial^2}{\partial y^2}$ is the Laplace operator.

The quantitative theory for solutions of equations (1.2) and (1.3) is developed essentially worse than the theory for plane harmonic functions (see [4], p. 18). Therefore, a development of the mentioned quantitative theory and methods for solving boundary problems for axial-symmetric potential fields is a topical problem.

In the paper [9], integral expressions for solutions of equations (1.2) and (1.3) were obtained on the basis of Green's formulas for the Euler – Poisson – Darboux operators in terms of the values of these functions and their normal derivatives on the boundary of the domain. Foloowing this approach, solutions of the Dirichlet problem for the axial-symmetric potential and for the Stockes flow function in a disk were obtained via analogs of the Poisson integral.

In the paper [15] the Dirichlet problem for the generalized equation (1.2) in a domain with the Lyapunov boundary was reduced to the Fredholm integral equation by means of the double layer potential associated with this equation.

We also note that the solution of the Dirichlet problem for the Stokes flow function is known in certain particular cases for steady streamline along an axial-symmetric body (see [4, 5, 1]).

In the paper [6, 7, 8] we have constructed explicit analytic functions of hypercomplex variable and have proved that components of these functions generate the axial-symmetric potentials and the Stockes flow functions (see also [11]). On this way new integral expressions for solutions of system (1.1) and equations (1.2), (1.3) were obtained and a new method for investigation of spatial potential solenoid fields with axial symmetry was found. These integral expressions for the functions φ and ψ are analogous to those for harmonic functions in the plane via the Cauchy integral, and the mentioned method to investigate axial-symmetric potential fields is analogous to analytic function method in the complex plane for plane potential fields.

Using these integral expressions for the functions φ and ψ, we develop methods which reduce Dirichlet problems for these functions to Cauchy singular integral equations and Fredholm integral equations. In this paper this is done for domains more general than in the papers [4, 9, 15, 5, 1].

2. Preliminary notes and notation

Let D be a bounded simply connected domain in the meridian plane xOy symmetric with respect to the axis Ox. Let ∂D denote its boundary. The domain of the complex plane $\mathbf{C}$ congruent to the domain D under the correspondence $z = x + iy, (x,y) \in D$ is denoted by D_z, its closure and boundary are denoted by $\overline{D_z}$ and ∂D_z, respectively.

By b_1 and b_2 we denote the points at which the boundary ∂D_z crosses the real axis $\mathbf{R}$. We assume that $b_1 < b_2$. In what follows, $(x,y) \in D$ and $z = x+iy$.

For every $z \in D_z$ with $\operatorname{Im} z \neq 0$, we fix an arbitrary Jordan rectifiable curve $\Gamma_{z\bar{z}}$ in the domain D_z which is symmetric with respect to the real axis $\mathbf{R}$ and connects the points z and $\bar{z}$. For every $z \in \partial D_z$ with $\operatorname{Im} z \neq 0$ by $\Gamma_{z\bar{z}}$ we denote the Jordan subarc of the boundary ∂D_z with the end points z and $\bar{z}$ which contains the point b_1.

For $z \in \overline{D_z}, \operatorname{Im} z \neq 0$, let $\sqrt{(t-z)(t-\bar{z})}$ be that continuous branch of the analytic function $G(t) = \sqrt{(t-z)(t-\bar{z})}$ outside of the cut along $\Gamma_{z\bar{z}}$ for which $G(b_2) > 0$. Let $(\sqrt{(\tau-z)(\tau-\bar{z})})^- := \lim_{t\to\tau, t\in\mathbf{C}\setminus\overline{D_z}} \sqrt{(t-z)(t-\bar{z})}$ at $\tau \in \partial D_z \setminus \{z, \bar{z}\}$.

The following theorem is a generalization of Theorem 5 from [6], and it can be proved in a similar way.

Theorem 2.1 *If F is an analytic function in the domain D_z, then the functions*

$$\varphi(x,y) = \begin{cases} \frac{1}{2\pi i}\int_\gamma \frac{F(t)}{\sqrt{(t-z)(t-\bar{z})}}dt, & \text{when } y \neq 0, \\ F(t), & \text{when } y = 0, \end{cases} \tag{2.1}$$

$$\psi(x,y) = \begin{cases} -\frac{1}{2\pi i}\int_\gamma \frac{F(t)(t-x)}{\sqrt{(t-z)(t-\bar{z})}}dt, & \text{when } y \neq 0, \\ 0, & \text{when } y = 0 \end{cases} \tag{2.2}$$

are solutions in D of system (1.1), *where γ is an arbitrary closed Jordan rectifiable curve in D_z which embraces $\Gamma_{z\bar{z}}$. Functions* (2.1) *and* (2.2) *are also solutions in D of equations* (1.2) *and* (1.3), *respectively.*

It is evident that if the boundary ∂D_z is a Jordan rectifiable curve and the function F belongs to the Smirnov class E_1 (see [14], p. 205) in the domain D_z, then formulas (2.1) and (2.2) can be transformed to the form

$$\varphi(x,y) = \begin{cases} \frac{1}{2\pi i}\int_{\partial D_z} \frac{F(t)}{\sqrt{(t-z)(t-\bar{z})}}dt, & \text{when } y \neq 0, \\ F(t), & \text{when } y = 0, \end{cases} \tag{2.3}$$

$$\psi(x,y) = \begin{cases} -\frac{1}{2\pi i}\int_{\partial D_z} \frac{F(t)(t-x)}{\sqrt{(t-z)(t-\bar{z})}}dt, & \text{when } y \neq 0, \\ 0, & \text{when } y = 0 \end{cases} \tag{2.4}$$

for all $(x,y) \in D$, where $F(t)$ is the angular boundary value of the function F which is known to exist almost everywhere on ∂D_z.

3. Dirichlet boundary problem for the axial-symmetric potential

Consider the following *Dirichlet boundary problem for the axial-symmetric potential*: to find the solution $\varphi(x,y)$ of equation (1.2) in D when its boundary values $\varphi_{\partial D}(x,y)$ are given on the boundary ∂D, i.e. $\varphi(x,y) = \varphi_{\partial D}(x,y)$ for all $(x,y) \in \partial D$.

To solve this Dirichlet boundary problem, we introduce and consider the following *auxiliary problem* for the given function $\varphi_{\partial D}$: to find a function Φ analytic in D_z and continuous in $\overline{D_z}$ under the condition

$$\Phi(\bar{z}) = \overline{\Phi(z)} \qquad \forall\, z \in D_z$$

and boundary values $\Phi(t)$ satisfying the integral equation

$$\frac{1}{2\pi i}\int_{\partial D_z} \frac{\Phi(t)}{\left(\sqrt{(t-z)(t-\bar{z})}\right)^-}\, dt = \varphi_{\partial D}(x,y). \tag{3.1}$$

For equation (3.1) we develop a method for transition to the Cauchy singular integral equation. Before to describe such a transition for equation (3.1), we introduce some notations.

Denote by $\mathcal{H}^u_\alpha(\partial D)$ the set of functions $h : \partial D \to \mathbf{R}$ satisfying the following conditions:

$$h(x,-y) = h(x,y) \quad \forall (x,y) \in \partial D,$$

$$\begin{gathered} |h(x_1,y_1) - h(x_2,y_2)| \\ \le c\left(\max\{|z_1-b_1||z_1-b_2|, |z_2-b_1||z_2-b_2|\}\right)^{-\nu} |z_1-z_2|^\alpha \\ \forall (x_1,y_1),(x_2,y_2) \in \partial D, \end{gathered}$$

where $z_1 = x_1 + iy_1$, $z_2 = x_2 + iy_2$, $\alpha \in (1/2;1]$, $\nu \in [0;\alpha)$ and the constant c does not depend on x_1, y_1, x_2, y_2. In what follows, we suppose that $\varphi_{\partial D} \in \mathcal{H}^u_\alpha(\partial D)$.

Let $\sigma(Z)$ be the conformal mapping which maps the unit disk $\{Z \in \mathbf{C} : |Z| < 1\}$ onto the domain D_z that $\sigma(-1) = b_1$, $\sigma(1) = b_2$ and $\mathrm{Im}\,\sigma(i) > 0$. Note that $\sigma(\bar{Z}) = \overline{\sigma(Z)}$ for all $Z \in \{Z \in \mathbf{C} : |Z| \le 1\}$.

We introduce the function

$$M(Z,T) := \sqrt{\frac{(T-Z)(T-\bar{Z})}{(\sigma(T)-\sigma(Z))(\sigma(T)-\sigma(\bar{Z}))}}.$$

For each $Z \ne -1$ it is understood as a continuous branch of the function analytic with respect to the variable T in the unit disk and satisfying the condition $M(Z,-1) > 0$.

Now, consider the conformal mapping $Z = \frac{\xi - i}{\xi + i}$ of the complex plane. This mapping assigns to the points ξ and $-\xi$ of the real axis $\mathbf{R}$ the points Z and

$\bar{Z}$ of the unit circle. Let us introduce the function

$$m(\xi,\tau) := M(Z,T)$$

of real variables ξ and τ, where $T = \frac{\tau-i}{\tau+i}$. In what follows, we use the functions

$$A(\xi,\tau) := 2\,\mathrm{Re}\,m(\xi,\tau), \qquad B(\xi,\tau) := 2\,\mathrm{Im}\,m(\xi,\tau).$$

Theorem 3.1 *Suppose that $\varphi_{\partial D} \in \mathcal{H}^u_\alpha(\partial D)$ and the conformal mapping $\sigma(Z)$ has the nonvanishing continuous derivative on the unit circle. If Φ is a solution of the auxiliary problem for the function $\varphi_{\partial D}$, then the function*

$$U_p(\xi) := \mathrm{Re}\,\frac{i\left(\Phi(\sigma(\frac{\xi-i}{\xi+i})) - \varphi_{\partial D}(b_2,0)\right)\sigma'(\frac{\xi-i}{\xi+i})}{2(\xi+i)}$$

is a solution of the Cauchy singular integral equation

$$A(\xi,\xi)U_p(\xi) + \frac{B(\xi,\xi)}{\pi}\int_{-\infty}^{\infty}\frac{U_p(\tau)}{\tau-\xi}\,d\tau$$
$$-\frac{2\xi}{\pi^2}\int_0^{\xi}\left(\int_\tau^{\xi}\frac{s(B(s,\tau)-B(\xi,\tau))}{(\xi^2-s^2)^{3/2}\sqrt{s^2-\tau^2}}\,ds\int_{-\infty}^{\infty}\frac{U_p(\eta)}{\eta-\tau}\,d\eta\right)d\tau$$
$$-\frac{2\xi}{\pi}\int_0^{\xi}U_p(\tau)\int_\tau^{\xi}\frac{s(A(s,\tau)-A(\xi,\tau))}{(\xi^2-s^2)^{3/2}\sqrt{s^2-\tau^2}}\,ds\,d\tau = f_*(\xi) \quad \forall\,\xi>0, \qquad (3.2)$$

where

$$f_*(\xi) := \varphi_*(\xi) - \xi\int_0^{\xi}\frac{s(\varphi_*(s)-\varphi_*(\xi))}{(\xi^2-s^2)^{3/2}}\,ds$$

and the function φ_* is expressed via the given function $\varphi_{\partial D}$ in the following form:

$$\varphi_*(\xi) := \frac{1}{\sqrt{\xi^2+1}}\left(\varphi_{\partial D}(x,y) - \varphi_{\partial D}(b_2,0)\right),$$

where $x+iy = \sigma(\frac{\xi-i}{\xi+i})$.

Proof. Using the Cauchy theorem, we transform (3.1) into the equation

$$\frac{1}{\pi i}\int_{\Gamma_{z\bar z}}\frac{\Phi(t)-\varphi_{\partial D}(b_2;0)}{(\sqrt{(t-z)(t-\bar z)})^-}\,dt = \varphi_{\partial D}(x,y) - \varphi_{\partial D}(b_2;0). \qquad (3.3)$$

Now, we perform the conformal mapping $z = \sigma\left(\frac{\xi+i}{\xi-i}\right)$ of the half-plane $\{\xi \in \mathbf{C} : \mathrm{Im}\,\xi > 0\}$ onto the domain D_z. Using this conformal mapping, we reduce (3.3) to the integral equation

$$\frac{2}{\pi}\int_{-|\xi|}^{|\xi|}\frac{\Phi_*(\tau)m(\xi,\tau)}{\sqrt{\xi^2-\tau^2}}\,d\tau = \varphi_*(\xi) \qquad \forall\,\xi \in \mathbf{R}\setminus\{0\}, \qquad (3.4)$$

where

$$\Phi_*(\xi) := \frac{i\left(\Phi(\sigma(\frac{\xi-i}{\xi+i})) - \varphi_{\partial D}(b_2,0)\right)\sigma'(\frac{\xi-i}{\xi+i})}{2(\xi+i)}.$$

By virtue of the evenness of the function φ_*, (3.4) is equivalent to the equation

$$\frac{2}{\pi}\int_0^\xi \frac{\Phi_*(\tau)m(\xi,\tau)+\Phi_*(-\tau)m(\xi,-\tau)}{\sqrt{\xi^2-\tau^2}}\,d\tau = \varphi_*(\xi) \qquad \forall\,\xi>0. \tag{3.5}$$

Setting now $\xi = s$ in the equality (3.5) and, further, multiplying both sides of the obtained equality by $s(\xi^2-s^2)^{-1/2}$ and, finally, integrating it with respect to s over the segment $[0,\xi]$, we get

$$\frac{2}{\pi}\int_0^\xi \frac{s}{\sqrt{\xi^2-s^2}}\int_0^s \frac{\Phi_*(\tau)m(s,\tau)+\Phi_*(-\tau)m(s,-\tau)}{\sqrt{s^2-\tau^2}}\,d\tau\,ds = \int_0^\xi \frac{s\,\varphi_*(s)}{\sqrt{\xi^2-s^2}}\,ds.$$

Changing the order of integration in the repeated integrals, we obtain the equality

$$\frac{2}{\pi}\int_0^\xi\int_\tau^\xi \frac{s\,(\Phi_*(\tau)m(s,\tau)+\Phi_*(-\tau)m(s,-\tau))}{\sqrt{(\xi^2-s^2)(s^2-\tau^2)}}\,ds\,d\tau = \int_0^\xi \frac{s\,\varphi_*(s)}{\sqrt{\xi^2-s^2}}\,ds.$$

Further, by differentiating this equality with respect to ξ, we obtain the relation

$$-\frac{2\xi}{\pi}\int_0^\xi\int_\tau^\xi \frac{s[\Phi_*(\tau)(m(s,\tau)-m(\xi,\tau))+\Phi_*(-\tau)(m(s,-\tau)-m(\xi,-\tau))]}{(\xi^2-s^2)^{3/2}\sqrt{s^2-\tau^2}}\,ds\,d\tau$$

$$+\Phi_*(\xi)m(\xi,\xi)+\Phi_*(-\xi)m(\xi,-\xi) = f_*(\xi). \tag{3.6}$$

We introduce the functions $U_p(\xi) := \operatorname{Re}\Phi_*(\xi)$ and $V_p(\xi) := \operatorname{Im}\Phi_*(\xi)$. Taking into account the equalities

$$m(\xi,-\tau)=\overline{m(\xi,\tau)}, \quad \Phi_*(-\tau)=\overline{\Phi_*(\tau)} \qquad \forall\,\tau\in\mathbf{R}\setminus\{0\},$$

we rewrite the relation (3.6) in the form

$$A(\xi,\xi)U_p(\xi)-B(\xi,\xi)V_p(\xi)-\frac{2\xi}{\pi}\int_0^\xi U_p(\tau)\int_\tau^\xi \frac{s(A(s,\tau)-A(\xi,\tau))}{(\xi^2-s^2)^{3/2}\sqrt{s^2-\tau^2}}\,ds\,d\tau$$

$$+\frac{2\xi}{\pi}\int_0^\xi V_p(\tau)\int_\tau^\xi \frac{s(B(s,\tau)-B(\xi,\tau))}{(\xi^2-s^2)^{3/2}\sqrt{s^2-\tau^2}}\,ds\,d\tau = f_*(\xi). \tag{3.7}$$

Finally, using the Hilbert inversion formula for the Cauchy singular integral (see [2], p. 93), we reduce (3.7) to the Cauchy singular integral equation (3.2) to determine the function U_p. The theorem is proved. □

Thus, equation (3.1) is reduced to equation (3.2). If the boundary ∂D is a smooth curve satisfying certain additional requirements, then (3.2) is reduced to a Fredholm integral equation. Before to formulate the theorem on such a reduction of equation (3.2), we introduce some notations.

Let $C(\mathbf{R})$ be the Banach space of complex valued continuous functions $h_* : (\mathbf{R} \cup \{\infty\}) \to \mathbf{C}$ with the norm $\|h_*\|_{C(\mathbf{R})} := \sup_{\tau \in \mathbf{R}} |h_*(\tau)|$ and let $C_e(\mathbf{R})$ be the subspace of $C(\mathbf{R})$ containing all even continuous functions. For a function $h_* \in C(\mathbf{R})$, we consider the modulus of continuity

$$\omega_{\mathbf{R}}(h_*, \varepsilon) := \sup_{\tau_1, \tau_2 \in \mathbf{R}, |\tau_1 - \tau_2| \le \varepsilon} |h_*(\tau_1) - h_*(\tau_2)|$$

and the local centralized (with respect to infinite point) modulus of continuity

$$\omega_{\mathbf{R},\infty}(h_*, \varepsilon) := \sup_{\tau \in \mathbf{R}, |\tau| \ge 1/\varepsilon} |h_*(\tau) - h_*(\infty)|.$$

Now, denote by $\mathcal{D}(\mathbf{R})$ the set of functions $h_* \in C(\mathbf{R})$ such that its modulus of continuity satisfies the conditions

$$\int_0^1 \frac{\omega_{\mathbf{R}}(h_*, \eta)}{\eta}\, d\eta < \infty, \qquad \int_0^1 \frac{\omega_{\mathbf{R},\infty}(h_*, \eta)}{\eta}\, d\eta < \infty.$$

We denote also by $\mathcal{D}_e(\mathbf{R})$ the set of even functions from $\mathcal{D}(\mathbf{R})$.

For a function $g(Z)$ defined on the unit circle, we introduce its modulus of continuity as follows

$$\omega(g, \varepsilon) := \sup_{|Z_1| = |Z_2| = 1, |Z_1 - Z_2| \le \varepsilon} |g(Z_1) - g(Z_2)|.$$

Consider the functions

$$\tilde{m}(\xi, \tau) := \begin{cases} \frac{2\xi}{\pi} \int_\tau^\xi \frac{s(m(s,\tau) - m(\xi,\tau))}{(\xi^2 - s^2)^{3/2} \sqrt{s^2 - \tau^2}}\, ds, & \text{when } \xi\tau > 0, |\tau| < |\xi|, \\ 0, & \text{when } \xi\tau < 0 \text{ or } \xi\tau > 0, |\tau| > |\xi|, \end{cases}$$

$$\widetilde{A}(\xi, \tau) := 2 \operatorname{Re} \tilde{m}(\xi, \tau), \qquad \widetilde{B}(\xi, \tau) := 2 \operatorname{Im} \tilde{m}(\xi, \tau),$$

$$k_p(\xi, \tau) := -\frac{\xi}{|\xi|} \widetilde{A}(\xi, \tau) + \frac{1}{\pi} \int_0^\xi \frac{\widetilde{B}(\xi, \eta)}{\eta - \tau}\, d\eta,$$

$$P(\xi) := \sqrt{\sigma'\left(\frac{\xi - i}{\xi + i}\right)} - \sqrt{\sigma'\left(\frac{\xi + i}{\xi - i}\right)}$$

and the integral operators

$$(k_p f)(\xi) := \int_{-\infty}^{\infty} \frac{k_p(\xi,\tau)}{\sqrt{\tau^2+1}}\, f(\tau)\, d\tau,$$

$$(Rf)(\xi) := \sqrt{\xi^2+1}\left(\frac{A(\xi,\xi)}{(A(\xi,\xi))^2+(B(\xi,\xi))^2} f(\xi)\right.$$

$$\left. + \frac{P(\xi)}{4\pi i}\int_{-\infty}^{\infty} f(\tau)\sqrt{\frac{(\tau^2+1)\,|\mathrm{Im}\,\sigma(\frac{\tau-i}{\tau+i})|}{2\,|\tau|}}\;\frac{d\tau}{\tau-\xi}\right).$$

The following theorem establishes sufficient conditions for regularization of equation (3.2).

Theorem 3.2 *Suppose that $\varphi_{\partial D} \in \mathcal{H}_\alpha^u(\partial D)$ and the conformal mapping $\sigma(Z)$ has the nonvanishing continuous derivative on the unit circle and its modulus of continuity satisfies the following condition:*

$$\int_0^1 \frac{\omega(\sigma',\eta)}{\eta}\, \ln^3\frac{1}{\eta}\, d\eta < \infty. \tag{3.8}$$

Then the operator $R\,k_p$ is compact in the space $C_e(\mathbf{R})$ and the Fredholm integral equation

$$U_0(\xi) + (R(k_p U_0))(\xi) = (Rf_*)(\xi) \quad \forall \xi \in \mathbf{R} \tag{3.9}$$

has the unique solution $U_0 \in C_e(\mathbf{R})$ which belongs necessarily to the set $\mathcal{D}_e(\mathbf{R})$.

If the function $U_0 \in \mathcal{D}_e(\mathbf{R})$ is the solution of equation (3.9), then the function

$$U_p(\xi) = \frac{U_0(\xi)}{\sqrt{\xi^2+1}}, \tag{3.10}$$

is a solution of equation (3.2). Moreover, equation (3.2) has the unique solution of the form (3.10), where $U_0 \in \mathcal{D}_e(\mathbf{R})$.

Taking into account that the index of equation (3.2) is equal to

$$\kappa = \frac{1}{2\pi}\int_{-\infty}^{\infty} d\arg\frac{\overline{m(\xi,\xi)}}{m(\xi,\xi)} = 0, \tag{3.11}$$

Theorem 3.2 is proved by means of the classical Carleman – Vekua method of regularization of the Cauchy singular integral equation (see [2]).

We observe that if the boundary ∂D is the Lyapunov curve, then the condition (3.8) is satisfied, which follows from the Kellog theorem. The condition (3.8) is also satisfied in a more general case where the boundary ∂D is a smooth curve and, furthermore, the tangent angle $\theta(s)$ as function of the arc length s has the modulus of continuity $\omega(\theta,\varepsilon)$ satisfying the following condition

$$\int_0^1 \frac{\omega(\theta,\eta)}{\eta} \ln^4 \frac{1}{\eta}\, d\eta < \infty.$$

The last statement follows from an estimate for the modulus of continuity of conformal mapping of the unit disk given in the papers [3, 16].

The following theorem establishes sufficient conditions for solvability of the auxiliary problem for the function $\varphi_{\partial D}$.

Theorem 3.3 *If the conditions of Theorem* 3.2 *are satisfied, then the auxiliary problem for the function* $\varphi_{\partial D}$ *has the unique solution which is given by the formula*

$$\Phi(z) = \varphi_{\partial D}(b_2, 0) - 2i\frac{\xi+i}{\sigma'(\frac{\xi-i}{\xi+i})}\Phi_*(\xi), \quad z = \sigma\left(\frac{\xi-i}{\xi+i}\right) \in D_z, \tag{3.12}$$

where

$$\Phi_*(\xi) = \frac{1}{\pi i}\int_{-\infty}^{\infty} \frac{U_p(\tau)}{\tau-\xi}\, d\tau \qquad \forall \xi \in \mathbf{C} : \operatorname{Im}\xi > 0, \tag{3.13}$$

in which the function U_p *is expressed by* (3.10), *where* U_0 *is the solution of the Fredholm integral equation* (3.9) *in the space* $C_e(\mathbf{R})$.

Proof. Under the conditions of the theorem, the function U_p is a solution of equation (3.2). From the proof of Theorem 3.1 it follows that equation (3.2) is equivalent to equation (3.4), in which the function Φ_* is expressed via the function U_p as the solution of the Schwarz problem for a half-plane in the form (3.13). Now, to complete the proof, it suffices to note that the function (3.12) is the solution of equation (3.1). □

From Theorem 3.3 there follows immediately the next theorem on solvability of the Dirichlet boundary problem for the axial-symmetric potential.

Theorem 3.4 *If the conditions of Theorem* 3.2 *are satisfied, then the solution of the Dirichlet problem for the axial-symmetric potential is given by formula* (2.3), *in which* $F = \Phi$, *where* Φ *is the solution* (3.12) *of the auxiliary problem for the function* $\varphi_{\partial D}$.

In the case where the domain D is the disk, the solutions of both the auxiliary problem for the function $\varphi_{\partial D}$ and the Dirichlet problem for the axial-symmetric potential were obtained explicitly (see [10, 12]). Indeed, we observe that if $D = \{(x,y) : x^2+y^2 < 1\}$, then $m(\xi,\tau) \equiv 1$ and the function $U_p(\xi) = f_*(\xi)/2$ is the unique solution of equation (3.2). Furthermore, in this case the mentioned solutions are determined in the way shown in Theorem 3.3 and Theorem 3.4, respectively.

The next result shows that under some natural condition the axial-symmetric potential $\varphi(x, y)$ is expressed in the domain D by formula (2.1).

Theorem 3.5 *Suppose that the function $\varphi(x, y)$ satisfies equation (1.2) in the domain D and the following condition*

$$\varphi(x, -y) = \varphi(x, y) \quad \forall (x, y) \in D. \tag{3.14}$$

In this case, there exists the unique function F analytic in the domain D_z and satisfying the condition

$$F(\bar{z}) = \overline{F(z)} \qquad \forall\, z \in D_z \tag{3.15}$$

that equality (2.1) is fulfilled for all $(x, y) \in D$.

Proof. Let γ_ρ be the image of the circle $\{Z \in \mathbf{C} : |Z| = \rho < 1\}$ under the conformal mapping $\sigma(Z)$ and $D_{z,\rho}$ be the domain bounded by the curve γ_ρ. For every $z \in \gamma_\rho$ with $\operatorname{Im} z \neq 0$, we take a Jordan subarc of the curve γ_ρ with the end points z and $\bar{z}$ as the curve $\Gamma_{z\bar{z}}$.

For $z \in \gamma_\rho, y \neq 0$, taking into account the Cauchy theorem, we transform the equality (2.1) into the form

$$\frac{1}{2\pi i} \int_{\gamma_\rho} \frac{F(t)}{(\sqrt{(t-z)(t-\bar{z})})^-}\, dt = \varphi(x, y), \tag{3.16}$$

where $\quad (\sqrt{(t-z)(t-\bar{z})})^- := \lim\limits_{\tau \to t, \tau \in \mathbf{C} \setminus \overline{D_{z,\rho}}} \sqrt{(\tau - z)(\tau - \bar{z})}.$

Let us show that there exists the unique function F analytic in the domain D_z which satisfies condition (3.15) and equation (3.16) for all $\rho \in (0; 1)$ and for all $z \in \gamma_\rho$ with $y \neq 0$. Actually, in accordance with Theorem 3.3, there exists the unique function F_ρ analytic in the domain $D_{z,\rho}$ which satisfies the condition of type (3.15) in this domain and equation (3.16) with $F = F_\rho$. Now, it makes sense to remark that the equality $F_\rho(x) = \varphi(x, 0)$ is fulfilled for all $x \in \overline{D_{z,\rho}} \cap \mathbf{R}$. In view of this remark it becomes obvious that for any $\rho \in (0; 1)$ the function F_ρ is a restriction on $D_{z,\rho}$ of the unique function F analytic in the domain D_z, which proves Theorem 3.5. □

We note that if certain additional requirements are fulfilled, then for every solution $\varphi(x, y)$ of equation (1.2) in D, the function F has been determined explicitly by formula (2.1) (see Theorem 6 from [6]).

4. Dirichlet boundary problem for the Stokes flow function

Consider the following *Dirichlet boundary problem for the Stokes flow function*: to find the solution $\psi(x, y)$ of equation (1.3) in D when its boundary values $\psi_{\partial D}(x, y)$ are given on the boundary ∂D, i.e. $\psi(x, y) = \psi_{\partial D}(x, y)$ for all $(x, y) \in \partial D$.

To solve this Dirichlet boundary problem, we introduce and consider the following *auxiliary problem* for the given function $\psi_{\partial D}$: to find a function Ψ analytic in D_z and continuous in $\overline{D_z} \setminus \{b_1, b_2\}$ under the condition that

$$\Psi(\bar{z}) = \overline{\Psi(z)} \qquad \forall\, z \in D_z \tag{4.1}$$

and boundary values $\Psi(t), t \in \partial D_z \setminus \{b_1, b_2\}$, satisfy the following integral equation

$$-\frac{1}{2\pi i} \int_{\partial D_z} \frac{\Psi(t)(t-x)}{(\sqrt{(t-z)(t-\bar{z})})^-}\, dt = \psi_{\partial D}(x, y). \tag{4.2}$$

We consider first the homogeneous auxiliary problem in the case where $\psi_{\partial D} \equiv 0$.

Theorem 4.1 *Suppose that the boundary ∂D_z is a Jordan rectifiable curve satisfying the condition*

$$\sup_{z \in \partial D_z} \operatorname{mes}\{t \in \partial D_z : |t - z| \leq \varepsilon\} = O(\varepsilon), \quad \varepsilon \to 0, \tag{4.3}$$

where mes *denotes the linear Lebesgue measure on ∂D_z.*

If a solution Ψ of the auxiliary problem for the function $\psi_{\partial D} \equiv 0$ satisfies the estimate

$$|\Psi(z)| \leq c\ \left(|z - b_1|^{-\beta} + |z - b_2|^{-\beta}\right) \qquad \forall\, z \in \overline{D_z} \setminus \{b_1, b_2\}, \tag{4.4}$$

in which $\beta \in (0; 1)$ and the constant c does not depend on z, then $\Psi(z) \equiv C$, where C is a real constant.

Proof. First we note that if under the conditions of the theorem the equality

$$\int_{\partial D_z} \frac{\Psi(t)\ (t - \operatorname{Re} z)}{(\sqrt{(t-z)(t-\bar{z})})^-}\, dt = 0 \tag{4.5}$$

is fulfilled for all $z \in \partial D_z \setminus \{b_1, b_2\}$, then it is also fulfilled for all $z \in D_z$. The last statement follows from the uniqueness of solution of the Dirichlet boundary problem for the Stokes flow function.

Now, let us consider a disk centered at a point of the real axis and contained in the domain D_z. Applying Theorem 7 from [6] to this disk and taking into account the uniqueness theorem for analytic functions, we establish that the equality (4.5) is fulfilled in the disk if and only if the function Ψ is identical constant C. This constant is real because of condition (4.1). The theorem is proved. □

In the same way as for equation (3.1), we develop a method for transition of equation (4.2) to the Cauchy singular integral equations. To describe such a transition for equation (4.2), we introduce the function

$$n(\xi, \tau) := M(Z, T)\ (\sigma(T) - \operatorname{Re} \sigma(Z))$$

of real variables ξ and τ, where $Z = \frac{\xi-i}{\xi+i}$ and $T = \frac{\tau-i}{\tau+i}$. In what follows, we use the functions

$$C(\xi,\tau) := 2\,\mathrm{Re}\,n(\xi,\tau), \qquad D(\xi,\tau) := 2\,\mathrm{Im}\,n(\xi,\tau).$$

Theorem 4.2 *Suppose that the function $\psi_{\partial D} \in \mathcal{H}_\alpha^u(\partial D)$ satisfies the following condition:*

$$\psi_{\partial D}(b_1;0) = \psi_{\partial D}(b_2;0) = 0. \tag{4.6}$$

Suppose also that the conformal mapping $\sigma(Z)$ has the nonvanishing continuous derivative on the unit circle.

If a solution Ψ of the auxiliary problem for the function $\psi_{\partial D}$ satisfies estimate (4.4), then the function

$$V_f(\xi) = \mathrm{Im}\,\frac{i\Psi(\sigma(\frac{\xi-i}{\xi+i}))\sigma'(\frac{\xi-i}{\xi+i})}{2(\xi+i)}$$

is a solution of the Cauchy singular integral equation

$$D(\xi,\xi)V_f(\xi) - \frac{C(\xi,\xi)}{\pi}\int_{-\infty}^{\infty}\frac{V_f(\tau)}{\tau-\xi}\,d\tau$$
$$+\frac{2\xi}{\pi^2}\int_0^\xi\left(\int_\tau^\xi\frac{s(C(s,\tau)-C(\xi,\tau))}{(\xi^2-s^2)^{3/2}\sqrt{s^2-\tau^2}}\,ds\int_{-\infty}^{\infty}\frac{V_f(\eta)}{\eta-\tau}\,d\eta\right)d\tau$$
$$-\frac{2\xi}{\pi}\int_0^\xi V_f(\tau)\int_\tau^\xi\frac{s(D(s,\tau)-D(\xi,\tau))}{(\xi^2-s^2)^{3/2}\sqrt{s^2-\tau^2}}\,ds\,d\tau = g_*(\xi) \quad \forall\,\xi>0, \tag{4.7}$$

where

$$g_*(\xi) := \psi_*(\xi) - \xi\int_0^\xi\frac{s(\psi_*(s)-\psi_*(\xi))}{(\xi^2-s^2)^{3/2}}\,ds$$

and the function ψ_* is expressed via the given function $\psi_{\partial D}$ in the following form:

$$\psi_*(\xi) := \frac{\psi_{\partial D}(x,y)}{\sqrt{\xi^2+1}}$$

with $x+iy = \sigma(\frac{\xi-i}{\xi+i})$.

Proof. Consider the functions

$$\Psi_*(\xi) := \frac{i\;\Psi(\sigma(\frac{\xi-i}{\xi+i}))\;\sigma'(\frac{\xi-i}{\xi+i})}{2(\xi+i)}$$

and $U_f(\xi) := \mathrm{Re}\,\Psi_*(\xi)$, $\quad V_f(\xi) := \mathrm{Im}\,\Psi_*(\xi)$.

By analogy with the reduction of to (3.7), we reduce (4.2) to the following equation:

$$D(\xi,\xi)V_f(\xi) - C(\xi,\xi)U_f(\xi) - \frac{2\xi}{\pi}\int_0^\xi V_f(\tau)\int_\tau^\xi \frac{s(D(s,\tau)-D(\xi,\tau))}{(\xi^2-s^2)^{3/2}\sqrt{s^2-\tau^2}}\,ds\,d\tau$$
$$+\frac{2\xi}{\pi}\int_0^\xi U_f(\tau)\int_\tau^\xi \frac{s(C(s,\tau)-C(\xi,\tau))}{(\xi^2-s^2)^{3/2}\sqrt{s^2-\tau^2}}\,ds\,d\tau = g_*(\xi). \tag{4.8}$$

Now, using the Hilbert inversion formula for the Cauchy singular integral (see [2], p. 93), we reduce (4.8) to the Cauchy singular integral equation (4.7) for the determination of the function V_f. The theorem is proved. □

Denote by $C_u(\mathbf{R})$ the subspace of $C(\mathbf{R})$ including all odd continuous functions. Denote also by $\mathcal{D}_u(\mathbf{R})$ the set of odd functions from $\mathcal{D}(\mathbf{R})$.

Consider the function $\Pi_*(\xi) := |\xi|^{\beta_0}\,(|\xi|+1)^{\beta_\infty-\beta_0}$, where the numbers β_0, β_∞ satisfy the inequalities $1-\alpha+\nu<\beta_0<1$ and $0<\beta_\infty<\min\{\alpha-\nu,1/2\}$, in which α and ν are the same numbers as in the definition of the set $\widetilde{\mathcal{H}}^u_\alpha(\partial D)$.

Consider also the functions

$$\widetilde{n}(\xi,\tau) := \begin{cases} \frac{2\xi}{\pi}\int_\tau^\xi \frac{s(n(s,\tau)-n(\xi,\tau))}{(\xi^2-s^2)^{3/2}\sqrt{s^2-\tau^2}}\,ds, & \text{when } \xi\tau>0, |\tau|<|\xi|, \\ 0, & \text{when } \xi\tau<0 \text{ or } \xi\tau>0, |\tau|>|\xi|, \end{cases}$$

$$\widetilde{C}(\xi,\tau) := 2\,\mathrm{Re}\,\tilde{n}(\xi,\tau), \qquad \widetilde{D}(\xi,\tau) := 2\,\mathrm{Im}\,\tilde{n}(\xi,\tau),$$

$$k_f(\xi,\tau) := -\frac{\xi}{|\xi|}\,\widetilde{D}(\xi,\tau) - \frac{1}{\pi}\int_0^\xi \frac{\widetilde{C}(\xi,\eta)}{\eta-\tau}\,d\eta$$

and the integral operators

$$(k_f f)(\xi) := \int_{-\infty}^{\infty} \frac{k_f(\xi,\tau)}{\Pi_*(\tau)}\,f(\tau)\,d\tau,$$

$$(Qf)(\xi) := \Pi_*(\xi)\left(\frac{A(\xi,\xi)}{(A(\xi,\xi))^2+(B(\xi,\xi))^2}\,\frac{f(\xi)}{\mathrm{Im}\,\sigma(\frac{\xi-i}{\xi+i})}\right.$$
$$\left.+\frac{P(\xi)}{4\pi i}\int_{-\infty}^{\infty}\frac{f(\tau)}{\mathrm{Im}\,\sigma(\frac{\tau-i}{\tau+i})}\sqrt{\frac{(\tau^2+1)\,|\mathrm{Im}\,\sigma(\frac{\tau-i}{\tau+i})|}{2\,|\tau|}}\,\frac{d\tau}{\tau-\xi}\right).$$

The following theorem establishes sufficient conditions for regularization of equation (4.7).

Theorem 4.3 *Suppose that the function $\psi_{\partial D}\in\mathcal{H}^u_\alpha(\partial D)$ satisfies the condition (4.6) and the conformal mapping $\sigma(Z)$ has the nonvanishing continuous derivative on the unit circle and its modulus of continuity satisfies the condition (3.8).*

Then the operator $Q\, k_f$ *is compact in the space* $C_u(\mathbf{R})$ *and the Fredholm integral equation*

$$V_0(\xi) + (Q(k_f V_0))(\xi) = (Qg_*)(\xi) \quad \forall \xi \in \mathbf{R} \tag{4.9}$$

has the unique solution $V_0 \in C_u(\mathbf{R})$ *that belongs necessarily to the set* $\mathcal{D}_u(\mathbf{R})$ *and satisfies the equality*

$$V_0(0) = V_0(\infty) = 0. \tag{4.10}$$

If the function $V_0 \in \mathcal{D}_u(\mathbf{R})$ *is the solution of* (4.9), *then the function*

$$V_f(\xi) = \frac{V_0(\xi)}{\Pi_*(\xi)}, \tag{4.11}$$

is a solution of (4.7). *Moreover, equation* (4.7) *has the unique solution of the form* (4.11), *where the function* $V_0 \in \mathcal{D}_u(\mathbf{R})$ *satisfies the equality* (4.10).

Taking into account that the index of equation (4.7) is determined by formula (3.11), equation (4.2) is reduced to Fredholm integral equation (4.9) in the same way as equation (3.1) was reduced to equation (3.9) and Theorem 4.3 is proved similarly to Theorem 3.2.

Thus, we have reduced (4.2) to Fredholm integral equation (4.9). Therefore, the following analogues of Theorems 3.3 and 3.4 are valid in the case of the auxiliary problem for the function $\psi_{\partial D}$ and in the case of the Dirichlet problem for the Stokes flow function, respectively.

Theorem 4.4 *If the conditions of Theorem* 4.3 *are satisfied, then the auxiliary problem for the function* $\psi_{\partial D}$ *has a solution* Ψ_0 *which is given by the formula*

$$\Psi_0(z) = 2\frac{\xi + i}{\sigma'\left(\frac{\xi - i}{\xi + i}\right)}\Psi_*(\xi), \quad z = \sigma\left(\frac{\xi - i}{\xi + i}\right) \in D_z, \tag{4.12}$$

where

$$\Psi_*(\xi) = \frac{1}{\pi i}\int_{-\infty}^{\infty} \frac{V_f(\tau)}{\tau - \xi}\, d\tau \qquad \forall \xi \in \mathbf{C} : \operatorname{Im}\xi > 0,$$

in which the function V_f *is expressed in the form* (4.11), *where* V_0 *is the solution of the Fredholm integral equation* (4.9) *in the space* $C_u(\mathbf{R})$. *Moreover, any solution of the auxiliary problem for the function* $\psi_{\partial D}$ *is expressed in the form* $\Psi(z) = \Psi_0(z) + C$, *where* C *is a real constant.*

From Theorems 4.1 and 4.4 there follows immediately the next theorem on solvability of the Dirichlet boundary problem for the Stokes flow function.

Theorem 4.5 *If the conditions of Theorem* 4.3 *are satisfied, then the solution of the Dirichlet problem for the Stokes flow function is given by the formula*

(2.4) *with* $F = \Psi_0$, *where* Ψ_0 *is the solution* (4.12) *of the auxiliary problem for the function* $\psi_{\partial D}$.

In the case where the domain D is the disk, the solutions of both the auxiliary problem for the function $\psi_{\partial D}$ and the Dirichlet problem for the Stokes flow function were obtained explicitly (see [10, 13]).

The following analogue of Theorem 3.5 is proved in a similar way for the Stokes flow function $\psi(x, y)$.

Theorem 4.6 *Suppose that the function* $\psi(x, y)$ *satisfies equation* (1.3) *in the domain* D, *the condition*

$$\psi(x, -y) = \psi(x, y) \quad \forall (x, y) \in D$$

and the additional assumption

$$\psi(x, 0) \equiv 0 \quad \forall (x, y) \in D. \tag{4.13}$$

Then there exists a function F_0 *analytic in the domain* D_z *such that the equality* (2.2) *is fulfilled with* $F = F_0$ *for all* $(x, y) \in D$. *Moreover, any analytic function* F *which satisfies the condition* (3.15) *and the equality* (2.2) *for all* $(x, y) \in D$ *is expressed in the form* $F(z) = F_0(z) + C$, *where* C *is a real constant.*

Proof. Let γ_ρ and $D_{z,\rho}$ be the same as in the proof of Theorem 3.5. For $z \in \gamma_\rho$ with $y \neq 0$, by analogy with the proof of equality (3.16), we transform equality (2.2) into the form

$$-\frac{1}{2\pi i} \int_{\gamma_\rho} \frac{F(t)\,(t - x)}{(\sqrt{(t - z)(t - \bar{z})})^-}\, dt = \psi(x, y). \tag{4.14}$$

Using the conformal mapping $z = \sigma\left(\rho\,\frac{\xi - i}{\xi + i}\right)$ of the half-plane $\{\xi \in \mathbf{C} : \operatorname{Im}\xi > 0\}$ onto the domain $D_{z,\rho}$, in accordance with Theorems 4.2 and 4.3, we reduce (4.14) to a Fredholm integral equation of the type (4.9) which has the unique solution $V_\rho \in C_u(\mathbf{R})$. Therefore, in accordance with Theorem 4.3, boundary values on $\gamma_\rho \setminus \mathbf{R}$ of the analytic in $D_{z,\rho}$ function

$$F_\rho(z) := \frac{2(\xi + i)}{\pi\rho i\, \sigma'\left(\rho\frac{\xi - i}{\xi + i}\right)} \int_{-\infty}^{\infty} \frac{V_\rho(\tau)}{\tau - \xi} \frac{d\tau}{\Pi_*(\tau)} - \frac{4}{\pi\rho\, \sigma'(0)} \int_{-\infty}^{\infty} \frac{\tau\, V_\rho(\tau)}{\tau^2 + 1} \frac{d\tau}{\Pi_*(\tau)}$$

$$z = \sigma\left(\rho\,\frac{\xi - i}{\xi + i}\right) \in D_{z,\rho}, \operatorname{Im}\xi > 0,$$

satisfy (4.14) with $F = F_\rho$. It is important that the equality $F_\rho(\sigma(0)) = 0$ is valid.

Now, let us show that there exists a function F_0 analytic in D_z which satisfies equation (4.14) with $F = F_0$ for all $\rho \in (0; 1)$ and for all $z \in \gamma_\rho$ with $y \neq 0$. Keeping this purpose in mind, we consider two equations of the type (4.14) with $\rho = \rho_1$ and $\rho = \rho_2$, where $0 < \rho_1 < \rho_2 < 1$. Let F_{ρ_1} and F_{ρ_2} be their solutions, respectively.

Taking into account the uniqueness of solution of the Dirichlet boundary problem for the Stokes flow function for the domain $D_{\rho_2} := \{(x,y) : z = x+iy \in D_{z,\rho_2}\}$ with given boundary values $\psi(x,y)$, $(x,y) \in \partial D_{\rho_2}$, we establish that the function F_{ρ_2} is also a solution of equation (4.14) with $\rho = \rho_1$. Further, taking into account Theorem 4.1, we establish that for all $z \in D_{z,\rho_1}$, the equality $F_{\rho_2}(z) = F_{\rho_1}(z) + C_0$ is fulfilled with a real constant C_0. Inasmuch as the equality $F_{\rho_1}(\sigma(0)) = F_{\rho_2}(\sigma(0)) = 0$ is fulfilled, the equality $C_0 = 0$ is true. Now, it becomes obvious that for any $\rho \in (0;1)$, the function F_ρ is restricted on $D_{z,\rho}$ of the unique function F_0 analytic in the domain D_z.

Finally, for any function F analytic in D_z satisfying condition (3.15) and equality (2.2) for all $(x,y) \in D$, using Theorem 4.1, we conclude that $F(z) = F_0(z) + C$, C being real.

Theorem 4.6 is proved. □

Let us note that if certain additional assumptions are fulfilled, then for every solution $\psi(x,y)$ of equation (1.3) in D, the function F is determined explicitly in formula (2.2) (see Theorem 7 from [6]).

The requirement (4.13) is natural. For example, for the model of steady flow of an ideal incompressible fluid without sources and vortexes it means that the axis Ox is a line of flow. Furthermore, under the conditions of Theorem 2.1, function (2.2) satisfies identity (4.13). So, the requirement (4.6) is also natural.

In conclusion, it may be noted that the outer Dirichlet problems for the axial-symmetric potential and the Stockes flow function can be also investigated in exactly the same way as the inner Dirichlet problems considered above.

References

[1] Batchelor, G. K., *An Introduction to Fluid Dynamics.* Cambridge University Press, Cambridge, 1967.

[2] Gakhov, F. D., *Boundary Value Problems.* Pergamon Press, Oxford, New York, Paris, 1966. Extended Russian edition: Nauka, Moscow, 1977.

[3] Heronimus, J. L., On some properties of function continues in the closed disk, *Dokl. Akad. Nauk SSSR*, **98** (1954), no. 6, 889–891 (in Russian).

[4] Lavrentyev, M. A. and Shabat, B. V., *Problems of Hydrodynamics and Theirs Mathematical Models.* Nauka, Moscow, 1977 (in Russian).

[5] Loitsyanskii, L. G., *Mechanics of Liquids and Gases.* Nauka, Moscow, 1987 (in Russian).

[6] Mel'nichenko, I. P. and Plaksa, S. A., Potential fields with axial symmetry and algebras of monogenic functions of vector variable, I, *Ukr. Math. J.*, **48**, (1996), no. 11, 1717–1730.

[7] Mel'nichenko, I. P. and Plaksa, S. A., Potential fields with axial symmetry and algebras of monogenic functions of vector variable, II, *Ukr. Math. J.*, **48**, (1996), no. 12, 1916–1926.

[8] Mel'nichenko, I. P. and Plaksa, S. A., Potential fields with axial symmetry and algebras of monogenic functions of vector variable, III, *Ukr. Math. J.*, **49**, (1997), no. 2, 253–268.

[9] Mikhailov, L. G. and Radzhabov, N., An analog of the Poisson formula for certain second order equations with a singular curve, *Dokl. Akad. Nauk Tadzh. SSR*, **15** (1972), no. 11, 6–9 (in Russian).

[10] Plaksa, S., Boundary properties of axial-symmetrical potential and Stokes flow function. In: *Finite or Infinite Dimensional Complex Analysis: Proceedings of the Seventh International Colloquium*, Lecture Notes in Pure and Applied Mathematics, **214**, 443–455. Marcel Dekker Inc., New York, 2000.

[11] Plaksa, S., Algebras of hypercomplex monogenic functions and axial-symmetrical potential fields, In: *Proceedings of the Second ISAAC Congress*, Fukuoka, August 16 – 21, 1999, 613–622. Kluwer Academic Publishers, Dordrecht, 2000.

[12] Plaksa, S. A., Dirichlet problem for axisymmetric potential fields in a disk of the meridian plane, I. *Ukr. Math. J.*, **52** (2000), no. 4, 564–585.

[13] Plaksa, S. A., Dirichlet problem for axisymmetric potential fields in a disk of the meridian plane, II. *Ukr. Math. J.*, **52** (2000), no. 6, 748–757.

[14] Privalov, I. I., *Boundary Properties of Analytic Functions.* Gostekhizdat, Moscow, Leningrad, 1950 (in Russian).

[15] Radzhabov, N., Construction of potentials and investigation of interior and exterior boundary value problems of Dirichlet and Neumann types for the Euler–Poisson–Darboux equations on the plane, *Dokl. Akad. Nauk Tadzh. SSR*, **17** (1974), no. 8, 7–11 (in Russian).

[16] Warschawski, S. E., On differentiability at the boundary in conformal mapping, *Proc. Amer. Math. Soc.*, **12** (1961), no. 4, 614–620.

ON THE ANALYTICITY OF THE SCHWARZ OPERATOR WITH RESPECT TO A CURVE*

Luca Preciso
University of Padova, Italy
preciso@math.unipd.it

Sergei Rogosin
Belarusian State University, Belarus
rogosin@bsu.by

To Professor G. S. Litvinchuk on the occasion of his 70th birthday

Abstract The classical Schwarz operator $\mathbf{T}[\cdot,\cdot,\cdot]$ assigns to each triple (ϕ, f, w), where ϕ is a plane closed curve enclosing a simply connected domain D, f is a real-valued function of the boundary of D and $w \in D$, the unique holomorphic function F of D satisfying $\operatorname{Re} F = f$ on the boundary of D and $\operatorname{Im} F(w) = 0$. We consider the modified Schwarz operator $\mathbf{T}_*$, $\mathbf{T}_*[\phi,p,w] \equiv \mathbf{T}[\phi, p \circ \phi^{(-1)}, w] \circ \phi$, which assigns to each triple (ϕ, p, w) the composition of the classical Schwarz operator $\mathbf{T}$ valued in $(\phi, p\circ\phi^{(-1)}, w)$ with the boundary curve ϕ. We show that $\mathbf{T}_*$ depends real analytically on its variables and compute the first differential of $\mathbf{T}_*$. Regularity of another type of modifications of the Schwarz operator $\mathbf{T}$ is studied too.

Keywords: Schwarz boundary value problem, nonlinear operators, simply connected domain, Schauder spaces, perturbation analysis.

1. Introduction

The classical Schwarz boundary value problem (see [8]) consists of searching for a holomorphic function F in a (bounded or unbounded) plane domain D by given boundary values f of the real part $u = \operatorname{Re} F$ of the function F. It

*The authors are indebted to the 'Gruppo Nazionale per l'Analisi Mat., la Probabilità e loro Applicaz.' of the Italian 'Istituto Nazionale di Alta Matematica' "F.Severi" and to the 'Belarusian Fund for Fundamental Scientific Research' for their financial support.

is known, see [8, p. 210], [19, §30] that the solution of such a problem is determined uniquely up to a pure imaginary constant whenever f belongs to a suitable function space (*e.g.*, f is Hölder-continuous), and the boundary ∂D is a smooth simple curve. Uniqueness can be achieved by adding, *e.g.*, the condition $\mathrm{Im}\, F(w) = 0$, where w is a prescribed point of D. An analogous result takes place for D being a multiply connected domain. The operator which assigns to each given function f the solution of the Schwarz problem (having prescribed imaginary part in a given point) is called the Schwarz operator, see [8, p. 208].

The Schwarz operator $\mathbf{T}$ depends in fact not only on the boundary data f but also on the curve $\phi : \mathbb{T} \to \partial D$ encircling the domain D, and on the point $w \in D$.

The aim of our paper is to carry out the perturbation analysis of the Schwarz problem, *i.e.*, to study the regularity of the dependence of $\mathbf{T}[\phi, f, w]$ on the functional variables ϕ, f and w, in a Schauder space setting.

In the case of some simple curves the Schwarz operator can be represented in the explicit form. If, for instance, $\partial D = \mathbb{T}$, *i.e.*, ∂D is the unit circle with the standard parametrization, then the Schwarz operator has the following representation:

$$(\mathbf{T}[\phi, f, w])(z) \equiv \frac{1}{2\pi i} \int_{\mathbb{T}} f(\sigma) \frac{\sigma + z}{\sigma - z} \frac{d\sigma}{\sigma} - \frac{1}{2\pi} \int_{\mathbb{T}} f(\sigma)\, \mathrm{Im} \left[\frac{\sigma + w}{\sigma - w} \right] \frac{d\sigma}{\sigma}, \tag{1.1}$$

where $\phi \equiv id_{\mathbb{T}}$, *i.e.*, ϕ is the identity map on the unit circle $\mathbb{T}$, $w \in \mathbb{U}$, and $\mathrm{Im}\, \mathbf{T}[\phi, f, w](w) = 0$, in particular

$$(\mathbf{T}[\phi, f, 0])(z) \equiv \frac{1}{2\pi i} \int_{\mathbb{T}} f(\sigma) \frac{\sigma + z}{\sigma - z} \frac{d\sigma}{\sigma}, \tag{1.2}$$

For a general simply connected domain $D \equiv \mathbb{I}[\phi]$ enclosed by a Lyapunov curve ϕ (*i.e.* of class $C^{1,\alpha}$), the Schwarz operator is determined up to an imaginary constant by the formula

$$(\mathbf{T}[\phi, f, w])(z) \equiv \frac{1}{2\pi} \int_{\phi(\mathbb{T})} f(\sigma) \frac{\partial}{\partial n_\sigma} \left[log \frac{\omega(z) - \omega(\sigma)}{1 - \overline{\omega(\sigma)}\omega(z)} \right] |d\sigma|, \tag{1.3}$$

(see, *e.g.*, [21, p. 30]) where $\omega(\cdot)$ is a conformal mapping of the domain $\mathbb{I}[\phi]$ onto the unit disc $\mathbb{U}$ such that $\omega(w) = 0$, where $w \in \mathbb{I}[\phi]$. Thus the explicit representation can be found for those domains which are mapped explicitly onto the unit disc.

The study of the regularity of certain nonlinear operators with respect to functional variables is quite intensive in the recent years. The questions we are discussing here are close to that on the regularity of certain families of curves [27]. The latter are motivated by the application in fluid mechanics (see, e.g., [26], [9]).

Among the results similar to ours we have to mention the classical theorem by Radó [25], which asserts the continuity of the dependence of the Riemann map of a simply connected Jordan domain upon the boundary curve in the topology of the uniform convergence. More recently, Coifman and Meyer [6] have proved the analyticity of a nonlinear operator associated to the conformal representation of an unbounded simply connected domain having arc-length parametrized boundary with the direction of the tangent vector described by a function of class BMO. Later Wu [28] with advice of Coifman and with ideas of [6] has obtained two analyticity statements for bounded domains with arc-length parametrized boundary having certain symmetries. By using a PDE approach and in the frame of Schauder spaces, Lanza [12] has shown the analyticity of the operator $\mathbf{h} : (\phi, w) \mapsto g_{\phi,w}^{(-1)} \circ \phi$, where $g_{\phi,w}$ is the conformal mapping of the unit disc $\mathbb{U}$ onto the bounded domain $\mathbb{I}[\phi]$ enclosed by ϕ, which is normalized by the conditions $g_{\phi,w}(0) = w \in \mathbb{I}[\phi]$, $g'_{\phi,w}(0) > 0$. Later it has appeared that the integral equation approach is more suitable to the study of the operator $\mathbf{h}$ even in the case of simply connected domains. Indeed Lanza and Rogosin have discovered a system of integral-functional equations for the operator $\mathbf{h}$ in the case of simply connected domains [16] and in the case of doubly connected domains [17] and have shown on the base of these systems that the operator $\mathbf{h}$ is real analytic in a Schauder space setting.

The systems of integral-functional equations mentioned above contain the Cauchy integral operator

$$(\mathbf{C}[\phi, f])(s) \equiv \frac{\text{p.v.}}{2\pi i} \int_{\mathbb{T}} \frac{f(\sigma)\phi'(\sigma)}{\phi(\sigma) - \phi(s)} d\sigma. \tag{1.4}$$

In connection with the study of the regularity of operators of the type (1.4) we mention the contribution by Calderón, Coifman, Meyer, McIntosh, David. Calderón [3, Thm. 1] has shown that if ϕ is a graph of a Lipschitz function ψ, i.e., if $\phi(x) = x + i\psi(x)$ with $\psi' \in L^\infty(\mathbb{R})$ and if $\|\psi'\|_{L^\infty(\mathbb{R})} < \varepsilon$ for some $\varepsilon > 0$, then the linear integral operator with singular kernel $\frac{\phi'(y)}{\phi(y)-\phi(x)}$ is an element of the space $\mathcal{L}\left(L^2(\mathbb{R}, \mathbb{C}), L^2(\mathbb{R}, \mathbb{C})\right)$ of the linear and continuous operators of $L^2(\mathbb{R}, \mathbb{C})$ to itself. Then, by using a standard argument of truncated kernel, one can deduce the analytic dependence of the integral operator with kernel $\frac{1}{\phi(y)-\phi(x)}$ upon ψ' with $\|\psi'\|_{L^\infty(\mathbb{R})} < \sigma$ and σ possibly less than 1 (cf., *e.g.*, [18, p. 438]). Later, Coifman, McIntosh and Meyer [4, Thm. 1], and, by different method, David [7, p. 178], have extended the validity of the same analyticity result to the case in which $\|\psi'\|_{L^\infty(\mathbb{R})} < 1$.

To study the regularity of the dependence of $\mathbf{T}[\phi, f, w]$ on ϕ, f, and w we need to deal with functional variables having domains independent of ϕ. Then we study the regularity of some variants of the Schwarz operator. First we consider the modified Schwarz operator which maps a simple closed curve ϕ with nonvanishing derivative, a real valued function p of $\mathbb{T}$, and $w \in \mathbb{I}[\phi]$

to $\mathbf{T}[\phi, p \circ \phi^{(-1)}, w] \circ \phi$. We represent the modified Schwarz operator as a composition of operators whose regularity is known. Then we prove the real analyticity of the modified Schwarz operator by using the corresponding results for the singular (Cauchy) integral operator (as stated in [13, Thm 3.16]), and for the operator

$$\mathbf{h} : (\phi, w) \mapsto h[\phi, w] = g_{\phi,w}^{(-1)} \circ \phi \tag{1.5}$$

(as stated in [16, Thm. 5.8]), where $g_{\phi,w} : \mathbb{U} \to \mathbb{I}[\phi]$ is the Riemann map of the unit disc $\mathbb{U}$ onto the domain $\mathbb{I}[\phi]$, which is normalized by the conditions $g_{\phi,w}(0) = w, g'_{\phi,w}(0) > 0$. Moreover we calculate explicitly the first differential of the modified Schwarz operator. We study also the regularity of another modifications, namely

$$\mathbf{T}_{K,\epsilon}[\phi, p, w] \equiv \mathbf{T}[\phi, p \circ \phi^{(-1)}, w]_{|\mathrm{cl}\, K_\epsilon},$$

with K compact subset of $\mathbb{I}[\phi_0]$, $K_\epsilon \equiv K + \epsilon\mathbb{U}$, $\epsilon \in \mathbb{R}^+$ and $\mathrm{cl}\, K_\epsilon \subset \mathbb{I}[\phi_0]$, and

$$\mathbf{T}_c[\phi, p, w] \equiv \mathbf{T}[\phi, p \circ \phi^{(-1)}, w] \circ g_{\phi,w|\mathbb{T}},$$

where $g_{\phi,w}$ is the Riemann map of $\mathbb{I}[\phi]$.

The paper is organized as follows. In Section 2 we present basic notation and auxiliary results. In Section 3 we introduce the modified Schwarz operator. We show here the real analyticity of the modified Schwarz operator and calculate its first differential. Section 4 is devoted to the study of the regularity of another variants of the classical Schwarz operator.

2. Preliminaries and notation

For standard definitions of calculus on normed spaces we refer, e.g., to Prodi and Ambrosetti [24] and to Berger [2]. In particular, a finite product of normed spaces is endowed with the sup-norm of the norms of its components. Further, any complex normed space can be viewed as a real normed space. Accordingly, we will say that a map between two complex normed spaces is real differentiable, real analytic or real linear, respectively, if such a map has the corresponding property as a map between the underlined real normed spaces. In a contrary, if we are retaining the complex structure, then a map with the corresponding property will be called complex differentiable, complex analytic or complex linear.

The inverse function of a function f is denoted $f^{(-1)}$ as opposed to the reciprocal of a real or complex-valued function g, which is denoted by g^{-1}. We denote by $\mathbb{U}$ the open unit disc in $\mathbb{C}$ (or in $\mathbb{R}^2$), by $\mathbb{T}$ the boundary of $\mathbb{U}$. The unit circle $\mathbb{T}$ is as usual counter-clockwise oriented. For any set $D \subseteq \mathbb{R}^n$ we denote by $\mathrm{cl}\, D$ its closure, and by $\mathrm{int}\, D$ its interior. The symbol Re (Im) denotes the function which takes a complex number z to its real (imaginary) part. If

$\mathbf{T}$ is an operator into a subset of complex-valued functions, then $\mathrm{Re}\,\mathbf{T}$ ($\mathrm{Im}\,\mathbf{T}$) is the operator which maps f to $\mathrm{Re}\,\mathbf{T}[f] \equiv \mathrm{Re} \circ \mathbf{T}[f]$ ($\mathrm{Im}\,\mathbf{T}[f] \equiv \mathrm{Im} \circ \mathbf{T}[f]$).

Let Ω be an open subset of $\mathbb{R}^n$. Let $\mathbb{N}$ be the set of nonnegative integers including 0 and let $m \in \mathbb{N}$. We denote by $\mathcal{C}^m(\Omega, \mathbb{C})$ the set of m-times continuously real differentiable complex-valued functions on Ω and by $\mathcal{C}^m(\mathrm{cl}\,\Omega, \mathbb{C})$ the subspace of those functions $f \in \mathcal{C}^m(\Omega, \mathbb{C})$ such that, for all $\eta \in \mathbb{N}^n$ with $|\eta| \equiv \eta_1 + \cdots + \eta_n \leq m$, the functions $D^\eta f \equiv \frac{\partial^{|\eta|} f}{\partial^{\eta_1} x_1 \ldots \partial^{\eta_n} x_n}$ can be continuously extended to $\mathrm{cl}\,\Omega$. If Ω is a bounded open subset of $\mathbb{R}^n$, we endow $\mathcal{C}^m(\mathrm{cl}\,\Omega, \mathbb{C})$ with the norm $\|f\|_{\mathcal{C}^m(\mathrm{cl}\,\Omega, \mathbb{C})} \equiv \sum_{|\eta| \leq m} \sup_{\mathrm{cl}\,\Omega} |D^\eta f|$. The subspace of $\mathcal{C}^m(\mathrm{cl}\,\Omega, \mathbb{C})$ of those functions whose m-th order derivatives are α-Hölder continuous with exponent $\alpha \in]0,1]$ is denoted by $\mathcal{C}^{m,\alpha}(\mathrm{cl}\,\Omega, \mathbb{C})$. We endow $\mathcal{C}^{m,\alpha}(\mathrm{cl}\,\Omega, \mathbb{C})$ by the usual norm

$$\|f\|_{\mathcal{C}^{m,\alpha}(\mathrm{cl}\,\Omega, \mathbb{C})}$$

$$\equiv \sum_{|\eta| \leq m} \sup_{\mathrm{cl}\,\Omega} |D^\eta f| + \sum_{|\eta| = m} \sup \left\{ \frac{|D^\eta f(s) - D^\eta f(t)|}{|s-t|^\alpha} : s, t \in \mathrm{cl}\,\Omega, s \neq t \right\}.$$

The spaces $\left(\mathcal{C}^m(\mathrm{cl}\,\Omega, \mathbb{C}); \|\cdot\|_{\mathcal{C}^m(\mathrm{cl}\,\Omega, \mathbb{C})}\right)$, $\left(\mathcal{C}^{m,\alpha}(\mathrm{cl}\,\Omega, \mathbb{C}); \|\cdot\|_{\mathcal{C}^{m,\alpha}(\mathrm{cl}\,\Omega, \mathbb{C})}\right)$ are complex Banach spaces. If it is clear on which domain these spaces are defined then we will use the following short notation $\|\cdot\|_m$ and $\|\cdot\|_{m,\alpha}$ for the corresponding norms. If $n = 2$ we denote by $\mathcal{H}(\Omega)$ the set of holomorphic functions of Ω. Let $c[\Omega] \equiv \sup\{\frac{\lambda(x,y)}{|x-y|} : x, y \in \Omega, x \neq y\}$, where $\lambda(x,y) \equiv \inf\{\text{length of } \gamma \in \mathcal{C}^1([0,1], \Omega) : \gamma(0) = x, \gamma(1) = y\}$. The open subset Ω of $\mathbb{R}^n$ is said to be regular in the sense of Whitney (or Whitney regular set), if Ω is bounded and connected, and $c[\Omega] < +\infty$. It is well known (cf., *e.g.*, Jones [10, p. 73]) that if Ω is a bounded, connected, open subset of $\mathbb{R}^n$ of class $\mathcal{C}^1$, then it is a Whitney regular set.

Lemma 2.1 *Let $m, n, r, h \in \mathbb{N}$, let $n, r, h \geq 1$, and $\alpha, \beta \in]0,1]$. Let Ω be an open Whitney regular subset of $\mathbb{R}^n$. Then the following holds:*

(i) *$\mathcal{C}^{m+1}(\mathrm{cl}\,\Omega, \mathbb{C})$ is continuously imbedded in $\mathcal{C}^{m,\alpha}(\mathrm{cl}\,\Omega, \mathbb{C})$.*

(ii) *The pointwise product in $\mathcal{C}^{m,\alpha}(\mathrm{cl}\,\Omega, \mathbb{C})$ is continuous and the space $\mathcal{C}^{m,\alpha}(\mathrm{cl}\,\Omega, \mathbb{C})$ with this product is a commutative Banach algebra with unity.*

(iii) *The reciprocal map in $\mathcal{C}^{m,\alpha}(\mathrm{cl}\,\Omega, \mathbb{C})$, which maps a nonvanishing function f to its reciprocal f^{-1}, is complex analytic as a map from the open subset $\{f \in \mathcal{C}^{m,\alpha}(\mathrm{cl}\,\Omega, \mathbb{C}) : f(x) \neq 0, \text{ for all } x \in \mathrm{cl}\,\Omega\}$ of $\mathcal{C}^{m,\alpha}(\mathrm{cl}\,\Omega, \mathbb{C})$ to itself.*

(iv) *Let Ω_1 be an open subset of $\mathbb{R}^r$, regular in the sense of Whitney. If $F \in \mathcal{C}^{m,\alpha}(\mathrm{cl}\,\Omega_1, \mathbb{R}^h)$ and $G \in \mathcal{C}^{m,\beta}(\mathrm{cl}\,\Omega, \mathrm{cl}\,\Omega_1)$, then $F \circ G \in \mathcal{C}^{m,\gamma_m(\alpha,\beta)}(\mathrm{cl}\,\Omega, \mathbb{R}^h)$, with $\gamma_0(\alpha, \beta) = \alpha\beta$, and $\gamma_m(\alpha, \beta) = \min\{\alpha, \beta\}$ if $m > 0$.*

(v) *Let $m \in \mathbb{N}, m \geq 1$. If $G \in \mathcal{C}^{m,\alpha}(\mathrm{cl}\,\Omega, \mathbb{R}^n)$ is injective and satisfies the condition $\det DG(x) \neq 0$ for all $x \in \mathrm{cl}\,\Omega$, then $G(\Omega)$ is a bounded connected open subset of $\mathbb{R}^n$, $G(\mathrm{cl}\,\Omega) = \mathrm{cl}\,G(\Omega)$, $c[G(\Omega)] < +\infty$, and $G^{(-1)} \in \mathcal{C}^{m,\alpha}(\mathrm{cl}\,G(\Omega), \mathrm{cl}\,\Omega)$.*

For the proof of this Lemma we refer to [13, Lem. 2.3].

We now define the Schauder spaces on a general compact subset K of $\mathbb{C}$ with no isolated points (in particular, on plane Jordan curves). We say that a function f of K to $\mathbb{C}$ is complex differentiable at $z_0 \in \mathbb{C}$ if there exists the finite limit $\lim_{K \ni z \to z_0} \frac{f(z)-f(z_0)}{z-z_0}$. We denote such a limit by $f'(z_0)$. The higher order derivatives are defined by induction. Let $m \in \mathbb{N}$. We denote by $C_*^m(K, \mathbb{C})$ the complex normed space of the m-times continuously complex differentiable functions f from K to $\mathbb{C}$ endowed with the norm $\|f\|_{C_*^m(K,\mathbb{C})} = \sum_{j=0}^m \sup_K |f^{(j)}|$. If $\alpha \in]0,1]$, we denote by $C_*^{m,\alpha}(K, \mathbb{C})$ the subspace of $C_*^m(K, \mathbb{C})$ of those functions having α-Hölder continuous m-th order derivative in K. If $f \in C_*^{0,\alpha}(K, \mathbb{C})$, then we set $|f : K|_\alpha \equiv \sup\left\{\frac{|f(z_1)-f(z_2)|}{|z_1-z_2|^\alpha} : z_1, z_2 \in K,\ z_1 \neq z_2\right\}$. We endow $C_*^{m,\alpha}(K, \mathbb{C})$ with the norm $\|f\|_{C_*^{m,\alpha}(K,\mathbb{C})} \equiv \|f\|_{C_*^m(K,\mathbb{C})} + |f^{(m)} : K|_\alpha$. If $B \subseteq \mathbb{C}$, we set $C_*^{m,\alpha}(K, B) \equiv \{f \in C_*^{m,\alpha}(K, \mathbb{C}) : f(K) \subseteq B\}$. If B is an open subset of $\mathbb{C}$ a straightforward compactness argument implies that $C_*^{m,\alpha}(K, B)$ is an open subset of $C_*^{m,\alpha}(K, \mathbb{C})$. We denote by $C_*^{m,\alpha,0}(K, \mathbb{C})$ the closure of $C_*^\infty(K, \mathbb{C})$ in $C_*^{m,\alpha}(K, \mathbb{C})$.

Then the following variant of [11, Cor. 4.24, Prop. 4.29] holds (cf. [13, Lem. 2.5].)

Lemma 2.2 *The following statements hold.*

(i) *Let* $\phi \in C_*^1(\mathbb{T}, \mathbb{C})$. *Then* $l_{\mathbb{T}}[\phi] \equiv \inf\left\{\frac{|\phi(x)-\phi(y)|}{|x-y|} : x, y \in \mathbb{T},\ x \neq y\right\} > 0$ *if and only if* ϕ *is injective and* $\phi'(\xi) \neq 0$ *for all* ξ *in* $\mathbb{T}$.

(ii) *The function from* $C_*^1(\mathbb{T}, \mathbb{C})$ *to* $\mathbb{R}$ *which maps* ϕ *to* $l_{\mathbb{T}}[\phi]$ *is continuous, and in particular, the set*

$$\mathcal{Z} \equiv \{\phi \in C_*^1(\mathbb{T}, \mathbb{C}) : l_{\mathbb{T}}[\phi] > 0\} \tag{2.1}$$

is open in $C_*^1(\mathbb{T}, \mathbb{C})$.

We have to note that a smooth curve is often defined as an equivalence class of smooth parametrizations. However, for our purposes, it is necessary to distinguish different parametrizations. Thus we define a (closed) curve of class $\mathcal{C}^{m,\alpha}$ of the complex plane $\mathbb{C}$ to be a map ϕ belonging to $C_*^{m,\alpha}(\mathbb{T}, \mathbb{C})$. By a simple curve of class $\mathcal{C}^{m,\alpha}$ we understand a curve ϕ of class $\mathcal{C}^{m,\alpha}$ which belongs to $\mathcal{Z}$. A curve ϕ should not be confused with its locus $\phi(\mathbb{T})$. Our perturbation analysis of the Schwarz problem will deal with domains enclosed by a simple curve ϕ of class $\mathcal{C}^{m,\alpha}$ with $m \geq 1, \alpha \in]0,1[$ (*i.e.* a Lyapunov curve). Let $\phi \in \mathcal{Z}$. By $\mathbb{I}[\phi]$ we denote the bounded connected component of $\mathbb{C} \setminus \phi(\mathbb{T})$. Let $w \in \mathbb{I}[\phi]$. Let $g_{\phi,w}$ be the Riemann map of $\mathbb{U}$ to $\mathbb{I}[\phi]$ normalized by the conditions

$$g_{\phi,w}(0) = w,\ g'_{\phi,w}(0) > 0. \tag{2.2}$$

The same symbol $g_{\phi,w}$ will be used also for the unique continuous extension to $\mathrm{cl}\,\mathbb{U}$. It is well known that $g_{\phi,w}$ is a homeomorphism between $\mathrm{cl}\,\mathbb{U}$ and $\mathrm{cl}\,\mathbb{I}[\phi]$

(cf., *e.g.*, [23, Thm. 2.6]). A classical Warschawski result (cf. *e.g.* [23, Thms. 3.5, 3.6]) implies that

$$g_{\phi,w} \in C^{m,\alpha}(\mathrm{cl}\,\mathbb{U},\mathbb{C}),\ g'_{\phi,w}(z) \neq 0\ \forall z \in \mathrm{cl}\,\mathbb{U} \tag{2.3}$$

whenever $\phi \in C^{m,\alpha}_*(\mathbb{T},\mathbb{C}) \cap \mathcal{Z}$ with $m \in \mathbb{N} \setminus \{0\}$ and $\alpha \in]0,1[$.

We are now ready to state the following Lemma, which collects a few facts which we need on the spaces $C^{m,\alpha}_*(K,\mathbb{C})$. For a proof and for appropriate references, we refer to [13, Lems. 2.7, 2.8].

Lemma 2.3 *Let $m \in \mathbb{N}$, $\alpha, \beta \in]0,1]$, $\phi \in \mathcal{Z}$, $L = \phi(\mathbb{T})$. Then the following statements hold:*

(i) *$C^{m+1}_*(L,\mathbb{C})$ is continuously imbedded in $C^{m,\alpha}_*(L,\mathbb{C})$. If $\alpha < \beta$, then $C^{m,\beta}_*(L,\mathbb{C})$ is continuously imbedded in $C^{m,\alpha}_*(L,\mathbb{C})$. In particular, $C^{m,\alpha}_*(\mathbb{T},\mathbb{C})\cap \mathcal{Z}$ is an open subset of $C^{m,\alpha}_*(\mathbb{T},\mathbb{C})$.*

(ii) *The pointwise product is continuous in the Banach space $C^{m,\alpha}_*(L,\mathbb{C})$.*

(iii) *The reciprocal map $\mathbf{R}$ in $C^{m,\alpha}_*(L,\mathbb{C})$, which maps a nonvanishing function f to its reciprocal f^{-1}, is complex analytic from the open subset $C^{m,\alpha}_*(L,\mathbb{C}\setminus\{0\})$ of $C^{m,\alpha}_*(L,\mathbb{C})$ to itself. Moreover $D\mathbf{R}[f_0][\mu] = -\frac{\mu}{f_0^2}$ for all $f_0 \in C^{m,\alpha}_*(L,\mathbb{C} \setminus \{0\})$ and $\mu \in C^{m,\alpha}_*(L,\mathbb{C})$.*

(iv) *Let $\phi_1 \in \mathcal{Z}$, $L_1 = \phi_1(\mathbb{T})$. If $f \in C^{m,\alpha}_*(L_1,\mathbb{C})$ and $g \in C^{m,\beta}_*(L,L_1)$, then $f \circ g \in C^{m,\gamma_m(\alpha,\beta)}_*(L,\mathbb{C})$ with $\gamma_0(\alpha,\beta) = \alpha\beta$ and $\gamma_m(\alpha,\beta) = \min\{\alpha,\beta\}$ if $m > 0$.*

(v) *Let $m \geq 1$. If $g \in C^{m,\alpha}_*(L,\mathbb{C})$ is injective and satisfies the condition $g'(\xi) \neq 0$, for all $\xi \in L$, then $g^{(-1)} \in C^{m,\alpha}_*(g(L),L)$.*

(vi) *If $\mathbb{I}[\phi]$ denotes the open bounded and connected component of $\mathbb{C}\setminus\phi(\mathbb{T})$, then $\partial\mathbb{I}[\phi] = \phi(\mathbb{T})$.*

(vii) *Let $\phi \in C^{m,\alpha}_*(\mathbb{T},\mathbb{C}) \cap \mathcal{Z}$. Then the trace operator $\mathbf{R} : C^{m,\alpha}(\mathrm{cl}\,\mathbb{I}[\phi],\mathbb{C}) \to C^{m,\alpha}_*(\phi(\mathbb{T}),\mathbb{C})$ defined by $\mathbf{R}[F] = F_{|\phi(\mathbb{T})}$ is complex linear and continuous.*

The following Theorem collects known facts about the singular integral (with Cauchy kernel) and the Schwarz operator for the unit circle (cf., *e.g.*, [8, §3-5, pp. 44-46, 208-210, 248]).

Theorem 2.4 *Let $m \in \mathbb{N}$, $\alpha \in]0,1[$. Then the following statements hold.*

(i) *Let $\phi \in C^{1,\alpha}_*(\mathbb{T},\mathbb{C})\cap\mathcal{Z}$, $L = \phi(\mathbb{T})$. Then for all $f \in C^{m,\alpha}_*(L,\mathbb{C})$, the singular integral*

$$\mathbf{S}_\phi[f](\tau) \equiv \frac{\mathrm{p.v.}}{\pi i}\int_\phi \frac{f(\sigma)}{\sigma-\tau}\,d\sigma = \frac{\mathrm{p.v.}}{\pi i}\int_{\mathbb{T}} \frac{f(\phi(\eta))\phi'(\eta)}{\phi(\eta)-\tau}\,d\eta, \tag{2.4}$$

exists in the sense of the principal value for all $\tau \in L$, and $\mathbf{S}_\phi[f](\cdot) \in C^{m,\alpha}_*(L,\mathbb{C})$. The operator $\mathbf{S}_\phi$ defined by (2.4) is linear and continuous from $C^{m,\alpha}_*(L,\mathbb{C})$ to *itself and the equality* $(\mathbf{S}_\phi[g])' = \mathbf{S}_\phi[g']$ *holds for all* $g \in C_*^{1,\alpha}(L,\mathbb{C})$. *If* ϕ *coincides with the identity map* $\mathrm{id}_\mathbb{T}$, *then we set* $\mathbf{S} \equiv \mathbf{S}_\phi$.

(ii) *Let* $\phi \in C_*^{1,\alpha}(\mathbb{T},\mathbb{C}) \cap \mathcal{Z}$, $L = \phi(\mathbb{T})$. *Let* $\mathrm{ind}\,[\phi]$ *be the index of the curve* $\theta \mapsto \phi(e^{i\theta})$, $\theta \in [0, 2\pi]$ *with respect to any of the points of* $\mathbb{I}[\phi]$. *Then for all* $f \in C_*^{m,\alpha}(L,\mathbb{C})$, *the Cauchy type integral* $\mathbf{C}_\phi[f]$ *from* $\mathbb{C} \setminus \{L\}$ *to* $\mathbb{C}$ *defined by*

$$\mathbf{C}_\phi[f](z) \equiv \frac{\mathrm{ind}\,[\phi]}{2\pi i} \int_\phi \frac{f(\sigma)}{\sigma - z}\, d\sigma, \qquad \forall z \in \mathbb{C} \setminus L,$$

is holomorphic. The function $\mathbf{C}_\phi^+[f] \equiv \mathbf{C}_\phi[f]_{|\mathbb{I}[\phi]}$ *admits a continuous extension to* $\mathrm{cl}\,\mathbb{I}[\phi]$, *which we still denote by* $\mathbf{C}_\phi^+[f]$ *and we have* $\mathbf{C}_\phi^+[f] \in \mathcal{H}(\mathbb{I}[\phi]) \cap C^{m,\alpha}(\mathrm{cl}\,\mathbb{I}[\phi],\mathbb{C})$. *Moreover the Sokhotsky-Plemelj formula holds*

$$\mathbf{C}_\phi^+[f](t) = \frac{1}{2} f(t) + \frac{\mathrm{ind}\,[\phi]}{2} \mathbf{S}_\phi[f](t), \quad \forall t \in L,$$

and $\mathbf{C}_\phi^+[\cdot]$ *defines a linear and continuous operator from* $C_*^{m,\alpha}(L,\mathbb{C})$ *to* $C^{m,\alpha}(\mathrm{cl}\,\mathbb{I}[\phi],\mathbb{C})$. *If* ϕ *coincides with the identity map* $\mathrm{id}_\mathbb{T}$, *then we set* $\mathbf{C} \equiv \mathbf{C}_\phi$.

(iii) *For all* $f \in C_*^{m,\alpha}(\mathbb{T},\mathbb{C})$, *the function defined by*

$$\mathbf{T}[f](z) \equiv \frac{1}{2\pi i} \int\limits_{\mathbb{T}} f(\sigma) \frac{\sigma + z}{\sigma - z} \frac{d\sigma}{\sigma}, \quad \forall z \in \mathbb{U}, \tag{2.5}$$

and extended by continuity to $\mathrm{cl}\,\mathbb{U}$ *belongs to* $\mathcal{H}(\mathbb{U}) \cap C^{m,\alpha}(\mathrm{cl}\,\mathbb{U},\mathbb{C})$. *This operator is called the Schwarz operator for the unit disc, while the integral in formula* (2.5) *is called the Schwarz integral for the unit disc. Moreover the following formulas connect the Schwarz operator to the Cauchy type operator and the singular (Cauchy) integral operator*

$$\mathbf{T}[f](z) = 2\mathbf{C}[f](z) - \frac{1}{2\pi i} \int\limits_{\mathbb{T}} \frac{f(\sigma) d\sigma}{\sigma}, \quad \forall z \in \mathbb{U}, \tag{2.6}$$

$$\mathbf{T}[f](t) = f(t) + \mathbf{S}[f](t) - \frac{1}{2\pi i} \int\limits_{\mathbb{T}} \frac{f(\sigma) d\sigma}{\sigma} = f(t) - i\mathbf{H}[f](t), \ t \in \mathbb{T}, \tag{2.7}$$

for all $f \in C_*^{m,\alpha}(\mathbb{T},\mathbb{C})$ *where* $\mathbf{H}[f](\cdot)$ *is the singular integral operator with the Hilbert kernel*

$$\mathbf{H}[f](e^{is}) \equiv \frac{1}{2\pi} \int\limits_0^{2\pi} f(e^{i\theta}) \cot \frac{\theta - s}{2} d\theta, \quad \forall s \in [0, 2\pi[. \tag{2.8}$$

In particular, the operator $\mathbf{T}[\cdot]$ *is linear and continuos from* $C_*^{m,\alpha}(\mathbb{T},\mathbb{R})$ *to* $C^{m,\alpha}(\mathrm{cl}\,\mathbb{U},\mathbb{C})$.

(iv) *For all* $f \in C_*^{m,\alpha}(\mathbb{T}, \mathbb{R})$, *the function* $\mathbf{T}[f]$ *determined by formula* (2.5) *is a solution of the following boundary value problem*

$$\operatorname{Re} F(t) = f(t),\ t \in \mathbb{T}, \tag{2.9}$$

and satisfies $\operatorname{Im} \mathbf{T}[f](0) = 0$. *The general solution of* (2.9) *in* $\mathcal{H}(\Omega) \cap C^0(\operatorname{cl} \mathbb{U}, \mathbb{C})$ *is* $F = \mathbf{T}[f] + ic$, *where* $c \in \mathbb{R}$.

3. Real analyticity of the modified Schwarz operator

In this section we carry out the perturbation analysis of the Schwarz problem in simply connected domains in the case of domains enclosed by Lyapunov curves and Hölder continuous boundary data. In this Schauder space setting we obtain the real analyticity of a variant of the Schwarz operator and we compute the first differential of this operator.

Let ϕ be a simple (closed) curve of class $C^{1,\alpha}$ with $\alpha \in]0,1[$, $f \in C_*^{0,\alpha}(\phi(\mathbb{T}), \mathbb{R})$ and let $w \in \mathbb{I}[\phi]$. *The classical Schwarz problem* consists in searching for a complex-valued function of $\operatorname{cl} \mathbb{I}[\phi]$, holomorphic in $\mathbb{I}[\phi]$ such that

$$\begin{cases} F \in \mathcal{H}(\mathbb{I}[\phi]) \cap C^0(\operatorname{cl} \mathbb{I}[\phi], \mathbb{C}), \\ \operatorname{Re} F(t) = f(t),\ \forall t \in \phi(\mathbb{I}),\quad \operatorname{Im} F(w) = 0. \end{cases} \tag{3.1}$$

Let $g_{\phi,w}$ be the Riemann map of $\mathbb{U}$ onto $\mathbb{I}[\phi]$ normalized by (2.2). A function F satisfies problem (3.1) if and only if the function $\tilde{F} \equiv F \circ g_{\phi,w}$ satisfies the following problem in $\operatorname{cl} \mathbb{U}$

$$\begin{cases} \tilde{F} \in \mathcal{H}(\mathbb{U}) \cap C^0(\operatorname{cl} \mathbb{U}, \mathbb{C}), \\ \operatorname{Re} \tilde{F}_{|\mathbb{T}} = f \circ (g_{\phi,w|\mathbb{T}}),\quad \operatorname{Im} \tilde{F}(0) = 0. \end{cases} \tag{3.2}$$

By (2.3), Lemma 2.3 (iv) and by Theorem 2.4 (iv), problem (3.2) has a unique solution $\tilde{F}$ and then $F \equiv \tilde{F} \circ g_{\phi,w}^{(-1)}$ is the unique solution of (3.1). By (2.6) this solution F can be represented in the form

$$F(z) = \frac{1}{\pi i} \int_{\mathbb{T}} \frac{(f \circ g_{\phi,w})(\sigma)}{\sigma - g_{\phi,w}^{(-1)}(z)} d\sigma - \frac{1}{2\pi i} \int_{\mathbb{T}} \frac{(f \circ g_{\phi,w})(\sigma)}{\sigma} d\sigma. \tag{3.3}$$

Let $\mathbf{T}$ be the operator from $\{(\phi, f, w) \in (C_*^{1,\alpha}(\mathbb{T}, \mathbb{C}) \cap \mathcal{Z}) \times C_*^{0,\alpha}(\phi(\mathbb{T}), \mathbb{R}) \times \mathbb{C} : w \in \mathbb{I}[\phi]\}$ to $C^0(\operatorname{cl} \mathbb{I}[\phi], \mathbb{C})$ defined by

$$\mathbf{T}[\phi, f, w] \equiv F, \tag{3.4}$$

where F is the unique solution of (3.1). We call this operator the (classical) Schwarz operator. We should note that the Schwarz problem is well-posed for classes of curves and boundary data more general than those we consider here, but we are not going to deal with those classes. By (2.3), Lemma 2.3 (iv) and (vii), Theorem 2.4 (ii) and Lemma 2.1 (iv) and (v) the following holds

$$\mathbf{T}[\phi, f, w] \in \mathcal{H}(\mathbb{I}[\phi]) \cap C^{m,\alpha}(\operatorname{cl} \mathbb{I}[\phi], \mathbb{C}), \tag{3.5}$$

for all $\phi \in C_*^{m,\alpha}(\mathbb{T},\mathbb{C}) \cap \mathcal{Z}$, $f \in C_*^{m,\alpha}(\phi(\mathbb{T}),\mathbb{R})$, $w \in \mathbb{I}[\phi]$ with $m \in \mathbb{N} \setminus \{0\}$ and $\alpha \in]0,1[$.

To get the regularity results about the dependence of $\mathbf{T}$ on its variables (ϕ, f, w), together with a description of the first variation with respect to these variables, we have to tackle two difficulties. First, we cannot study directly the operator $\mathbf{T}$ because the functions f and $\mathbf{T}[\phi, f, w]$ are defined in domains depending on ϕ. Then we introduce the modified Schwarz operator $\mathbf{T}_*$ by the following formula

$$\mathbf{T}_*[\phi, p, w] \equiv \mathbf{T}[\phi, p \circ \phi^{(-1)}, w] \circ \phi, \tag{3.6}$$

which is the composition with ϕ of the classical Schwarz operator corresponding to $(\phi, p \circ \phi^{(-1)}, w)$. Second, we need to choose a suitable representation of the modified Schwarz operator involving operators which have known regularity and whose first differentials can be explicitly calculated.

In order to prove an analyticity result for the modified Schwarz operator and to compute its first differential, we need the following auxiliary results.

Lemma 3.1 *Let*

$$\mathcal{A}_*^{m,\alpha} \equiv \{(\phi, p, w) \in (C_*^{m,\alpha}(\mathbb{T},\mathbb{C}) \cap \mathcal{Z}) \times C_*^{m,\alpha}(\mathbb{T},\mathbb{R}) \times \mathbb{C} : w \in \mathbb{I}[\phi]\}. \tag{3.7}$$

Then the operator $\mathbf{T}_*$ *maps* $\mathcal{A}_*^{m,\alpha}$ *into* $C_*^{m,\alpha}(\mathbb{T},\mathbb{C})$ *and is represented in the following form*

$$\begin{aligned} \mathbf{T}_*[\phi, p, w](t) &\equiv \mathbf{T}[\phi, p \circ \phi^{(-1)}, w](\phi(t)) \\ &= p(t) + \frac{\text{p.v.}}{\pi i} \int_{\mathbb{T}} \frac{p \circ h[\phi, w]^{(-1)}(\sigma)}{\sigma - h[\phi, w](t)} d\sigma - \frac{1}{2\pi i} \int_{\mathbb{T}} \frac{p \circ h[\phi, w]^{(-1)}(\sigma)}{\sigma} d\sigma \end{aligned} \tag{3.8}$$

for all $t \in \mathbb{T}$ *and for all* $(\phi, p, w) \in \mathcal{A}_*^{m,\alpha}$, *where* $h[\phi, w]$ *is the value in* (ϕ, w) *of the operator* $\mathbf{h}$ *defined by* (1.5), *i.e.*,

$$\mathbf{h} : (\phi, w) \mapsto h[\phi, w] = g_{\phi,w}^{(-1)} \circ \phi \tag{3.9}$$

for each pair (ϕ, w) *of the set*

$$\mathcal{E}^{m,\alpha} \equiv \{(\phi, w) \in (C_*^{m,\alpha}(\mathbb{T},\mathbb{C}) \cap \mathcal{Z}) \times \mathbb{C} : w \in \mathbb{I}[\phi]\}. \tag{3.10}$$

Moreover $h[\phi, w] \in C_*^{m,\alpha}(\mathbb{T},\mathbb{C})$ *whenever* $(\phi, w) \in \mathcal{E}^{m,\alpha}$.

Proof. By Lemma 2.3 (iv), (v) and (vii) and (3.5), $\mathbf{T}_*[\phi, p, w] \in C_*^{m,\alpha}(\mathbb{T},\mathbb{C})$ whenever $(\phi, p, w) \in \mathcal{A}_*^{m,\alpha}$. By formula (3.3) and Theorem 2.4 (ii) one can get representation (3.8). By (2.3), Lemma 2.1 (v), Lemma 2.3 (vii) and (iv), the operator $\mathbf{h}$ takes its values in $C_*^{m,\alpha}(\mathbb{T},\mathbb{C})$. □

Lemma 3.2 ([16, Prop. 3.5, 3.7]) *Let $m \in \mathbb{N} \setminus \{0\}$, $\alpha \in]0,1[$. Let the set $\mathcal{A}$ be defined by the relation*

$$\mathcal{A} \equiv \Big\{ (\phi, w, h) \in C_*^{m,\alpha}(\mathbb{T}, \mathbb{C}) \times \mathbb{C} \times C_*^{m,\alpha}(\mathbb{T}, \mathbb{C}) : \tag{3.11}$$
$$\phi \in \mathcal{Z},\ w \in \mathbb{I}[\phi], \mathrm{Re}\left\{ \frac{\mathrm{ind}[h]}{2\pi i} \int_{\mathbb{T}} \frac{\phi(s)h'(s)}{h^2(s)}\, ds \right\} > 0 \Big\}.$$

Then $\mathcal{A}$ is an open subset of $C_*^{m,\alpha}(\mathbb{T}, \mathbb{C}) \times \mathbb{C} \times C_*^{m,\alpha}(\mathbb{T}, \mathbb{C})$, and the triple $(\phi, w, h) \in \mathcal{A}$ satisfies the operator equation

$$\mathbf{P}[\phi, w, h] = 0, \tag{3.12}$$

with

$$\mathbf{P}[\phi, w, h] \equiv (\mathbf{P}_l[\phi, w, h])_{l=1,2,3,4} \tag{3.13}$$
$$\equiv \Bigg(\mathrm{Re}\left\{ \phi(\cdot) - \frac{\mathrm{ind}[h]}{\pi i}\, \mathrm{p.v.}\!\int_{\mathbb{T}} \frac{\phi(s)h'(s)}{h(s) - h(\cdot)}\, ds \right\}, \frac{\mathrm{ind}[h]}{2\pi i} \int_{\mathbb{T}} \frac{\phi(s)h'(s)}{h(s)}\, ds - w,$$
$$\mathrm{Im}\left\{ \frac{\mathrm{ind}[h]}{2\pi i} \int_{\mathbb{T}} \frac{\phi(s)h'(s)}{h^2(s)}\, ds \right\}, h(\cdot)\overline{h(\cdot)} - 1 \Bigg),$$

if and only if $h = h[\phi, w] = g_{\phi,w}^{(-1)} \circ \phi$.

The operator $\mathbf{P}$ is real analytic from $\mathcal{A}$ into $C_*^{m,\alpha}(\mathbb{T}, \mathbb{R}) \times \mathbb{C} \times \mathbb{R} \times C_*^{m,\alpha}(\mathbb{T}, \mathbb{C})$.

Lemma 3.3 ([16, Thm. 5.4]) *Let $m \in \mathbb{N} \setminus \{0\}$, $\alpha \in]0,1[$. Let the set $\mathcal{E}^{m,\alpha}$ and the operator $\mathbf{h}$ be defined by (3.10) and (3.9). Then $\mathcal{E}^{m,\alpha}$ is an open subset of $C_*^{m,\alpha}(\mathbb{T}, \mathbb{C}) \times \mathbb{C}$ and the operator $\mathbf{h}$ of $\mathcal{E}^{m,\alpha}$ to $C_*^{m,\alpha}(\mathbb{T}, \mathbb{T})$ is real analytic on $\mathcal{E}^{m,\alpha}$.*

Lemma 3.4 *Let $m \in \mathbb{N} \setminus \{0\}$, $\alpha \in]0,1[$. Let the set $\mathcal{E}^{m,\alpha}$ and the operator $\mathbf{h}$ be defined by (3.10) and (3.9). Let $(\phi_0, w_0) \in \mathcal{E}^{m,\alpha}$. The first order derivatives of the operator $\mathbf{h}$ are given by the following formulas*

$$\frac{\partial \mathbf{h}}{\partial \phi}[\phi_0, w_0][\mu](t) = 2ih[\phi_0, w_0](t)\, \mathrm{Im}\Bigg\{ -\frac{\omega}{h[\phi_0, w_0](t) g_0'(0)} \tag{3.14}$$
$$- \frac{\mathrm{ind}[h[\phi_0, w_0]]}{4\pi i} \int_{\mathbb{T}} \frac{\omega (h[\phi_0, w_0]')^2(\tau)}{\phi_0'(\tau)} \frac{d\tau}{h^2[\phi_0, w_0](\tau)}$$
$$+ \frac{\mathrm{ind}[h[\phi_0, w_0]]}{2\pi i} \int_{\mathbb{T}} \frac{f(\tau)(h[\phi_0, w_0]')^2(\tau)}{\phi_0'(\tau) h[\phi_0, w_0](\tau)} \frac{d\tau}{h[\phi_0, w_0](\tau) - h[\phi_0, w_0](t)}$$
$$- \frac{\mathrm{ind}[h[\phi_0, w_0]]}{4\pi i} \int_{\mathbb{T}} \frac{f(\tau)(h[\phi_0, w_0]')^2(\tau)}{\phi_0'(\tau)} \frac{d\tau}{h^2[\phi_0, w_0](\tau)}$$
$$- \frac{f(t) h[\phi_0, w_0]'(t)}{2\phi_0'(t) h[\phi_0, w_0](t)} + \frac{1}{g_0'(0)} \frac{\mathrm{ind}[h[\phi_0, w_0]]}{4\pi i} \int_{\mathbb{T}} \frac{f(\tau) h[\phi_0, w_0]'(\tau)}{h^2[\phi_0, w_0](\tau)} d\tau$$
$$+ \frac{1}{g_0'(0)} \frac{\mathrm{ind}[h[\phi_0, w_0]]}{4\pi i} \int_{\mathbb{T}} \frac{\mu(\tau) h[\phi_0, w_0]'(\tau)}{h^2[\phi_0, w_0](\tau)} d\tau \Bigg\},$$

$$\frac{\partial \mathbf{h}}{\partial w}[\phi_0, w_0](\xi)(t) = 2ih[\phi_0, w_0](t)\,\mathrm{Im}\left\{-\frac{\xi}{h[\phi_0, w_0](t)g_0'(0)} - \frac{\mathrm{ind}[h[\phi_0, w_0]]}{4\pi i}\int_{\mathbb{T}} \frac{\xi(h[\phi_0, w_0]')^2(\tau)}{\phi_0'(\tau)}\frac{d\tau}{h^2[\phi_0, w_0](\tau)}\right\}, \tag{3.15}$$

for all $\mu \in C_*^{m,\alpha}(\mathbb{T}, \mathbb{C})$, $t \in \mathbb{T}$, $\xi \in \mathbb{C}$, *where* $g_0 \equiv g_{\phi_0, w_0}$,

$$g_0'(0) = \frac{\mathrm{ind}[h[\phi_0, w_0]]}{2\pi i}\int_{\mathbb{T}} \frac{\phi_0'(\tau)}{h[\phi_0, w_0](\tau)}d\tau,$$

$$\omega \equiv -\frac{\mathrm{ind}[h[\phi_0, w_0]]}{2\pi i}\int_{\mathbb{T}} \frac{\mu(\tau)h[\phi_0, w_0]'(\tau)}{h[\phi_0, w_0](\tau)}d\tau,$$

$$f(t) \equiv -\mathrm{Re}\left\{\mu(t) - \frac{\mathrm{ind}[h[\phi_0, w_0]]}{\pi i}\int_{\mathbb{T}} \frac{\mu(\tau)h[\phi_0, w_0]'(\tau)}{h[\phi_0, w_0](\tau) - h[\phi_0, w_0](t)}d\tau\right\}.$$

Proof. The proof is based on the identity

$$\mathbf{P}[\phi, w, h[\phi, w]] \equiv 0.$$

Therefore

$$\frac{\partial \mathbf{P}}{\partial h}[\phi_0, w_0, h_0][q] = -\frac{\partial \mathbf{P}}{\partial \phi}[\phi_0, w_0, h_0][\mu] \tag{3.16}$$

$$= \left(-\mathrm{Re}\,(\mu - \mathbf{S}[\mu \circ k_0] \circ h_0), -\frac{1}{2\pi i}\int_{\mathbb{T}} \frac{\mu \circ k_0(\eta)}{\eta}d\eta, -\mathrm{Im}\left(\frac{1}{2\pi i}\int_{\mathbb{T}} \frac{\mu \circ k_0(\eta)}{\eta^2}d\eta\right), 0\right),$$

where $q \equiv \frac{\partial \mathbf{h}}{\partial \phi}[\phi_0, w_0][\mu]$, $h_0 \equiv h[\phi_0, w_0]$ and $k_0 \equiv h_0^{(-1)}$. Then formula (3.4) can be derived by using the explicit solution of $\frac{\partial \mathbf{P}}{\partial h}[\phi_0, w_0, h_0][q] = d$ given in [16, Props. 4.3 and 5.3] and by (3.16).

To simplify the formula for $\frac{\partial \mathbf{h}}{\partial \phi}$ we apply also the following consideration. Let $g_0 \equiv g_{\phi_0, w_0}$. By (2.3), $\frac{1}{g_0'}$ is the boundary function of a function holomorphic in the unit disc $\mathbb{U}$. Then from the mean value theorem it follows

$$\frac{1}{g_0'(0)} = \frac{\mathrm{ind}[h[\phi_0, w_0]]}{2\pi i}\int_{\mathbb{T}} \frac{h[\phi_0, w_0]'(\tau)}{g_0'(h[\phi_0, w_0](\tau))}\frac{d\tau}{h[\phi_0, w_0](\tau)},$$

and from the Sokhotsky-Plemelj formula

$$\frac{1}{2g_0'(h[\phi_0, w_0](t))} = \frac{\mathrm{ind}[h[\phi_0, w_0]]}{2\pi i}\int_{\mathbb{T}} \frac{h[\phi_0, w_0]'(\tau)}{g_0'(h[\phi_0, w_0](\tau))}\frac{d\tau}{h[\phi_0, w_0](\tau) - h[\phi_0, w_0](t)}.$$

The formula for $\frac{\partial \mathbf{h}}{\partial w}$ can be obtained by a similar argument. □

Lemma 3.5 ([13, Prop. 4.1]) *Let* $m \in \mathbb{N} \setminus \{0\}$, $\alpha \in]0, 1[$. *Let* $\mathbf{C}$ *be the Cauchy integral operator from the open subset* $(C_*^{m,\alpha}(\mathbb{T}, \mathbb{C}) \cap \mathcal{Z}) \times C_*^{m,\alpha}(\mathbb{T}, \mathbb{C})$

of $(C_*^{m,\alpha}(\mathbb{T},\mathbb{C}))^2$ *to* $C_*^{m,\alpha}(\mathbb{T},\mathbb{C})$ *defined by* (1.4). *Then* $\mathbf{C}$ *is a complex analytic operator and the following formulas for the first order derivatives of* $\mathbf{C}$ *at* $(\psi_0, f_0) \in (C_*^{m,\alpha}(\mathbb{T},\mathbb{C}) \cap \mathcal{Z}) \times C_*^{m,\alpha}(\mathbb{T},\mathbb{C})$ *hold:*

$$\frac{\partial \mathbf{C}}{\partial \psi}[\psi_0, f_0][\nu](t) = -\frac{1}{2\pi i}\int_{\mathbb{T}} f_0'(\tau)\frac{\nu(\tau)-\nu(t)}{\psi_0(\tau)-\psi_0(t)}d\tau, \tag{3.17}$$

for each $\nu \in C_*^{m,\alpha}(\mathbb{T},\mathbb{C})$ *and* $t \in \mathbb{T}$,

$$\frac{\partial \mathbf{C}}{\partial f}[\psi_0, f_0][\theta](t) = \mathbf{C}[\psi_0, \theta](t) = \frac{1}{2\pi i}\int_{\mathbb{T}} \frac{\theta(\tau)\psi_0'(\tau)}{\psi_0(\tau)-\psi_0(t)}d\tau, \tag{3.18}$$

for each $\theta \in C_*^{m,\alpha}(\mathbb{T},\mathbb{C})$ *and* $t \in \mathbb{T}$.

Lemma 3.6 *Let* $m \in \mathbb{N}\setminus\{0\}$, $\alpha \in]0,1]$. *Then* $(C_*^{m,\alpha}(\mathbb{T},\mathbb{C}\setminus\{0\}) \cap \mathcal{Z}) \times C_*^{m,\alpha}(\mathbb{T},\mathbb{C})$ *is an open subset of* $(C_*^{m,\alpha}(\mathbb{T},\mathbb{C}))^2$ *and the operator* $\mathbf{G}$ *from* $(C_*^{m,\alpha}(\mathbb{T},\mathbb{C}\setminus\{0\}) \cap \mathcal{Z}) \times C_*^{m,\alpha}(\mathbb{T},\mathbb{C})$ *to* $\mathbb{C}$ *defined by*

$$\mathbf{G}[\psi, f] \equiv \frac{\mathrm{ind}[\psi]}{2\pi i}\int_{\mathbb{T}} \frac{f(\tau)\psi'(\tau)}{\psi(\tau)}d\tau \tag{3.19}$$

is complex analytic. Moreover, its first order derivatives at the pair $(\psi_0, f_0) \in (C_*^{m,\alpha}(\mathbb{T},\mathbb{C}\setminus\{0\}) \cap \mathcal{Z}) \times C_*^{m,\alpha}(\mathbb{T},\mathbb{C})$ *satisfies*

$$\frac{\partial \mathbf{G}}{\partial \psi}[\psi_0, f_0][\nu] = -\frac{\mathrm{ind}[\psi_0]}{2\pi i}\int_{\mathbb{T}} \frac{f_0'(\tau)\nu(\tau)}{\psi_0(\tau)}d\tau, \tag{3.20}$$

$$\frac{\partial \mathbf{G}}{\partial f}[\psi_0, f_0][\theta] = \mathbf{G}[\psi_0, \theta] = \frac{\mathrm{ind}[\psi_0]}{2\pi i}\int_{\mathbb{T}} \frac{\theta(\tau)\psi_0'(\tau)}{\psi_0(\tau)}d\tau, \tag{3.21}$$

for all $\mu \in C_*^{m,\alpha}(\mathbb{T},\mathbb{C})$.

Proof. Since $\mathbb{C}\setminus\{0\}$ is an open subset of $\mathbb{C}$, then $C_*^{m,\alpha}(\mathbb{T},\mathbb{C}\setminus\{0\})$ is an open subset of $C_*^{m,\alpha}(\mathbb{T},\mathbb{C})$. Hence by Lemma 2.2 (ii) and Lemma 2.3 (i), $(C_*^{m,\alpha}(\mathbb{T},\mathbb{C}) \cap \mathcal{Z}) \times C_*^{m,\alpha}(\mathbb{T},\mathbb{C})$ and $(C_*^{m,\alpha}(\mathbb{T},\mathbb{C}\setminus\{0\}) \cap \mathcal{Z}) \times C_*^{m,\alpha}(\mathbb{T},\mathbb{C})$ are open subsets of $(C_*^{m,\alpha}(\mathbb{T},\mathbb{C}))^2$. Then by Lemma 2.3 (ii) and (iii) the operator $(\psi, f) \mapsto \frac{f(\cdot)\psi'(\cdot)}{\psi(\cdot)}$ is complex analytic from $(C_*^{m,\alpha}(\mathbb{T},\mathbb{C}\setminus\{0\}) \cap \mathcal{Z}) \times C_*^{m,\alpha}(\mathbb{T},\mathbb{C})$ to $C_*^{m-1,\alpha}(\mathbb{T},\mathbb{C})$ and then $\mathbf{G}$ is complex analytic. By the chain rule and Lemma 2.3 (iii) we obtain formulas (3.20), (3.21). □

Now we are at the position to formulate the result on the regularity of the modified Schwarz operator, as well as to calculate the first order derivatives of this operator.

Theorem 3.7 *Let* $m \in \mathbb{N}\setminus\{0\}, \alpha \in]0,1[$. *Let* $\mathbf{T}_*$ *be the operator from* $\mathcal{A}_*^{m,\alpha}$ *to* $C_*^{m,\alpha}(\mathbb{T},\mathbb{C})$ *defined by* $\mathbf{T}_*[\phi, p, w] \equiv \mathbf{T}[\phi, p\circ\phi^{(-1)}, w]\circ\phi$ *for all* $(\phi, p, w) \in \mathcal{A}_*^{m,\alpha}$. *Then the following statements hold.*

(i) *The set* $\mathcal{A}_*^{m,\alpha}$ *is an open subset of* $C_*^{m,\alpha}(\mathbb{T},\mathbb{C}) \times C_*^{m,\alpha}(\mathbb{T},\mathbb{R}) \times \mathbb{C}$ *and the operator* $\mathbf{T}_*$ *is real analytic on its domain* $\mathcal{A}_*^{m,\alpha}$.

(ii) *Let* $(\phi_0, p_0, w_0) \in \mathcal{A}_*^{m,\alpha}$, $h_0 \equiv g_{\phi_0,w_0}^{(-1)} \circ \phi_0$, *Then the following formulas for the partial derivatives with respect to* ϕ, p, *and* w *hold.*

$$\frac{\partial \mathbf{T}_*}{\partial \phi}[\phi_0, p_0, w_0][\mu](t) = 2\,\mathrm{ind}[h[\phi_0, w_0]]\frac{\partial \mathbf{C}}{\partial \psi}[h[\phi_0, w_0], p_0][\frac{\partial \mathbf{h}}{\partial \phi}[\phi_0, w_0][\mu]](t) - \frac{\partial \mathbf{G}}{\partial \psi}[h[\phi_0, w_0], p_0][\frac{\partial \mathbf{h}}{\partial \phi}[\phi_0, w_0][\mu]] \tag{3.22}$$

for all $\mu \in C_*^{m,\alpha}(\mathbb{T},\mathbb{C})$ *and for all* $t \in \mathbb{T}$,

$$\frac{\partial \mathbf{T}_*}{\partial p}[\phi_0, p_0, w_0][\eta](t) = \eta(t) + 2\,\mathrm{ind}[h[\phi_0, w_0]]\mathbf{C}[h[\phi_0, w_0], \eta](t) - \mathbf{G}[h[\phi_0, w_0], \eta] \tag{3.23}$$

for all $\eta \in C_*^{m,\alpha}(\mathbb{T},\mathbb{C})$ *and for all* $t \in \mathbb{T}$,

$$\frac{\partial \mathbf{T}_*}{\partial w}[\phi_0, p_0, w_0](\xi)(t) = 2\,\mathrm{ind}[h[\phi_0, w_0]]\frac{\partial \mathbf{C}}{\partial \psi}[h[\phi_0, w_0], p_0][\frac{\partial \mathbf{h}}{\partial w}[\phi_0, w_0](\xi)](t) - \frac{\partial \mathbf{G}}{\partial \psi}[h[\phi_0, w_0], p_0][\frac{\partial \mathbf{h}}{\partial w}[\phi_0, w_0][\xi]] \tag{3.24}$$

for all $\xi \in \mathbb{C}$ *and for all* $t \in \mathbb{T}$, *where the derivatives of the operator* $\mathbf{h}$ *are given in Lemma* 3.4, *the derivatives of the operator* $\mathbf{C}$ *are given in Lemma* 3.5, *and the derivatives of the operator* $\mathbf{G}$ *are given in Lemma* 3.6.

Proof. Let $\mathcal{E}^{m,\alpha}$ be as in (3.10). By Lemma 3.3, $\mathcal{E}^{m,\alpha}$ is an open subset of $C_*^{m,\alpha}(\mathbb{T},\mathbb{C}) \times \mathbb{C}$. Then the first part of statement (i) follows. The real analyticity of the operator $\mathbf{T}_*$ follows from the real analyticity of the operator $\mathbf{h}$, the (complex) analyticity of the operators $\mathbf{C}$ and $\mathbf{G}$, and from representation (3.8).

The formulas for the first order derivatives (3.22), (3.23), (3.24) can be obtained by applying Lemmas 3.4, 3.5, 3.6, and representation (3.8). □

Remark 3.8 If the function $p_0 \in C_*^{m,\alpha}(\mathbb{T},\mathbb{R})$ is fixed, one may wonder whether the function $\mathbf{T}_*[\phi, p_0, w]$ depends complex analytically (besides real analytical dependence) on (ϕ, w). The answer is negative even if we choose a different normalization of the solution of the Schwarz problem. Let $\phi_0 \in C_*^{m,\alpha}(\mathbb{T},\mathbb{C}) \cap \mathcal{Z}$, $p_0 \in C_*^{m,\alpha}(\mathbb{T},\mathbb{R})$ and let $w_0 \in \mathbb{I}[\phi_0]$. We assume that the operator $\phi \mapsto \mathbf{T}_*[\phi, p_0, w_0]$ is complex analytic in an open connected neighborhood $\mathcal{U}_{\phi_0}$ of ϕ_0. Let $t \in \mathbb{T}$. Then $\phi \mapsto \mathbf{T}_*[\phi, p_0, w_0](t)$ is a complex analytic map of $\mathcal{U}_{\phi_0}$ to $\mathbb{C}$. By definition, $\mathrm{Re}(\mathbf{T}_*[\phi, p_0, w_0](t)) = p_0(t)$ for all $\phi \in C_*^{m,\alpha}(\mathbb{T},\mathbb{C}) \cap \mathcal{Z}$. Then the image of the map $\phi \mapsto \mathbf{T}_*[\phi, p_0, w_0](t)$ is contained in a line. A standard fact about holomorphic maps implies that $\mathbf{T}_*[\phi, p_0, w_0](t)$ is a complex constant independent of $\phi \in \mathcal{U}_{\phi_0}$ and then $\mathbf{T}_*[\cdot, p_0, w_0]$ is constant in $\mathcal{U}_{\phi_0}$. This yields a contradiction. Indeed let $\phi_0 \equiv \mathrm{id}_{\mathbb{T}}$, $p_0 \equiv \mathrm{Re} \circ \mathrm{id}_{\mathbb{T}}$, $w_0 \equiv 0$ and $g \in C_*^{m,\alpha}(\mathbb{T}, \mathbb{C})$. It is not difficult to verify that $\mathbf{T}_*[\phi_0 + g, p_0, w_0]$ is different from $\mathbf{T}_*[\phi_0, p_0, w_0] = \mathrm{id}_{\mathbb{T}}$ whenever g is not the trace of a holomorphic map of $\mathbb{U}$.

4. Regularity of another variants of the Schwarz operator

In this section we study the regularity of another variants of the Schwarz operator. First, we consider the classical Schwarz operator in $\mathrm{cl}\, \mathbb{I}[\phi]$ associated to the boundary data $p \circ \phi^{(-1)}$ and restricted to a proper and fixed compact subset K of $\mathbb{I}[\phi]$. Second, we study the regularity of the operator which maps the curve ϕ and the boundary data $p \circ \phi^{(-1)}$ to the composition of the solution of the Schwarz problem in $\mathrm{cl}\, \mathbb{I}[\phi]$ with the Riemann map of $\mathbb{I}[\phi]$.

Let us begin with the proof of the real analyticity for the classical Schwarz operator restricted to a fixed compact subset.

Proposition 4.1 *Let $\phi_0 \in C_*^{m,\alpha}(\mathbb{T}, \mathbb{C}) \cap \mathcal{Z}$. Let K be a compact subset of $\mathbb{I}[\phi_0]$ and $K_\epsilon \equiv K + \epsilon \mathbb{U}$ for all $\epsilon \in \mathbb{R}^+$. Let $\mathbf{T}$ be the Schwarz operator introduced in (3.4) and let $\mathcal{A}_*^{m,\alpha}$ be as in (3.7). Then there exists $\epsilon \in \mathbb{R}^+$ and an open neighborhood $\mathcal{U}_{\phi_0}$ of ϕ_0 in $C_*^{m,\alpha}(\mathbb{T}, \mathbb{C}) \cap \mathcal{Z}$ such that $\mathrm{cl}\, K_\epsilon \subset \mathbb{I}[\phi]$ for all $\phi \in \mathcal{U}_{\phi_0}$ and the operator $\mathbf{T}_{K,\epsilon}$ from $(\mathcal{U}_{\phi_0} \times C_*^{m,\alpha}(\mathbb{T}, \mathbb{R}) \times \mathbb{C}) \cap \mathcal{A}_*^{m,\alpha}$ to $C^0(\mathrm{cl}\, K_\epsilon, \mathbb{C})$ defined by*

$$\mathbf{T}_{K,\epsilon}[\phi, p, w] \equiv \mathbf{T}[\phi, p \circ \phi^{(-1)}, w]_{|\mathrm{cl}\, K_\epsilon}, \tag{4.1}$$

for all $(\phi, p, w) \in (\mathcal{U}_{\phi_0} \times C_*^{m,\alpha}(\mathbb{T}, \mathbb{R}) \times \mathbb{C}) \cap \mathcal{A}_*^{m,\alpha}$, is real analytic. Moreover, by possibly shrinking ϵ, $\mathbf{T}_{K,\epsilon}$ is a real analytic operator from $(\mathcal{U}_{\phi_0} \times C_*^{m,\alpha}(\mathbb{T}, \mathbb{R}) \times \mathbb{C}) \cap \mathcal{A}_*^{m,\alpha}$ to $C^h(\mathrm{cl}\, K_\epsilon, \mathbb{C})$ for all $h \in \mathbb{N}$.

Proof. A standard compactness argument shows the existence of $\epsilon \in \mathbb{R}^+$ such that $\mathrm{dist}(\phi_0(\mathbb{T}), \mathrm{cl}\, K_\epsilon) > 0$. We choose such an ϵ. Then the ball $\mathcal{U}_{\phi_0}$ in $C_*^{m,\alpha}(\mathbb{T}, \mathbb{C}) \cap \mathcal{Z}$ centered at ϕ_0 with radius $\rho < \mathrm{dist}(\phi_0(\mathbb{T}), \mathrm{cl}\, K_\epsilon)$ satisfies $\mathrm{cl}\, K_\epsilon \subset \mathbb{I}[\phi]$ for all $\phi \in \mathcal{U}_{\phi_0}$. Let $\mathbf{T}_*$ be the operator introduced in Theorem 3.7. By the definition of $\mathbf{T}_*$ and by the Cauchy formula the following equality holds

$$\begin{aligned} \mathbf{T}_{K,\epsilon}[\phi, p, w](z) &= \frac{\mathrm{ind}[\phi]}{2\pi i} \int_\phi \frac{\mathbf{T}_*[\phi, p, w](\phi^{(-1)}(\eta))}{\eta - z} d\eta \\ &= \frac{\mathrm{ind}[\phi]}{2\pi i} \int_{\mathbb{T}} \frac{\mathbf{T}_*[\phi, p, w](\sigma)\phi'(\sigma)}{\phi(\sigma) - z} d\sigma, \end{aligned} \tag{4.2}$$

for all $z \in \mathrm{cl}\, K_\epsilon$ and $(\phi, p, w) \in (\mathcal{U}_{\phi_0} \times C_*^{m,\alpha}(\mathbb{T}, \mathbb{R}) \times \mathbb{C}) \cap \mathcal{A}_*^{m,\alpha}$. By Theorem 3.7 and Lemma 2.3 (ii), the operator $(\phi, p, w) \mapsto \mathbf{T}_*[\phi, p, w]\phi'$ is real analytic from $\mathcal{A}_*^{m,\alpha}$ to $C_*^{0,\alpha}(\mathbb{T}, \mathbb{C})$ and, in particular, to $C_*^0(\mathbb{T}, \mathbb{C})$. Since the composition of two analytic operators is analytic, to conclude it is sufficient to prove that the operator $\mathbf{K}$ from $\mathcal{U}_{\phi_0} \times C^0_*(\mathbb{T}, \mathbb{C})$ to $C^0(\mathrm{cl}\, K_\epsilon, \mathbb{C})$ defined by

$$\mathbf{K}[\phi, f](z) \equiv \int_{\mathbb{T}} \frac{f(\sigma)}{\phi(\sigma) - z} d\sigma, \tag{4.3}$$

for all $z \in \mathrm{cl}\, K_\epsilon$ and $(\phi, f) \in \mathcal{U}_{\phi_0} \times C^0_*(\mathbb{T}, \mathbb{C})$ is complex analytic.

We observe that the following equality holds in $C^0_*(\mathbb{T}, \mathbb{C})$

$$\frac{f}{\phi_0 + \mu - z} = \frac{f}{\phi_0 - z} \left[\sum_{n=0}^{+\infty} \left(-\frac{\mu}{\phi_0 - z} \right)^n \right], \tag{4.4}$$

for all $z \in \mathrm{cl}\, K_\epsilon$ and μ with $2\|\mu\|_0 \leq \mathrm{dist}(\phi_0(\mathbb{T}), \mathrm{cl}\, K_\epsilon)$. Moreover the series in (4.4) converges totally for $z \in \mathrm{cl}\, K_\epsilon$ and for μ such that $2\|\mu\|_0 \leq \mathrm{dist}(\phi_0(\mathbb{T}), \mathrm{cl}\, K_\epsilon)$. By integrating (4.4) on $\mathbb{T}$, the analyticity of $\mathbf{K}[\cdot, f]$ for a fixed f follows. Since $\mathbf{K}[\phi, \cdot]$ is linear and continuous, the operators $\mathbf{K}[\cdot, \cdot]$ and $\mathbf{T}_{K,\epsilon}$ are analytic. The second part of the statement follows in view of the Cauchy estimates. Indeed, the Cauchy estimates for holomorphic functions imply that the subspace $C^0(\mathrm{cl}\, K_\epsilon, \mathbb{C}) \cap \mathcal{H}(K_\epsilon)$ of $C^0(\mathrm{cl}\, K_\epsilon, \mathbb{C})$ is continuously embedded in $C^h(\mathrm{cl}\, K_{\epsilon/2}, \mathbb{C})$ for all $h \in \mathbb{N}$. □

In the following Proposition we prove a sharp regularity statement for the composition of the Schwarz operator with the Riemann map of $\mathbb{I}[\phi]$.

Proposition 4.2 *Let* $m, r \in \mathbb{N} \setminus \{0\}$, $\alpha \in]0, 1[$. *Let* $\mathbf{T}$ *be the Schwarz operator introduced in* (3.4) *and let* $\mathcal{A}^{m,\alpha}_*$ *be as in* (3.7).

Then the operator $\mathbf{T}_c$ *from* $(C^{m+r,\alpha,0}_*(\mathbb{T}, \mathbb{C}) \times C^{m+r,\alpha,0}_*(\mathbb{T}, \mathbb{R}) \times \mathbb{C}) \cap \mathcal{A}^{m+r,\alpha}_*$ *to* $C^{m,\alpha,0}_*(\mathbb{T}, \mathbb{C})$ *defined by*

$$\mathbf{T}_c[\phi, p, w] \equiv \mathbf{T}[\phi, p \circ \phi^{(-1)}, w] \circ g_{\phi,w|\mathbb{T}}, \tag{4.5}$$

where $g_{\phi,w}$ *is the Riemann map of* $\mathbb{I}[\phi]$ *as in* (2.2), *is of class* C^r *and the* C^r *regularity is sharp.*

Proof. Let $(\phi_0, p_0, w_0) \in (C^{m+r,\alpha,0}_*(\mathbb{T}, \mathbb{C}) \times C^{m+r,\alpha,0}_*(\mathbb{T}, \mathbb{R}) \times \mathbb{C}) \cap \mathcal{A}^{m+r,\alpha}_*$. By the arbitrariness of (ϕ_0, p_0, w_0), it is sufficient to prove that $\mathbf{T}_c$ is of class C^r in a neighborhood of (ϕ_0, p_0, w_0) in $(C^{m+r,\alpha,0}_*(\mathbb{T}, \mathbb{C}) \times C^{m+r,\alpha,0}_*(\mathbb{T}, \mathbb{R}) \times \mathbb{C}) \cap \mathcal{A}^{m+r,\alpha}_*$ to $C^{m,\alpha,0}_*(\mathbb{T}, \mathbb{C})$.

Let the operator $\mathbf{h}$ and $\mathcal{E}^{m,\alpha}$ be as in (3.9) and (3.10), respectively. Let $\mathbf{T}_*$ be the modified Schwarz operator (3.6). Clearly, the following equality holds,

$$\mathbf{T}_c[\phi, p, w] = \mathbf{T}_*[\phi, p, w] \circ \mathbf{h}[\phi, w]^{(-1)}, \tag{4.6}$$

for all $(\phi, p, w) \in \mathcal{A}^{m,\alpha}_*$. By Lemma 3.3, $\mathbf{h}$ is analytic and, in particular, continuous from $\mathcal{E}^{m,\alpha}$ to $C^{m,\alpha}_*(\mathbb{T}, \mathbb{C})$ and $\mathbf{h}[\phi, w] \in C^\infty_*(\mathbb{T}, \mathbb{C})$ whenever $\phi \in C^\infty_*(\mathbb{T}, \mathbb{C})$. Then $\mathbf{h}$ restricts to a real analytic operator from $(C^{m+r,\alpha,0}_*(\mathbb{T}, \mathbb{C}) \times \mathbb{C}) \cap \mathcal{E}^{m+r,\alpha}$ to $C^{m+r,\alpha,0}_*(\mathbb{T}, \mathbb{C})$. By Theorem 3.7, $\mathbf{T}_*$ is analytic and, in particular continuous, from $\mathcal{A}^{m+r,\alpha}_*$ to $C^{m+r,\alpha}_*(\mathbb{T}, \mathbb{C})$ and $\mathbf{T}_*[\phi, p, w] \in C^\infty_*(\mathbb{T}, \mathbb{C})$

whenever $\phi, p \in C_*^\infty(\mathbb{T},\mathbb{C})$. Then the operator $\mathbf{T}_*$ restricts to an analytic operator from the set $(C_*^{m+r,\alpha,0}(\mathbb{T},\mathbb{C}) \times C_*^{m+r,\alpha,0}(\mathbb{T},\mathbb{R}) \times \mathbb{C}) \cap \mathcal{A}_*^{m+r,\alpha}$ to $C_*^{m+r,\alpha,0}(\mathbb{T},\mathbb{C})$. By [15, Thm. 2.11], the inversion operator $f \mapsto \mathbf{J}[f] \equiv f^{(-1)}$ from $C_*^{m+r,\alpha,0}(\mathbb{T},\mathbb{T}) \cap \mathcal{Z}$ to $C_*^{m,\alpha,0}(\mathbb{T},\mathbb{C})$ extends to an operator $\tilde{\mathbf{J}}$ of class C^r from an open neighborhood of $h[\phi_0, w_0]$ in $C_*^{m+r,\alpha,0}(\mathbb{T},\mathbb{C}) \cap \mathcal{Z}$ to $C_*^{m,\alpha,0}(\mathbb{T},\mathbb{C})$. By [15, Thm. 2.9], the ordinary composition operator $(f,g) \mapsto f \circ g$ from $C_*^{m+r,\alpha,0}(\mathbb{T},\mathbb{C}) \times C_*^{m,\alpha,0}(\mathbb{T},\mathbb{T})$ to $C_*^{m,\alpha,0}(\mathbb{T},\mathbb{C})$ extends to an operator of class C^r from an open subset of $C_*^{m+r,\alpha,0}(\mathbb{T},\mathbb{C}) \times C_*^{m,\alpha,0}(\mathbb{T},\mathbb{C})$ to $C_*^{m,\alpha,0}(\mathbb{T},\mathbb{C})$. Since the composition of operators of class C^r is an operator of class C^r, the first part of the statement follows.

To prove the sharpness of the result, we argue by contradiction and assume that $\mathbf{T}_c$ is of class C^{r+1}. Let $\phi_0 \in C_*^{m+r,\alpha,0}(\mathbb{T},\mathbb{C}) \cap \mathcal{Z}$ and $p_0 \in C_*^{m+r,\alpha,0}(\mathbb{T},\mathbb{R})$. We assume that $0 \in \mathbf{I}[\phi_0]$ and let $\mathcal{U}_{\phi_0}$ be an open neighborhood of ϕ_0 in $C_*^{m+r,\alpha,0}(\mathbb{T},\mathbb{C}) \cap \mathcal{Z}$ such that $0 \in \mathbf{I}[\phi]$ for all $\phi \in \mathcal{U}_{\phi_0}$. Then $\mathbf{L}[\phi] \equiv \operatorname{Re} \mathbf{T}_c[\phi, p_0, 0] = p_0 \circ \mathbf{h}[\phi, 0]^{(-1)}$ is an operator of the class C^{r+1} from $\mathcal{U}_{\phi_0}$ to $C_*^{m,\alpha,0}(\mathbb{T},\mathbb{C})$. Let $\sigma \in \mathbb{T}$ and M_σ be the multiplication by σ from $\mathbb{T}$ to itself. It is not difficult to check that

$$
\begin{aligned}
g_{M_{\sigma^{-1}} \circ \phi_0, 0} &= M_{\sigma^{-1}} \circ g_{\phi_0,0} \circ M_\sigma, \\
\mathbf{L}[M_{\sigma^{-1}} \circ \phi_0] &= p_0 \circ \mathbf{h}[\phi_0, 0]^{(-1)} \circ M_\sigma,
\end{aligned}
\tag{4.7}
$$

for all $\sigma \in \mathbb{T}$ such that $M_{\sigma^{-1}} \circ \phi_0 \in \mathcal{U}_{\phi_0}$. In particular, $\theta \mapsto \mathbf{L}[M_{e^{-i\theta}} \circ \phi_0]$ is of class C^{r+1} from $]-\epsilon, \epsilon[$ to $C_*^{m,\alpha,0}(\mathbb{T},\mathbb{C})$ for some $\epsilon > 0$. Then by (4.7) and by [14, Lem. 2.5 (ii)], this implies that $p_0 \circ \mathbf{h}[\phi_0, 0]^{(-1)} \in C_*^{m+r+1,\alpha,0}(\mathbb{T},\mathbb{C})$ which in general is not true when p_0 and ϕ_0 belong to $C_*^{m+r,\alpha,0}(\mathbb{T},\mathbb{C})$. □

References

[1] Akhiezer, N. I., *Elements of the Theory of Elliptic Functions*, Translations of Mathematical Monographs, **79**. AMS, Providence, R.I., 1990.

[2] Berger, M. S., *Nonlinearity and Functional Analysis*. Academic Press, 1977.

[3] Calderón, A.-P., Cauchy integrals on Lipschitz curves and related operators, *Proc. Natl. Acad. Sci. U.S.A.*, **74** (1977), 1324–1327.

[4] Coifman, R. R., McIntosh, A., and Meyer, Y., L'integrale de Cauchy définit un operator borné sur L^2 pour les courbes lipschitziennes, *Ann. Math.*, **116** (1982), 361–387.

[5] Coifman, R. R. and Meyer, Y., L'analyse harmonique non linéaire. In: *Topics in Modern Harmonic Analysis*, Vol. I, II (Turin/Milan, 1982), 707–721. Ist. Naz. Alta Mat. Francesco Severi, Rome, 1983.

[6] Coifman, R. R. and Meyer, Y., Lavrentiev's curves and conformal mappings, Mittag-Leffler Institute, Stockholm, Report No. 5, 1983.

[7] David, G., Opérateurs intégraux singuliers sur certaines courbes du plan complexe, *Ann. Sci. École Norm. Sup.*, **17** (1984), 157–189.

[8] Gakhov, F. D., *Boundary Value Problems*. Pergamon Press, Oxford, New York, Paris, 1966.

[9] Gustafsson, B. and Sakai, M., Some geometric properties of a Hele-Shaw flow moving boundary problem. In: *Free boundary value problems in fluid flow with applications*, (Montreal, PQ 1990), Pitman Research Notes Math. Ser., **282**, 53-57. Longman Sci. Tech., Harlow, 1993.

[10] Jones, P., Quasiconformal mappings and extendability of functions in Sobolev spaces, *Acta Math.*, **147** (1981), 71–88.

[11] Lanza de Cristoforis, M., Properties and pathologies of the composition and inversion operators in Schauder spaces, *Acc. Naz. delle Sci. detta dei XL*, **15** (1991), fasc. 5, 93–109.

[12] Lanza de Cristoforis, M., A functional decomposition theorem for the conformal representation, *J. Math. Soc. Japan*, **49** (1997), 759–780.

[13] Lanza de Cristoforis, M. and Preciso, L., On the analyticity of the Cauchy integral in Schauder spaces, *J. Int. Equat. Appl.*, **11** (1999), no. 3, 363–391.

[14] Lanza de Cristoforis, M. and Preciso, L., Regularity properties of a nonlinear operator associated to the conformal welding, In: *Proceedings on Analysis and Geometry*, 300–317. Sobolev Institute Press, Novosibirsk, 2000.

[15] Lanza de Cristoforis, M. and Preciso, L., Differentiability properties of some nonlinear operators associated to the conformal welding of Jordan curves in Schauder spaces, *Hiroshima Math. J.*, to appear.

[16] Lanza de Cristoforis, M. and Rogosin, S. V., Analyticity of a nonlinear operator associated to the conformal representation in Schauder spaces. An integral equation approach, *Math. Nachr.*, **220** (2000), 59–77.

[17] Lanza de Cristoforis, M. and Rogosin, S. V., Analyticity of a nonlinear operator associated to the conformal representation of a doubly connected domain in Schauder spaces, *Complex Variables*, **44** (2001), 193–223.

[18] Meyer, Y. and Coifman, R. R., *Ondelettes et Opérateurs, III. Opérateurs Multilinéaires.* Hermann, Paris, 1991.

[19] Mikhlin, S. G., *Integral Equations.* Pergamon Press, New York, 1964.

[20] Mityushev, V. V., Hilbert boundary value problem for multiply connected domains, *Complex Variables*, **35** (1998), 283–295.

[21] Mityushev, V. V. and Rogosin, S. V., *Constructive Methods for Linear and Nonlinear Boundary Value Problems for Analytic Functions*, Chapman & Hall/CRC Monographs and Surveys in Pure and Applied Mathematics, **108**. Chapman & Hall/CRC, Boca Raton, FL, 2000.

[22] Pommerenke, Ch., *Univalent Functions.* Vandenhoeck & Ruprecht, Göttingen, 1975.

[23] Pommerenke, Ch., *Boundary Behaviour of Conformal Maps.* Springer Verlag, Berlin, 1992.

[24] Prodi, G. and Ambrosetti, A., *Analisi Non Lineare.* Editrice Tecnico Scientifica, Pisa, 1973.

[25] Radó, T., Sur la représentation conforme de domains variables, *Acta Univ. Szeged*, **1** (1923), 180–186.

[26] Sakai, M., *Quadrature Domains*, Lecture Notes in Math., **934**. Springer Verlag, Berlin, 1982.

[27] Sakai, M., Regularity of boundaries of quadrature domains in two dimensions, *SIAM J. Math. Anal.*, **24** (1993), no. 2, 341–364.

[28] Wu, S., Analytic dependence of Riemann mappings for bounded domains and minimal surfaces, *Comm. Pure Appl. Math.*, **46** (1993), 1303–1326.

INTEGRAL OPERATORS WITH SHIFTS ON HOMOGENEOUS GROUPS

Vladimir S. Rabinovich*
Instituto Politechnico National, ESIME-Zacatenco, México
rabinov@maya.esimez.ipn.mx

Steffen Roch
Fachbereich Mathematik, Technische Universität Darmstadt, Germany
roch@mathematik.tu-darmstadt.de

Dedicated to G. S. Litvinchuk on the occasion of his 70th birthday

Abstract We study Fredholm properties of integral operators with shifts on homogeneous groups. This investigation is based on the limit operators method which allows us to reduce the problem of Fredholmness of convolution operators with variable coefficients and with variable shifts to the problem of invertibility of convolution operators with constant coefficients and constant shifts. For the invertibility of these operators, methods of harmonic analysis on noncommutative groups are available.

Keywords: Integral equations, slowly oscillating shifts, Fredholm properties

1. Introduction

Let $\mathbb{X}$ be a homogeneous group (see, for instance, [18], and also Section 3.1). We consider the C^*-algebra of operators acting on $L^2(\mathbb{X})$ which is generated by the operators of the form

$$\gamma I + \sum_{i=1}^{N} \prod_{j=1}^{M} a_{ij} K_{ij} T_{ij} \tag{1.1}$$

*Supported by the German Research Foundation (DFG) under Grant Nr. 436 RUS 17/24/01.

where $\gamma \in \mathbb{C}$, the a_{ij} are operators of multiplication by functions $a_{ij} \in L^\infty(\mathbb{X})$, the K_{ij} are operators of right convolution on the group $\mathbb{X}$ with kernels k_{ij} in $L^1(\mathbb{X})$, and the T_{ij} are operators of right shift by functions $g_{ij} : \mathbb{X} \to \mathbb{X}$,

$$(T_{ij})u(x) = u(x \cdot g_{ij}(x)), \quad x \in \mathbb{X}.$$

The functions g_{ij} will be specified later such that the operators $T_{ij} : L^2(\mathbb{X}) \to L^2(\mathbb{X})$ become bounded.

The goal of this paper is to examine the Fredholm and semi-Fredholm properties of operators of the form (1.1) and of their limits with respect to the norm topology of $L(L^2(\mathbb{X}))$.

A well-known and archetypical example of a noncommutative homogeneous group is the Heisenberg group. Singular integral operators and pseudodifferential operators on the Heisenberg group have been intensively studied by many authors (see, for example, the monographs [18, 12, 21, 22] which also contain extensive bibliographies). The Fredholm property of operators in some algebras generated by convolution operators and operators of multiplication by bounded functions on general noncommutative locally compact groups was the subject of [19, 20]. Integral operators with constant coefficients and constant shifts (i.e., $a_{ij} \in \mathbb{C}$ and $g_{ij} \in \mathbb{R}$) on the real half-line are considered in [4]. There is also an extensive bibliography devoted to singular integral and pseudodifferential operators with shifts (see, for instance, [1, 3, 6, 7, 5] and the references therein).

Our approach is essentially different from the approaches of the cited papers. It is based on the limit operators method which has been developed in [8, 9, 10, 13, 14, 16] to study the Fredholm properties of large classes of pseudodifferential operators and convolution operators on $\mathbb{R}^N$ and $\mathbb{Z}^N$. Here we apply this method to investigate the Fredholm and semi-Fredholm properties of integral operators with shifts. We employ an axiomatic scheme for the limit operators method which has been proposed in [15]. It should be mentioned that the results of this paper are new even for the operators with nonconstant shifts on the group $\mathbb{R}^N$.

2. The limit operators method

We start with recalling the axiomatic scheme for the application of the limit operators method developed in [15]. Let H be a Hilbert space and $L(H)$ the C^*-algebra of all bounded linear operators acting on H. Suppose that we are given

(A1) operators $P, \hat{P} \in L(H)$ with $P\hat{P} = \hat{P}P = P$.

(A2) a countable set $\{U_\alpha\}_{\alpha\in\Lambda}$ of unitary operators on H such that, with $P_\alpha := U_\alpha P U_\alpha^*$ and $\hat{P}_\alpha := U_\alpha \hat{P} U_\alpha^*$,

$$\sum_{\alpha\in\Lambda} \|P_\alpha u\|^2 = \|u\|^2 \quad \text{and} \quad \sum_{\alpha\in\Lambda} \|\hat{P}_\alpha u\|^2 \le C\|u\|^2 \tag{2.1}$$

for all $u \in H$ with a constant C independent of u.

(A3) a sequence $(W_k)_{k\in\mathbb{N}}$ of unitary operators on H and a related sequence $(D_k)_{k\in\mathbb{N}}$ of mappings from Λ into itself such that $W_k U_\alpha = U_{D_k(\alpha)} W_k$ for all $\alpha \in \Lambda$ and $k \in \mathbb{N}$, and such that the operators $\hat{P}^{(k)} := W_k \hat{P} W_k^*$ converge strongly to the identity operator on H. We also set $P^{(k)} := W_k P W_k^*$ and $P_{k,\alpha} := W_k P_\alpha W_k^*$ as well as $\hat{P}_{k,\alpha} := W_k \hat{P}_\alpha W_k^*$.

(A4) a bounded sequence $(Q_r)_{r\in\mathbb{N}}$ of operators in $L(H)$ such that

- there is a distinguished set $\mathfrak{B}$ of sequences in Λ which contains all sequences (β_m) for which there exist a $k \in \mathbb{N}$ and a sequence (r_m) in $\mathbb{N}$ tending to infinity such that

$$P_{k,\beta_m} Q_{r_m} \neq 0 \quad \text{for all } m \in \mathbb{N}, \tag{2.2}$$

- every subsequence of a sequence in $\mathfrak{B}$ belongs to $\mathfrak{B}$,
- the set $\mathfrak{B}$ is invariant with respect to each of the mappings D_k, i.e. if $(\beta_m) \in \mathfrak{B}$, then $(D_k\beta_m) \in \mathfrak{B}$ for every k,
- for each $r \in \mathbb{N}$ and each sequence $(\beta_m) \in \mathfrak{B}$,

$$\text{s-lim}_{m\to\infty} U_{\beta_m}^* Q_r U_{\beta_m} = I. \tag{2.3}$$

Since both the U_α and the W_k are unitary operators, one also has

$$\sum_{\alpha\in\Lambda} \|P_{k,\alpha}u\|^2 = \|u\|^2 \quad \text{and} \quad \sum_{\alpha\in\Lambda} \|\hat{P}_{k,\alpha}u\|^2 \leq C\|u\|^2$$

for all $u \in H$ and $k \in \mathbb{N}$ and

$$P_{k,\alpha}\hat{P}_{k,\alpha} = \hat{P}_{k,\alpha}P_{k,\alpha} = P_{k,\alpha}$$

for all $\alpha \in \Lambda$ and $k \in \mathbb{N}$.

Definition 2.1 We say that the operator A_β is the *limit operator of* $A \in L(H)$ with respect to the sequence $\beta = (\beta_j) \in \mathfrak{B}$ if, for every $k \in \mathbb{N}$,

$$\lim_{j\to\infty} \|(U_{\beta_j}^* A U_{\beta_j} - A_\beta)\hat{P}^{(k)}\| = \lim_{j\to\infty} \|(\hat{P}^{(k)})^*(U_{\beta_j}^* A U_{\beta_j} - A_\beta)\| = 0.$$

The set of all limit operators of A with respect to sequences in $\mathfrak{B}$ will be denoted by $\lim_{\mathfrak{B}}(A)$.

The following proposition describes some elementary properties of limit operators.

Proposition 2.2 *Let $\beta \in \mathfrak{B}$, and let A, $B \in L(H)$ be operators for which the limit operators A_β and B_β exist. Then* (*a*) $\|A_\beta\| \leq \|A\|$.

(*b*) $(A+B)_\beta$ *exists and* $(A+B)_\beta = A_\beta + B_\beta$.

(*c*) $(A^*)_\beta$ *exists and* $(A^*)_\beta = (A_\beta)^*$.

(*d*) *if* $C, C_n \in L(H)$ *are operators with* $\|C - C_n\| \to 0$, *and if the limit operators* $(C_n)_\beta$ *exist for all sufficiently large* n, *then* C_β *exists and* $\|C_\beta - (C_n)_\beta\| \to 0$.

Definition 2.3 Let $\mathcal{A}_0(H)$ denote the set of all operators $A \in L(H)$ with the following properties

(*a*) $\lim_{k\to\infty} \|[P_{k,\alpha}, A]\| = 0$ and $\lim_{k\to\infty} \|[P_{k,\alpha}, A^*]\| = 0$ uniformly with respect to $\alpha \in \Lambda$,

(*b*) every sequence in $\mathfrak{B}$ possesses a subsequence β for which the limit operator A_β exists,

(*c*) there is a $k_0 \in \mathbb{N}$ such that $P_{k,\alpha}A = P_{k,\alpha}A\hat{P}_{k,\alpha}$ for all $k \geq k_0$.

Further, let $\mathcal{A}(H)$ denote the closure of $\mathcal{A}_0(H)$ in $L(H)$.

It is easy to check that $\mathcal{A}_0(H)$ and $\mathcal{A}(H)$ are linear spaces. Moreover, every operator A in $\mathcal{A}(H)$ also satisfies conditions (*a*) and (*b*) (the latter follows from Proposition 2.2), and if A and B are operators which satisfy (*a*), then their product also satisfies this condition. On the other hand, condition (*c*) (which is the abstract analogue of the band property) is not stable with respect to norm limits and products of operators.

Let $\nu(A) := \inf_{\|f\|=1} \|Af\|$ refer to the lower norm of the operator $A \in L(H)$. It is well-known that A is invertible from the left if and only if $\nu(A) > 0$ and invertible from the right if and only if $\nu(A^*) > 0$. Thus, A is invertible if and only if both $\nu(A) > 0$ and $\nu(A^*) > 0$.

For every non-zero (but not necessarily closed) subspace L of H we also consider the lower norm of the restriction $A|_L$ of A onto L. If, in particular, L is the range of a non-zero operator $P \in L(H)$, then we call

$$\nu(A|_{P(H)}) = \inf_{\|Pf\|=1} \|APf\|$$

the lower norm of A relative to P. The lower norms of A relative to the Q_r are closely related to the Fredholm properties of A (see the proof of Theorem 4.5 below).

The following result has been proved in [15].

Theorem 2.4 *Let* $A \in \mathcal{A}(H)$. *Then*

$$\liminf_{r\to\infty} \nu(A|_{Q_r(H)}) = \inf_{A_\beta \in \lim_{\mathfrak{B}}(A)} \nu(A_\beta). \tag{2.4}$$

3. Operators on homogeneous groups

3.1 Homogeneous groups

Following [18], Chapter XIII, Section 5, we cite some facts on homogeneous groups which are needed in what follows. Homogeneous groups $\mathbb{X}$ arise by equipping $\mathbb{R}^m$ with a Lie group structure and with a family of dilations that act as group automorphisms on $\mathbb{X}$. To be precise, assume we are given smooth mappings $(x, y) \mapsto x \cdot y$ and $x \mapsto x^{-1}$ from $\mathbb{R}^m$ to $\mathbb{R}^m$ which provide $\mathbb{R}^m$ with a Lie group structure, and assume that the origin of $\mathbb{R}^m$ is the identity element of the associated Lie group. Further we suppose that $a_1 \leq \cdots \leq a_m$ are positive integers such that the dilations

$$D_\delta : (x_1, \ldots, x_m) \mapsto (\delta^{a_1} x_1, \ldots, \delta^{a_m} x_m)$$

are group automorphisms for every $\delta > 0$, i.e. that

$$D_\delta(x \cdot y) = D_\delta x \cdot D_\delta y \quad \text{for all } x, y \in \mathbb{R}^m.$$

It follows from these assumptions that the group operation is necessarily of the form

$$x \cdot y = x + y + Q(x, y)$$

where $Q : \mathbb{R}^m \to \mathbb{R}^m$ satisfies

$$Q(0,0) = Q(x,0) = Q(0,x) = 0 \quad \text{for every } x \in \mathbb{R}^m.$$

Moreover, if one writes $Q = (Q_1, \ldots, Q_m)$, then each Q_r is a polynomial in $2m$ real variables which is homogeneous of degree a_r. Thus, Q contains no pure monomials in x or y.

The Euclidean measure dx is both left and right invariant with respect to the group multiplication, hence, it is the Haar measure on $\mathbb{X}$. Note also that $d(D_\delta x) = \delta^a dx$ where $a := a_1 + \cdots + a_m$.

An archetypical example of a homogeneous non-commutative group is the Heisenberg group $\mathbb{H}^n$ which can be identified with the product $\mathbb{C}^n \times \mathbb{R}$, provided with the group operation

$$(w, s) \cdot (z, t) := (w + z, s + t + 2 \operatorname{Im}\langle w, z \rangle)$$

where $\langle w, z \rangle := \sum_{j=1}^{n} w_j \bar{z}_j$. Consider the norm function ρ on $\mathbb{R}^m$, defined by

$$\rho(x) := \max_{1 \leq j \leq m} \{|x_j|^{1/a_j}\}.$$

Note that $\rho(x) \geq 0$ and $\rho(x) = 0$ if and only if $x = 0$. Also, $\rho(D_\delta x) = \delta \rho(x)$, and there is a constant $c > 0$ such that

$$\rho(x \cdot y) \leq c(\rho(x) + \rho(y)) \quad \text{and} \quad \rho(x^{-1}) \leq c\rho(x).$$

Set $\rho(x, y) := \rho(x^{-1} \cdot y)$. Then the collection of all balls

$$B(x, \varepsilon) := \{y \in \mathbb{X} : \rho(x, y) < \varepsilon\}, \quad \varepsilon > 0,$$

forms an open neighborhood basis of the point $x \in \mathbb{X}$.

3.2 Multiplication operators on $\mathbb{X}$

Throughout what follows, let $\mathbb{X}$ be a homogeneous group. By $C_b(\mathbb{X})$ we denote the C^*-algebra of all continuous functions on $\mathbb{X}$ with $\|f\|_\infty := \sup_{x\in\mathbb{X}} |f(x)| < \infty$, and we let $BUC(\mathbb{X})$ stand for the C^*-subalgebra of $C_b(\mathbb{X})$ which consists of the uniformly continuous functions, i.e. $f \in C_b(\mathbb{X})$ belongs to $BUC(\mathbb{X})$ if, for each $\varepsilon > 0$, there is a $\delta > 0$ such that $|f(x) - f(y)| < \varepsilon$ whenever $\rho(x, y) < \delta$.

Further, we let $Q(\mathbb{X})$ refer to set of all measurable bounded functions a on $\mathbb{X}$ such that

$$\limsup_{y\to\infty} \int_\Omega |a(y^{-1} \cdot x)| \, dx = 0$$

for every compact $\Omega \subset \mathbb{X}$. It is easy to check that $BUC(\mathbb{X})$ is a C^*-subalgebra of $L^\infty(\mathbb{X})$ and that $Q(\mathbb{X})$ is a closed ideal of $L^\infty(\mathbb{X})$.

3.3 Convolution operators on $\mathbb{X}$

Given $k \in L^1(\mathbb{X})$ we define the operator of right convolution by k by

$$(C_{k,r}u)(x) := \int_{\mathbb{R}^m} k(x^{-1} \cdot y)u(y)dy = \int_{\mathbb{R}^m} k(z)u(x \cdot z)dz, \quad x \in \mathbb{R}^m.$$

It is well-known that $C_{k,r}$ is bounded on $L^2(\mathbb{R}^m)$ and invariant with respect to the left shift,

$$U_{l,g}C_{k,r} = C_{k,r}U_{l,g} \quad \text{where} \quad (U_{l,g}f)(x) := f(g \cdot x) \quad \text{for } g \in \mathbb{X}.$$

We denote by $V_r(\mathbb{X})$ the set of all operators $C_{k,r}$ of right convolution by a function $k \in L^1(\mathbb{R}^m)$. Note that, if $a \in Q(\mathbb{X})$ and $T \in V_r(\mathbb{X})$, then aT and TaI are compact operators on $L^2(\mathbb{X})$ (see [20]).

Let $\mathbb{Y}$ be a discrete subgroup of the group $\mathbb{X}$ which acts freely on $\mathbb{X}$ such that $\mathbb{X}/\mathbb{Y}$ is a compact manifold. Let M be a fundamental domain of $\mathbb{X}$ with respect to the action of $\mathbb{Y}$ on $\mathbb{X}$ by left shift, i.e., M is a bounded domain in $\mathbb{X}$ such that

$$\mathbb{X} = \bigcup_{\alpha\in\mathbb{Y}} \alpha \cdot \overline{M}.$$

Let M' be an open neighborhood of $\overline{M}$ such that the family $\{\alpha M'\}_{\alpha\in\mathbb{Y}}$ provides a covering of $\mathbb{X}$ of finite multiplicity. Let $f : \mathbb{X} \to [0, 1]$ be a continuous function with $f(x) = 1$ if $x \in \overline{M}$ and $f(x) = 0$ outside M', and let φ be the non-negative function which satisfies

$$\varphi^2(x) := \frac{f(x)}{\sum_{\beta \in \mathbb{Y}} f(\beta \cdot x)}.$$

For $\alpha \in \mathbb{Y}$, set $\varphi_\alpha(x) := \varphi(\alpha \cdot x)$. Evidently, $0 \le \varphi_\alpha(x) \le 1$ and

$$\sum_{\alpha \in \mathbb{Y}} \varphi_\alpha^2(x) = 1, \quad x \in \mathbb{X}. \tag{3.1}$$

In that sense, the family $\{\varphi_\alpha^2\}_{\alpha \in \mathbb{Y}}$ forms a partition of unity on $\mathbb{X}$. For $\delta > 0$, set $\varphi_{\delta,\,\alpha}(x) := \varphi_\alpha(D_\delta x)$. The following is proved in [15].

Proposition 3.1 *Let $K \in V_r(\mathbb{X})$. Then $\lim_{\delta \to 0} \|[\varphi_{\delta,\,\alpha} I,\, K]\| = 0$ uniformly with respect to $\alpha \in \mathbb{Y}$.*

3.4 Shift operators on $\mathbb{X}$

Let $g = (g_1, \dots, g_m) : \mathbb{X} \to \mathbb{X}$. We consider the shift operators of the form

$$(T_g u)(x) := u(x \cdot g(x))$$

where

(α) $g_j \in C_b^1(\mathbb{X})$ for all j.

(β) The mapping $F_g : \mathbb{X} \to \mathbb{X}$, $x \mapsto x \cdot g(x)$ is invertible.

(γ) $\lim_{x \to \infty} \det(dF_g(x)) = 1$ were df refers to the derivative of the function $f : \mathbb{R}^m \to \mathbb{R}^m$.

Proposition 3.2 *If g satisfies the conditions $(\alpha) - (\gamma)$, then the operator T_g is bounded on $L^2(\mathbb{X})$.*

Proof. We have

$$\|T_g u\|^2 = \int_{\mathbb{R}^m} |u(F_g(x))|^2 \, dx = \int_{\mathbb{R}^m} |u(y)|^2 |\det dF_g^{-1}(y)| \, dy \le C\|u\|^2$$

where $C := \sup_{y \in \mathbb{R}^m} |\det dF_g^{-1}(y)| < \infty$ due to conditions (β) and (γ). □

We call the function g *slowly oscillating* if, in addition to the conditions $(\alpha) - (\gamma)$,

(δ) $\lim_{x \to \infty} \|dg(x)\| = 0$.

The class of all shifts T_g with g slowly oscillating will be denoted by $\mathcal{R}(\mathbb{X})$.

Proposition 3.3 *Let $T_g \in \mathcal{R}(\mathbb{X})$. Then*

$$\lim_{\delta\to 0} \|[\varphi_{\delta,\,\alpha} I,\, T_g]\| = 0 \quad \textit{uniformly with respect to } \alpha \in \mathbb{Y}.$$

Proof. For every $u \in L^2(\mathbb{X})$, one has

$$\begin{aligned} \|[\varphi_{\delta,\,\alpha} I,\, T_g]\, u\| &\leq \sup_{x\in\mathbb{X}} |\varphi_{\delta,\,\alpha}(x) - \varphi_{\delta,\,\alpha}(x \cdot g(x))|\, \|T_g u\| \\ &\leq C \sup_{x\in\mathbb{X}} |\varphi_{\delta,\,\alpha}(x) - \varphi_{\delta,\,\alpha}(x \cdot g(x))|\, \|u\|. \end{aligned}$$

Let us estimate

$$|\varphi_{\delta,\,\alpha}(x) - \varphi_{\delta,\,\alpha}(x \cdot g(x))| = |\varphi(D_\delta(\alpha \cdot x)) - \varphi(D_\delta(\alpha \cdot x) \cdot D_\delta(g(x)))|.$$

The function g is bounded due to assumption (α). Hence,

$$\|D_\delta(g(x))\|_\infty \leq \max_{1\leq j\leq m} \delta^{a_j}\, \|g\|_\infty.$$

Since φ is uniformly continuous on $\mathbb{X}$ we obtain that, given $\varepsilon > 0$, one finds a $\delta_0 > 0$ such that, if $\max_{1\leq j\leq m} \delta^{a_j}\, \|g\|_\infty < \delta_0$, then

$$\sup_{x\in\mathbb{X}} |\varphi_{\delta,\,\alpha}(x) - \varphi_{\delta,\,\alpha}(x \cdot g(x))| < \varepsilon$$

uniformly with respect to $\alpha \in \mathbb{Y}$. This implies the assertion. □

Here are a few instances where the requirements $(\alpha) - (\delta)$ are satisfied.

Example 3.4 If g is a constant function then, evidently, T_g belongs to $\mathcal{R}(\mathbb{X})$.

Example 3.5 Let $\mathbb{X}$ be the commutative group $\mathbb{R}^m$ where

$$(T_g u)(x) = u(x + g(x)),$$

and let the conditions (α) and (δ) be fulfilled. If one of the conditions

$$\max_{1\leq j\leq m} \sum_{k=1}^{m} \sup_x \left|\frac{\partial g_j(x)}{\partial x_k}\right| < 1, \quad \max_{1\leq k\leq m} \sum_{j=1}^{m} \sup_x \left|\frac{\partial g_j(x)}{\partial x_k}\right| < 1 \tag{3.2}$$

is satisfied, then $T_g \in \mathcal{R}(\mathbb{X})$. Indeed, each of the conditions (3.2) implies that g is a contraction. Hence, by the Banach fixed point theorem, the mapping

$$F_g : \mathbb{R}^m \to \mathbb{R}^m, \quad x \mapsto x + g(x)$$

is invertible, and it follows from condition (δ) that

$$\lim_{x\to\infty} \det(dF_g(x)) = 1$$

whence condition (γ).

Example 3.6 Let $\mathbb{H}^n$ be the Heisenberg group with coordinates $(x,\, y,\, t) \in \mathbb{R}^n \times \mathbb{R}^n \times \mathbb{R}$. Consider the function $g(x,\, y,\, t) := (p(x,\, y),\, q(x,\, y),\, \tau(x,\, y,\, t))$ where the mapping

$$(x, y) \mapsto (p(x,\, y),\, q(x,\, y)) : \mathbb{R}^{2n} \to \mathbb{R}^{2n}$$

is subject to a condition analogous to (3.2) and where, consequently, the mapping

$$(x, y) \mapsto \Phi(x,\, y) := (x + p(x,\, y),\, y + q(x,\, y))$$

is invertible. Thus, the system

$$x' = x + p(x,\, y), \quad y' = y + q(x,\, y)$$

possesses a unique solution

$$x = f(x',\, y'), \quad y = \varphi(x',\, y').$$

Moreover we suppose that g is slowly oscillating in the sense that

$$\lim_{(x,y)\to\infty} d_x p(x,y) = \lim_{(x,y)\to\infty} d_y p(x,y) = 0, \tag{3.3}$$

$$\lim_{(x,y)\to\infty} d_x q(x,y) = \lim_{(x,y)\to\infty} d_y q(x,y) = 0, \tag{3.4}$$

$$\lim_{(x,y,t)\to\infty} d_x\tau(x,y,t) = \lim_{(x,y,t)\to\infty} d_y\tau(x,y,t) = \lim_{(x,y,t)\to\infty} d_t\tau(x,y,t) = 0. \tag{3.5}$$

Let, moreover,

$$\sup_{(x,y,t)\in\mathbb{R}^{2n+1}} |d_t\tau(x,y,t)| < 1. \tag{3.6}$$

Then the mapping $F_g : \mathbb{R}^{2n+1} \longrightarrow \mathbb{R}^{2n+1}$, which sends (x, y, t) to

$$\begin{aligned}(x,y,t)\cdot g(x,y,t) \quad &= \quad (x + p(x,\, y),\, y + q(x,\, y), \\ &\qquad t + \tau(x,\, y,\, t) + 2\left(\langle (x,\, q(x,\, y)\rangle - \langle y,\, p(x,y)\rangle\right)\end{aligned}$$

is invertible. Indeed, for arbitrary $(x', y', t') \in \mathbb{R}^{2n+1}$, the equation

$$t' = t + \tau(f(x',y'), \varphi(x',y'),\, t) + 2\,\Psi(x',y')$$

where

$$\Psi(x',y') = \langle f(x',y'),\, q(f(x',y'),\, \varphi(x',y'))\rangle - \langle \varphi(x',y'),\, p(f(x',y'),\, \varphi(x',y'))\rangle$$

has a unique solution t due to (3.6). This proves condition (β), and condition (γ) follows since (3.3) – (3.5) imply that

$$\lim_{(x,\, y,\, t)\to\infty} \det(d_{(x,\, y,\, t)} F_g(x,\, y,\, t)) = 1.$$

Consequently, under the above assumptions, $T_g \in \mathcal{R}(\mathbb{H}^{2n+1})$.

4. Fredholmness of convolution operators with shifts

Let $\mathcal{B}(BUC(\mathbb{X}), V_r(\mathbb{X}), \mathcal{R}(\mathbb{X}))$ denote the smallest C^*-subalgebra of $L(L^2(\mathbb{X}))$ which contains the compact operators and all operators of the form

$$A = \gamma I + \sum_{j=1}^{N} \prod_{k=1}^{L} a_{jk} K_{jk} b_{jk} T_{jk} \tag{4.1}$$

where $\gamma \in \mathbb{C}$, a_{jk}, $b_{jk} \in BUC(\mathbb{X})$, $K_{jk} \in V_r(\mathbb{X})$ and $T_{jk} \in \mathcal{R}(\mathbb{X})$. Further we write $\mathcal{B}_0(BUC(\mathbb{X}), V_r(\mathbb{X}), \mathcal{R}(\mathbb{X}))$ for the smallest symmetric (but not necessarily closed) subalgebra of $L(L^2(\mathbb{X}))$ which contains all operators of the form

$$\varphi_\alpha C \varphi_\alpha I \quad \text{where } C \text{ is compact and } \alpha \in \mathbb{Z}^n \tag{4.2}$$

as well as all operators of the form (4.1) where the kernel functions of the convolution operators K_{jk} are compactly supported.

To study the Fredholmness of operators in $\mathcal{B}(BUC(\mathbb{X}), V_r(\mathbb{X}), \mathcal{R}(\mathbb{X}))$ by the limit operators method, we specify the axioms (A1) – (A4) as follows.

(A1) Let $\hat{M}$ be an open set which contains the closure $\overline{M'}$ of M' and for which the covering $\{\alpha\hat{M}\}_{\alpha\in\mathbb{Y}}$ of $\mathbb{X}$ has a finite multiplicity. Then we let P be the operator of multiplication by φ and $\hat{P}$ be the operator of multiplication by the characteristic function of $\hat{M}$.

(A2) We choose $\Lambda := \mathbb{Y}$ and let U_α, $\alpha \in \Lambda$, be the operator of left shift by α,

$$(U_\alpha u)(x) := (U_{l,\,\alpha} u)(x) = u(\alpha \cdot x).$$

Observe that then P_α is the operator of multiplication by φ_α. Hence, the first condition in (2.1) follows from (3.1)

$$\sum \|P_\alpha u\|^2 = \sum \langle \varphi_\alpha u,\, \varphi_\alpha u\rangle = \sum \langle \varphi_\alpha^2 u,\, u\rangle = \langle u,\, u\rangle = \|u\|^2,$$

and the second one follows similarly due to the finite multiplicity of the covering $\{\alpha\hat{M}\}$.

(A3) We choose a sequence $(\delta_k)_{k\in\mathbb{N}}$ of positive numbers with $\delta_k \to 0$ as $k \to \infty$ and such that $D_{\delta_k^{-1}}\mathbb{Y} \subseteq \mathbb{Y}$, and we define W_k

$$(W_k u)(x) := \delta_k^{a/2} u(D_{\delta_k} x)$$

with $a := a_1 + \cdots + a_m$. Then

$$W_k U_\alpha = U_{D_{\delta_k^{-1}}\alpha} W_k \quad \text{for all } k \in \mathbb{N} \text{ and } \alpha \in \mathbb{Y},$$

and the operators $\hat{P}^{(k)}$ converge strongly to the identity.

(A4) For $r \in \mathbb{N}$, let Q_r be the operator of multiplication by the characteristic function of $\{x \in \mathbb{X} : \rho(x, 0) > r\}$, and let $\mathfrak{B}$ be the set of all sequences in $\mathbb{Y}$ which tend to infinity. Then conditions (2.2) and (2.3) are fulfilled.

We show that, under these assumptions, the algebra $\mathcal{B}_0(BUC(\mathbb{X}), V_r(\mathbb{X}), \mathcal{R}(\mathbb{X}))$ is a subset of $\mathcal{A}_0$ and, hence, its closure $\mathcal{B}(BUC(\mathbb{X}), V_r(\mathbb{X}), \mathcal{R}(\mathbb{X}))$ is a subset of $\mathcal{A}$.

Proposition 4.1 *Let* $A \in \mathcal{B}(BUC(\mathbb{X}), V_r(\mathbb{X}), \mathcal{R}(\mathbb{X}))$. *Then*

$$\lim_{k\to\infty} \|[A, P_{k,\alpha}]\| = 0 \quad \textit{uniformly with respect to } \alpha \in \mathbb{Y}. \tag{4.3}$$

Proof. For operators of the form (4.1), the proof follows immediately from Propositions 3.1 and 3.3 in connection with Proposition 2.2. If A is of the form (4.2), then the proof follows easily from the *-strong convergence of the $P_{k,\alpha}$ to the identity as $k \to \infty$. Since the set of all operators A which satisfy (4.3) is a C^*-subalgebra of $L(L^2(\mathbb{X}))$, the result holds for all operators in $\mathcal{B}(BUC(\mathbb{X}), V_r(\mathbb{X}), \mathcal{R}(\mathbb{X}))$. □

Proposition 4.2 *Let* $A \in \mathcal{B}_0(BUC(\mathbb{X}), V_r(\mathbb{X}), \mathcal{R}(\mathbb{X}))$. *Then, for all* $\alpha \in \mathbb{Y}$ *and for all sufficiently large* k,

$$P_{k,\alpha}A = P_{k,\alpha}A\hat{P}_{k,\alpha}.$$

Proof. Given an open set $N \subseteq \mathbb{X}$, let χ^N refer to the characteristic function of N and, for $k \in \mathbb{N}$ and $\alpha \in \mathbb{Y}$, define $\chi^N_{k,\alpha}$ in analogy to $\varphi_{k,\alpha}$. Thus, $\hat{P}_{k,\alpha}$ is the operator of multiplication by $\chi^{\hat{M}}_{k,\alpha}$. We consider the set $\mathcal{C}$ of all operators B on $L^2(\mathbb{X})$ with the following property. If N_1, N_2 are open sets with $\overline{N_1} \subset N_2$ then, for every $\alpha \in \mathbb{Y}$ and every sufficiently large k,

$$\chi^{N_1}_{k,\alpha}B = \chi^{N_1}_{k,\alpha}B\chi^{N_2}_{k,\alpha}I.$$

It is evident that $\mathcal{C}$ contains all operators of the form (4.2), all operators aI of multiplication by bounded functions on $\mathbb{X}$, all operators K of right convolution by compactly supported measurable functions on $\mathcal{X}$, and all shift operators $T \in \mathcal{R}(\mathbb{X})$. The set $\mathcal{C}$ also contains the adjoints of these operators. This is again evident for the adjoints of the operators of the form (4.2) and for the adjoints of aI and K (the adjoint of aI is the operator of multiplication by the complex conjugate of a, and K^* is the operator of right convolution by the function $x \mapsto \overline{k(x^{-1})}$ which is also compactly supported). Let now $T = T_g$ be a shift operator in $\mathcal{R}(\mathbb{X})$. The substitution rule shows that the adjoint of T_g is the operator $T_h bI$ where

$$h(y) = (g(F_g^{-1}(y)))^{-1} \quad \text{and} \quad b = |\det dF_g^{-1}|.$$

The mapping h is continuous and bounded, and b is a bounded and uniformly continuous function, which follows from property (γ). Thus, the inclusion $T^* \in \mathcal{C}$ will follow once we have shown that $\mathcal{C}$ is an algebra.

Let A, B in $\mathcal{C}$. Then, clearly, $A+B \in \mathcal{C}$. To prove that $AB \in \mathcal{C}$, let N_1, N_2 be open sets with $\overline{N_1} \subset N_2$, and choose an open set N' such that

$$\overline{N_1} \subset N' \subset \overline{N'} \subset N_2.$$

Then

$$\chi_{k,\alpha}^{N_1} AB = \chi_{k,\alpha}^{N_1} A \chi_{k,\alpha}^{N'} B \chi_{k,\alpha}^{N_2} I = \chi_{k,\alpha}^{N_1} AB \chi_{k,\alpha}^{N_2} I,$$

which implies that $AB \in \mathcal{C}$. Consequently, $\mathcal{B}_0(BUC(\mathbb{X}), V_r(\mathbb{X}), \mathcal{R}(\mathbb{X})) \subseteq \mathcal{C}$, whence the assertion. □

Proposition 4.3 *Let $A \in \mathcal{B}(BUC(\mathbb{X}), V_r(\mathbb{X}), \mathcal{R}(\mathbb{X}))$, and let $\beta = (\beta_k)$ be a sequence in $\mathfrak{B}$. Then there is a subsequence $\tilde{\beta}$ of β for which the limit operator $A_{\tilde{\beta}}$ exists.*

Proof. We start with verifying the assertion for the operators in the algebra $\mathcal{B}_0 := \mathcal{B}_0(BUC(\mathbb{X}), V_r(\mathbb{X}), \mathcal{R}(\mathbb{X}))$. Let $\mathcal{D}$ denote the set of all operators $A \in \mathcal{B}_0$ having the property that every sequence in $\mathfrak{B}$ has a subsequence $\tilde{\beta} = (\tilde{\beta}_k)$ such that the operators $U_{\tilde{\beta}_k}^{-1} A U_{\tilde{\beta}_k}$ converge *-strongly as $k \to \infty$.

We claim that $\mathcal{D}$ contains the generating operators of $\mathcal{B}_0$. Since $\mathcal{D}$ is an algebra (which can be easily shown), this implies that $\mathcal{B}_0 \subseteq \mathcal{D}$.

If C is compact, then the limit operator of C with respect to every sequence $\beta \in \mathfrak{B}$ exists and is equal to zero. If aI is the operator of multiplication by the function $a \in BUC(\mathbb{X})$, then $U_{\beta_k}^{-1} a U_{\beta_k}$ is the operator of multiplication by the function $x \mapsto a(\beta_k \cdot x)$. The functions in $BUC(\mathbb{X})$ are bounded and uniformly continuous by definition. Hence, by the Arzelà-Ascoli theorem, the sequence β possesses a subsequence $\tilde{\beta}$ such that the functions $x \mapsto a(\tilde{\beta}_k \cdot x)$ tend uniformly on compact subsets of $\mathbb{X}$ to a certain bounded function $a_{\tilde{\beta}}$ as $k \to \infty$. Consequently, the operators $U_{\tilde{\beta}_k}^{-1} a U_{\tilde{\beta}_k}$ converge strongly to $a_{\tilde{\beta}} I$, and the strong convergence of the adjoint sequence follows analogously.

If $A = K$ is a convolution operator, then there is nothing to prove because A commutes with the U_k.

Next we consider the operator $T = T_g \in \mathcal{R}(\mathbb{X})$ of shift by the function g. Then one has

$$(U_{\beta_k}^{-1} T_g U_{\beta_k} u)(x) = u(x \cdot g(\beta_k \cdot x)).$$

Since the functions $x \mapsto g(\beta_k \cdot x)$ are uniformly bounded with respect to $k \in \mathbb{N}$ and equicontinuous on compact subsets of $\mathbb{X}$, the Arzelà-Ascoli theorem implies the existence of a subsequence $\tilde{\beta}$ of β such that the functions $x \mapsto g(\tilde{\beta}_k \cdot x)$

converge uniformly on compacts in $\mathbb{X}$ to a certain bounded function $g_{\tilde{\beta}}$. Since g is slowly oscillating, the function $g_{\tilde{\beta}}$ is constant.

We proceed with showing that the strong limit of the operators $U_{\tilde{\beta}_k}^{-1} T_g U_{\tilde{\beta}_k}$ as $k \to \infty$ exists and that

$$\text{s-lim}_{k\to\infty} U_{\tilde{\beta}_k}^{-1} T_g U_{\tilde{\beta}_k} = T_{g_{\tilde{\beta}}}. \tag{4.4}$$

Let u be a compactly supported continuous function on $\mathbb{X}$. Thus, u is uniformly continuous on $\mathbb{X}$, and there exists a compact subset Ω of $\mathbb{X}$ such that

$$u(x \cdot g(\tilde{\beta}_k \cdot x)) - u(x \cdot g_{\tilde{\beta}}) = 0 \quad \text{whenever} \quad x \notin \Omega$$

(recall that g is bounded). It is further evident from the definition of $g_{\tilde{\beta}}$ that, for arbitrary $\delta > 0$, there exists a $k_0 \in \mathbb{N}$ such that, for all $k \geq k_0$ and all $x \in \Omega$, $\rho(g(\tilde{\beta}_k \cdot x), g_{\tilde{\beta}}) < \delta$. Since u is uniformly continuous, this implies that for each $\varepsilon > 0$, there exists a $k_0 \in \mathbb{N}$ such that

$$\sup_{x\in\Omega} |u(x \cdot g(\tilde{\beta}_k \cdot x)) - u(x \cdot g_{\tilde{\beta}})| < \varepsilon \quad \text{for all } k \geq k_0.$$

Thus,

$$\lim_{k\to\infty} U_{\tilde{\beta}_k}^{-1} T_g U_{\tilde{\beta}_k} u = T_{g_{\tilde{\beta}}} u$$

for every continuous and compactly supported function u on $\mathbb{X}$. Since these functions form a dense subset of $L^2(\mathbb{X})$, this implies (4.4). The strong convergence of the adjoint sequence follows from the representation of T_g^* derived in the previous proof, from the above results, and from the fact that $\mathcal{D}$ is an algebra.

The remaining part of the proof makes use of the elementary fact that, if $A_n \to A$ and $B_n^* \to B^*$ strongly, and if K is compact, then $\|A_n K B_n - AKB\| \to 0$.

Every operator in $\mathcal{B}_0$ can be written as a sum of operators of the form ACB where C is the operator of convolution by a compactly supported function, and where $A, B \in \mathcal{B}_0$. Let $\beta \in \mathfrak{B}$ and $m \in \mathbb{N}$. We choose a subsequence $\tilde{\beta}$ of β such that

$$U_{\tilde{\beta}_k}^{-1} A U_{\tilde{\beta}_k} \to A_{\tilde{\beta}} \quad \text{and} \quad U_{\tilde{\beta}_k}^{-1} B U_{\tilde{\beta}_k} \to B_{\tilde{\beta}}$$

*-strongly as $k \to \infty$. It follows from the proof of the previous proposition that, if N is sufficiently large,

$$\begin{aligned} U_{\tilde{\beta}_k}^{-1} ACB U_{\tilde{\beta}_k} \hat{P}^{(m)} &= U_{\tilde{\beta}_k}^{-1} A U_{\tilde{\beta}_k} C U_{\tilde{\beta}_k}^{-1} B U_{\tilde{\beta}_k} \hat{P}^{(m)} \\ &= U_{\tilde{\beta}_k}^{-1} A U_{\tilde{\beta}_k} C \hat{P}^{(N)} U_{\tilde{\beta}_k}^{-1} B U_{\tilde{\beta}_k} \hat{P}^{(m)}. \end{aligned}$$

Since the operator $C\hat{P}^{(N)}$ is compact, it follows from the fact just mentioned that

$$\|(U_{\tilde{\beta}_k}^{-1} ACBU_{\tilde{\beta}_k} - A_{\tilde{\beta}} CB_{\tilde{\beta}})\hat{P}^{(m)}\| \to 0$$

as $k \to \infty$. The dual condition follows analogously. Thus, $A_{\tilde{\beta}} CB_{\tilde{\beta}}$ is a limit operator of ACB. This yields the assertion for operators in $\mathcal{B}_0$ and, employing Proposition 2.2 (d), also for operators in $\mathcal{B}(BUC(\mathbb{X}),\, V_r(\mathbb{X}),\, \mathcal{R}(\mathbb{X}))$. □

The following theorem is a specification of the general Theorem 2.4.

Theorem 4.4 *Let $A \in \mathcal{B}(BUC(\mathbb{X}),\, V_r(\mathbb{X}),\, \mathcal{R}(\mathbb{X}))$. Then*

$$\liminf_{r\to\infty} \nu(A|_{Q_r L^2(\mathbb{X})}) = \inf_{A_\beta \in \sigma_{\mathfrak{B}}(A)} \nu(A_\beta).$$

As a corollary, we derive the desired criteria of semi-Fredholmness and Fredholmness.

Theorem 4.5 *Let $A \in \mathcal{B}(BUC(\mathbb{X}),\, V_r(\mathbb{X}),\, \mathcal{R}(\mathbb{X}))$. Then*
(a) A is a Φ_+-operator if and only if

$$\inf_{A_\beta \in \sigma_{\mathfrak{B}}(A)} \nu(A_\beta) > 0. \tag{4.5}$$

(b) A is a Φ_--operator if and only if

$$\inf_{A_\beta \in \sigma_{\mathfrak{B}}(A)} \nu(A_\beta^*) > 0. \tag{4.6}$$

(c) A is a Fredholm operator if and only if all limit operators of A are invertible and if the norms of their inverses are uniformly bounded.

Proof. We will show assertion (a) only. The proof of (b) proceeds similarly, and (c) is a consequence of (a) and (b).

Let (4.5) be satisfied. Then, by Theorem 4.4, there exist an $r \in \mathbb{N}$ and a constant $C > 0$ such that

$$|\langle Q_r A^* A Q_r f,\, Q_r f\rangle| \ge C\|Q_r f\|^2 \quad \text{for all } f \in L^2(\mathbb{X}). \tag{4.7}$$

This implies that the operator $Q_r A^* A Q_r$ is invertible on $L^2(Q_r\mathbb{X})$, i.e. there is an operator B such that

$$BQ_r A^* A Q_r = Q_r. \tag{4.8}$$

It follows from the inverse closeness of C^*-algebras that the operator B belongs to the smallest C^*-subalgebra $\mathcal{B}(BUC(\mathbb{X}),\, V_r(\mathbb{X}),\, \mathcal{R}(\mathbb{X}),\, Q_r)$ of $L(L^2(\mathbb{X}))$ which contains the algebra $\mathcal{B}(BUC(\mathbb{X}),\, V_r(\mathbb{X}),\, \mathcal{R}(\mathbb{X}))$ and the operator Q_r.

Let J_0 refer to the closed ideal of $\mathcal{B}(BUC(\mathbb{X}),\, V_r(\mathbb{X}),\, \mathcal{R}(\mathbb{X}),\, Q_r)$ which is generated by the operators $I - Q_r$, $r \in \mathbb{N}$, and let J_1 stand for the smallest closed ideal of that algebra which contains all operators in $V_r(\mathbb{X})$ and all compact operators. It is evident from the definition of $\mathcal{B}(BUC(\mathbb{X}),\, V_r(\mathbb{X}),\, \mathcal{R}(\mathbb{X}),\, Q_r)$ that for every operator G in this algebra, there is a (uniquely determined) complex number γ_G such that $G - \gamma_G I \in J_1$. Clearly, the mapping $G \mapsto \gamma_G$ is a continuous algebra homomorphism. Since $\gamma_{Q_r} = 1$, it follows from (4.8) that $\gamma_A \neq 0$.

The equality (4.8) further implies that there is an operator R' in the algebra $\mathcal{B}(BUC(\mathbb{X}),\, V_r(\mathbb{X}),\, \mathcal{R}(\mathbb{X}),\, Q_r)$ such that $R'A - I \in J_0$. If we set $R := \gamma_A R' - AR' + I$, then

$$RA - \gamma_A I = \gamma_A R'A - AR'A + A - \gamma_A I = (\gamma_A I - A)(R'A - I).$$

Since $T_0 := R'A - I \in J_0$ and $T_1 := \gamma_A I - A \in J_1$, the operator $RA - \gamma_A I = T_1 T_0$ is compact. Hence, and because of $\gamma_A \neq 0$, A is a Φ_+-operator.

Conversely, let A be a Φ_+-operator. Then there is a compact operator T as well as a positive constant C such that the a priori estimate

$$C\|Au\| \geq \|u\| - \|Tu\|, \quad u \in L^2(\mathbb{X}),$$

holds (see [11], I, Lemma 2.1). This estimate yields

$$C\|AQ_r u\| \geq \|Q_r u\| - \|TQ_r u\|$$

for all $u \in L^2(\mathbb{X})$ and $r \in \mathbb{N}$. Due to the strong convergence of the operators Q_r to 0, there is an $r_0 \in \mathbb{N}$ such that $\|TQ_{r_0}\| \leq C/2$. Thus,

$$\|AQ_{r_0} u\| \geq \frac{C}{2}\,\|Q_{r_0} u\| \quad \text{for all } u \in L^2(\mathbb{X})$$

whence $\liminf_{r\to\infty} \nu(A|_{Q_r L^2(\mathbb{X})}) > 0$. This implies (4.5) via Theorem 4.4. □

Finally, we are going to specialize the results of the previous theorem to a class of operators for which the invertibility of their limit operators can be effectively checked. A function $a \in C_b(\mathbb{X})$ is called *slowly oscillating at infinity* if, for every compact $\Omega \subset \mathbb{X}$,

$$\lim_{x\to\infty} \sup_{y\in\Omega} |a(x \cdot y) - a(x)| = 0.$$

For example, if $a \in C_b^1(\mathbb{X})$ and

$$\lim_{x\to\infty} \frac{\partial a(x)}{\partial x_j} = 0, \quad 1 \leq j \leq m,$$

then a is slowly oscillating at infinity. We write $SO(\mathbb{X})$ for the class of all slowly oscillating functions on $\mathbb{X}$ and set $W(\mathbb{X}) := SO(\mathbb{X}) + Q(\mathbb{X})$. Let further the algebra $\mathcal{B}(W(\mathbb{X}),\, V_r(\mathbb{X}),\, \mathcal{R}(\mathbb{X}))$ be defined in analogy to $\mathcal{B}(BUC(\mathbb{X}),\, V_r(\mathbb{X}),\, \mathcal{R}(\mathbb{X}))$.

We claim that all limit operators of operators in $\mathcal{B}(W(\mathbb{X}), V_r(\mathbb{X}), \mathcal{R}(\mathbb{X}))$ are invariant with respect to left shifts.

Let $a \in Q(\mathbb{X})$ and $K \in V_r(\mathbb{X})$. Then the operators aK and KaI are compact (see [20]). Hence, the limit operators of these operators exist with respect to every sequence $\beta \in \mathfrak{B}$, and they are equal to zero.

Further, let a be slowly oscillating, $\beta = (\beta_m)_{m\in\mathbb{N}} \in \mathfrak{B}$, and let $a_{\tilde{\beta}}$ be as in the proof of Proposition 4.3. Then, evidently,

$$a_{\tilde{\beta}}(x) - a_{\tilde{\beta}}(y) = \lim_{m\to\infty} (a(\beta_m \cdot x) - a((\beta_m \cdot x) \cdot (x^{-1} \cdot y))) = 0$$

for arbitrary $x, y \in \mathbb{X}$. Thus, $a_{\tilde{\beta}}$ is indeed a constant function.

In particular, if $A \in \mathcal{B}(W(\mathbb{X}), V_r(\mathbb{X}), \mathcal{R}(\mathbb{X}))$, then every limit operator of A belongs to the smallest C^*-subalgebra $\mathcal{B}(V_r(\mathbb{X}), \mathcal{R}^c(\mathbb{X}))$ of $L(L^2(\mathbb{X}))$ which contains all convolution operators in $V_r(\mathbb{X})$ and all shift operators in $\mathcal{R}(\mathbb{X})$ by a constant function (i.e. by an element of the group $\mathbb{X}$).

Thus, in this special setting, Theorem 4.5 reduces the problem of (semi-) Fredholmness for operators in $\mathcal{B}(W(\mathbb{X}), V_r(\mathbb{X}), \mathcal{R}(\mathbb{X}))$ to the problem of invertibility of operators in the algebra $\mathcal{B}(V_r(\mathbb{X}), \mathcal{R}^c(\mathbb{X}))$ which are invariant with respect to left shifts by elements in $\mathbb{X}$. To study this invertibility problem, methods of (noncommutative) harmonic analysis are available (cp. [22]). For example, in case of the commutative group $\mathbb{R}^n$, the operator

$$A := \gamma I + \sum_{j=1}^{N} K_j T_j$$

where $\gamma \in \mathbb{C}$, K_j is a convolution with kernel $k_j \in L^1(\mathbb{R}^n)$ and T_j is the shift by $g_j \in \mathbb{R}^n$, is invertible on $L^2(\mathbb{R}^n)$ if and only if

$$\inf_{\xi\in\mathbb{R}^n} |\gamma + \sum_{j=1}^{N} \hat{k}_j(\xi) e^{i\langle\xi,\, g_j\rangle}| > 0$$

where $\hat{k}_j$ refers to the Fourier transform of k_j.

References

[1] Antonevich, A., *Linear Functional Equations: Operator Approach*, Operator Theory: Advances and Applications, **83**. Birkhäuser, Basel, 1996.

[2] Antonevich, A., Belousov, M., and Lebedev, A., *Functional Differential Equations: 2. C*-applications. Parts 1, 2*, Pitman Monographs and Surveys in Pure and Applied Mathematics **94, 95**. Longman, Harlow, 1998.

[3] Antonevich, A. and Lebedev, A., *Functional Differential Equations: I. C*-theory*, Pitman Monographs and Surveys in Pure and Applied Mathematics, **70**. Longman Scientific & Technical, Harlow, 1994.

[4] Gohberg, I. and Feldman, I. A., *Convolution Equations and Projection Methods for Their Solution*, Nauka, Moscow, 1971 (in Russian). English translation: Translations of Mathematical Monographs, **41**. AMS, Providence, R.I., 1974.

[5] Karapetiants, N. and Samko, S., *Equations with Involutive Operators*. Birkhäuser Boston, Inc., Boston, MA, 2001.

[6] Kravchenko, V. G. and Litvinchuk, G. S., *Introduction to the Theory of Singular Integral Operators with Shift*, Mathematics and its Applications, **289**. Kluwer Academic Publishers, Dordrecht, 1994.

[7] Kurbatov, V. G., *Functional Differential Operators and Equations*, Mathematics and its Applications, **473**. Kluwer Academic Publishers, Dordrecht, 1999.

[8] Lange, B. V. and Rabinovich, V. S., On the Noether property of multi-dimensional discrete convolutions, *Mat. Zametki*, **37** (1985), no. 3, 407–421 (in Russian).

[9] Lange, B. V. and Rabinovich, V. S., On the Noether property of multi-dimensional operators of convolution type with measurable bounded coefficients, *Izv. Vyssh. Uchebn. Zaved., Mat.*, **6** (1985), 22–30 (in Russian).

[10] Lange, B. V. and Rabinovich, V. S., Pseudo-differential operators on $\mathbb{R}^n$ and limit operators, *Mat. Sb.*, **129** (1986), no. 2, 175–185 (in Russian). English translation: *Math. USSR Sbornik*, **577** (1987), no. 1, 183–194.

[11] Mikhlin, S. G. and Prössdorf, S., *Singular Integral Operators*. Springer-Verlag, Berlin, 1986.

[12] Nagel, A. and Stein, E. M., *Lectures on Pseudo-Differential Operators: Regularity Theorems and Applications to Non-Elliptic Problems*, Mathematical Notes, **24**. Princeton University Press, Princeton, N.J.; University of Tokyo Press, Tokyo, 1979.

[13] Rabinovich, V. S., Fredholmness of pseudo-differential operators on $\mathbb{R}^n$ in the scale of $L_{p,q}$-spaces, *Sib. Mat. Zh.*, **29** (1988), no. 4, 635–646 (in Russian). English translation: *Siberian Math. J.*, **29** (1988), no. 4, 635–646.

[14] Rabinovich, V. S., Criterion for local invertibility of pseudodifferential operators with operator symbols and some applications. In: *Proceedings of the St. Petersburg Mathematical Society*, Vol. V. Amer. Math. Soc. Transl. Ser. 2, **193**, 239–259. AMS, Providence, R.I., 1999.

[15] Rabinovich, V. S. and Roch, S., An axiomatic approach to the limit operators method. Preprint 2153 Technical University Darmstadt 2001. To appear in: Operator Theory: Advances and Applications, **142**. Birkhäuser Verlag, 2003.

[16] Rabinovich, V. S., Roch, S., and Silbermann, B., Fredholm theory and finite section method for band-dominated operators, *Integr. Equat. Oper. Theor.*, **30** (1998), 452–495.

[17] Rabinovich, V. S., Roch, S., and Silbermann, B., Band-dominated operators with operator-valued coefficients, their Fredholm properties and finite section, *Integr. Equat. Oper. Theor.*, **40** (2001), 342–381.

[18] Stein, E. M., *Harmonic Analysis: Real-Variable Methods, Orthogonality, and Oscillatory Integrals*, Princeton Mathematical Series, **43**. Princeton University Press, Princeton, NJ, 1993.

[19] Steinberg, B. Ya., On convolution operators on locally compact groups, *Funkts. Anal. Prilozh.*, **15** (1981), 95–96 (in Russian).

[20] Steinberg, B. Ya., Boundedness and compactness of convolution operators on locally compact groups, *Mat. Zametki*, **38** (1985), no. 2, 278–292 (in Russian).

[21] Taylor, M. E., Noncommutative Microlocal Analysis, I, *Mem. Amer. Math. Soc.*, **52** (1984), no. 313.

[22] Taylor, M. E., *Noncommutative Harmonic Analysis*, Mathematical Surveys and Monographs, **22**. AMS, Providence, R.I., 1986.

ON THE ALGEBRA GENERATED BY A POLY-BERGMAN PROJECTION AND A COMPOSITION OPERATOR

Josue Ramírez*
Universidad Autónoma Metropolitana, México
josue@math.cinvestav.mx

Ilya M. Spitkovsky†
College of William & Mary, Williamsburg, USA
ilya@math.wm.edu

Dedicated to Professor Georgii Litvinchuk on the occasion of his 70th birthday

Abstract Let G be a bounded domain in $\mathbb{C}$ with smooth boundary, and let α be a C^2-diffeomorphism of $\overline{G}$ such that $\alpha \circ \alpha = \mathrm{id}_{\overline{G}}$. A symbol algebra is described for the C^*-algebra generated by a poly-Bergman projection of G, all multiplication operators aI $(a \in C(\overline{G}))$ and the composition operator $Wf = f \circ \alpha$.

Keywords: Bergman projection, composition operator, symbol algebra, factorization, Fredholm operator.

1. Introduction

Let G be a domain in $\mathbb{C}$ and $d\mu = dxdy$ be the usual Lebesgue measure. Define the space $\mathcal{A}_n^2(G)$ of n-analytic functions as the subspace of $L_2(G)$ consisting of all functions $f = f(z) = f(x, y)$ which satisfy the equation

$$\left(\frac{\partial}{\partial \overline{z}}\right)^n f = 0.$$

*Supported by CONACyT during the stay at the College of William & Mary when the research was conducted.
†Supported in part by NSF grant DMS 9988579.

The set $\mathcal{A}^2_n(G)$ thus defined is a closed subspace of $L_2(G)$ [3, 13], and is called the poly-Bergman space of G. The orthogonal projection

$$B_n : L_2(G) \to \mathcal{A}^2_n(G)$$

is called n-poly-Bergman projection. The true-n-analytic function space was introduced in [13] as $\mathcal{A}^2_{(n)}(G) = \mathcal{A}^2_n(G) \ominus \mathcal{A}^2_{n-1}(G)$, where by convention $\mathcal{A}^2_0(G) = \{0\}$ (that is, $\mathcal{A}^2_{(1)}(G) = \mathcal{A}^2_1(G)$). We denote by $B_{(n)}$ the (true-n-poly-Bergman) orthoprojection from $L_2(G)$ onto $\mathcal{A}^2_{(n)}(G)$. Obviously, $B_n = \sum_1^n B_{(k)}$.

The n-anti-analytic and true-n-anti-analytic function spaces $\tilde{\mathcal{A}}^2_n(G)$ and $\tilde{\mathcal{A}}^2_{(n)}(G)$ are defined along the same lines [13], with $\partial/\partial\bar{z}$ changed to $\partial/\partial z$. The corresponding orthoprojections are denoted by $\tilde{B}_n$ and $\tilde{B}_{(n)}$.

For the upper half-plane Π, the explicit integral representations of the poly-Bergman projections were obtained in [13], thus extending the well known results for the standard Bergman space $\mathcal{A}^2(\Pi) := \mathcal{A}^2_1(\Pi)$ [3]. Another representation for B_n is given by the following

Theorem 1.1 *We have*

$$B_n = I - \chi_\Pi S^n \chi_\Pi (S^*)^n \chi_\Pi I, \tag{1.1}$$

where χ_Π *is the characteristic function of* Π *and* $S = S_{\mathbb{R}^2}$ *is the so-called two-dimensional Hilbert transform*

$$(S_{\mathbb{R}^2} f)(z) = -\frac{1}{\pi} \int_{\mathbb{R}^2} \frac{1}{(z-\zeta)^2} f(\zeta)\, d\mu(\zeta).$$

The proof of Theorem 1.1 is relegated to Section 5. We point out that the representation (1.1) is not totally new, it was established (module compact operators) for bounded domains in [3].

From now on $B(\mathfrak{H})$ shall stand for the C^*-algebra of all bounded linear operators on a Hilbert space $\mathfrak{H}$, and $\mathcal{R}(\mathcal{F}; A_1, ..., A_m)$ will denote its C^*-subalgebra generated by certain $\mathcal{F} \subset B(\mathfrak{H})$ and $A_1, ..., A_m \in B(\mathfrak{H})$.

One of the main properties of $S_{\mathbb{R}^2}$ is the equality $F_2 S_{\mathbb{R}^2} F_2^* = \Phi(z) I$, where $\Phi(z) = \bar{z}/z$ and F_2 is the Fourier transform on $L_2(\mathbb{R}^2)$. The operator

$$B_{n,A} = I - \chi_\Pi S_A^n \chi_\Pi (S_A^*)^n \chi_\Pi I, \text{ where } S_A = F_2^* \Phi(A\zeta) F_2,$$

is obtained from (1.1) by means of a nonsingular $\mathbb{R}$-linear change of variable $\zeta = Az$. In Section 2 we study the algebra $\mathcal{R}_{n,A} = \mathcal{R}(\mathbb{C}I; B_n, B_{n,A})$. Actually, $B_{n,A}$ is an orthogonal projection, thus $\mathcal{R}_{n,A}$ is isomorphic to an algebra of 2×2 matrix-valued functions continuous on $\Delta = \mathrm{sp}(B_n - B_{n,A})^2$ [14]. In case $\det A > 0$, the spectrum of $B_n - B_{n,A}$ can be obtained by solving a Riemann boundary value problem and it can also be described as the set of all singular values of the Hankel operator H_σ with a certain rational symbol σ depending on n and the matrix A.

Let α be a C^2-diffeomorphism of $\overline{G}$ satisfying the Carleman condition $\alpha^2 = \mathrm{id}_G$. Introduce the unitary operator

$$(Wf)(z) = \sqrt{|\det J_\alpha(z)|}\,f(\alpha(z)),$$

where J_α denotes the Jacobian matrix of α. Combining results of Section 2 with the local techniques [9, 11], in Section 3 we describe the Calkin algebra of $\mathcal{R}_n = \mathcal{R}(C(\overline{G})I; B_n, WB_nW)$ as the algebra of all continuous sections of a C^*-bundle. Each local algebra (fiber) of $\mathcal{R}_n$ is described via the isomorphic description of $\mathcal{R}_{n,A}$. Finally, a symbol algebra for $\mathcal{R}(C(\overline{G})I; B_n, W)$ is presented in Section 4.

Note that the role of singular values of Hankel operators in factorization theory and its applications to boundary value problems was observed twenty years ago by Professor Litvinchuk and one of the authors [6, 7], see also [5, Chapter 4]. It therefore gives us a great pleasure to publish our results in this special volume.

2. Symbol algebra of $\mathcal{R}_{n,A} = \mathcal{R}(\mathbb{C}I; B_n, B_{n,A})$

We will use $\mathbb{R}_+$ ($\mathbb{R}_-$) to denote the positive (negative) half-line, and $\chi_\pm$ to denote the characteristic functions of $\mathbb{R}_\pm$, respectively. Recall that the Laguerre polynomial $L_n(y)$ of degree n and type 0 is defined by

$$L_n(y) = \frac{e^y}{n!}\frac{d^n}{dy^n}(e^{-y}y^n), \quad y \in \mathbb{R}_+,$$

and that the system of functions $\{e^{-y/2}L_n(y)\}_0^\infty$ forms an orthonormal basis in the space $L_2(\mathbb{R}_+)$. Let L_n be the (one-dimensional) subspace of $L_2(\mathbb{R}_+)$ generated by $\ell_n(y) = e^{-y/2}L_n(y)$ and let P_n be the orthogonal projection of $L_2(\mathbb{R}_+)$ onto L_n.

Introduce $U = U_2(F \otimes I)$, where $F \in B(L_2(\mathbb{R}))$ is the standard Fourier transform:

$$(Ff)(x) = \frac{1}{\sqrt{2\pi}}\int_{\mathbb{R}} f(t)e^{-itx}\,dt,$$

and $U_2 \in B(L_2(\mathbb{R}^2))$ is defined by the formula

$$(U_2f)(x,y) = \frac{1}{\sqrt{2|x|}}f(x,\, y/2|x|).$$

Of course, the operator U is unitary along with F and U_2.

The following results were established in [13]:

- $L_2(\Pi) = \bigoplus_{n=1}^\infty \mathcal{A}^2_{(n)}(\Pi) \oplus \bigoplus_{n=1}^\infty \tilde{\mathcal{A}}^2_{(n)}(\Pi)$.
- The mapping $U\colon \mathcal{A}^2_{(n)}(\Pi) \to L_2(\mathbb{R}_+) \otimes L_{n-1}$ is an isomorphism, or equivalently,

$$UB_{(n)}U^* = \chi_+ I \otimes P_{n-1}.$$

As usual, let H stand for the Hilbert transform

$$(Hf)(y) = \frac{1}{\pi i} \int_{\mathbb{R}} \frac{f(t)}{t-y}\, dt,$$

$P_\pm$ for the Szegö projections $(I \pm H)/2$, and $H^2_\pm$ for the Hardy spaces in the upper/lower half-plane. Denote also $\dot{H}^2_\pm = H^2_\pm + \mathbb{C}$.

The system of functions

$$\left\{ \varphi_k(y) = \frac{1}{\sqrt{\pi}} \frac{[\bar{\mu}\varphi(y)]^k}{i+y} \right\}_{-\infty}^{\infty}, \quad \text{with} \quad \varphi(y) = \frac{i-y}{i+y},$$

forms an orthonormal basis in $L_2(\mathbb{R})$ for each $\mu \in \mathbb{R}$ of modulus 1.

Introduce the following operators on $L_2(\Pi)$:

$$B_{n,A} = I - \chi_\Pi S_A^n \chi_\Pi (S_A^*)^n \chi_\Pi I, \quad n \geq 0,$$
$$B_{(n),A} = B_{n,A} - B_{n-1,A}, \quad n \geq 1,$$

where $S_A = (F^* \otimes F^*)\Phi(A\zeta)I(F \otimes F)$, $\Phi(z) = \overline{z}/z$ and A is a 2×2 invertible matrix with real entries. In case $A = I$ we have $S_I = S_{\mathbb{R}^2}$, i.e., $(F \otimes F)S_{\mathbb{R}^2}(F^* \otimes F^*) = \Phi(z)I$ [9].

The system of functions

$$\left\{ \psi_k(y) = \sqrt{\frac{|\operatorname{Im} \omega|}{\pi}} \frac{[\bar{\nu}\psi(y)]^k}{\omega + y} \right\}_{-\infty}^{\infty}, \quad \text{with} \quad \psi(y) = \gamma \frac{\bar{\omega}+y}{\omega+y},$$

also forms an orthonormal basis in $L_2(\mathbb{R})$, where $\gamma = b/\bar{b}$, $\omega = \bar{a}/\bar{b}$, $a = A1$, $b = Ai$ and $|\nu| = 1$.

Let $P_{\psi,k}$ be the orthogonal projection of $L_2(\mathbb{R})$ onto the space generated by ψ_k.

Introduce the operator $P_\psi = P_- - P_-[\overline{\psi(t)}]^n P_-[\psi(t)]^n P_-$. Note that $\psi^n \in \dot{H}^2_+$ if and only if $\operatorname{Im} \omega > 0$. For $\operatorname{Im} \omega > 0$, the equality $P_\psi = \sum_{k=1}^n P_{\psi,-k}$ can be easily proved by applying P_ψ to the elements of the basis $\{\psi_k\}$. We point out that P_ψ acts on H^2_-.

Let $V = U_3(F \otimes F)$, where U_3 is the unitary operator defined by $(U_3 f)(z) = |x|^{1/2} f(x, |x|y)$.

Theorem 2.1 *Both operators $B_{n,A}$ and $B_{(n),A}$ are orthogonal projections. If $\det A > 0$, then $VB_{n,A}V^* = \chi_+ I \otimes P_\psi$. If $\det A < 0$, then $VB_{n,A}V^* = \chi_- I \otimes P_{\overline{\phi}}$, where $\phi(y) = (\bar{a} - \bar{b}y)/(a - by)$.*

Proof. We shall use $\bar{T}$ to denote VTV^*, but still write $\bar{P} := V(\chi_\Pi I)V^*$. From $F\chi_+ IF^* = P_-$ and the fact that H commutes with dilations we get $\bar{P} = I \otimes P_-$. Since $\bar{S}_A = \Phi(a \operatorname{sgn}(x) + by)I$ we have $\bar{S}_A = (\chi_- I \otimes \phi(y)I) \oplus (\chi_+ I \otimes \phi(-y)I)$.

If $\det A > 0$, then $0 < \arg(-\bar{\omega}) = \arg(Ai) - \arg(A1) < \pi$. Therefore $\operatorname{Im} \omega > 0$ and $\overline{\phi} \in \dot{H}^2_-$. From $P_- \overline{\phi}^n P_- = \overline{\phi}^n P_-$ it follows that $VB_{n,A}V^* = \chi_+ I \otimes$

$(P_- - P_-[\phi(-t)]^n P_-[\overline{\phi(-t)}]^n P_-)$. Since $\psi(y) = \overline{\phi(-y)}$, we have $VB_{n,A}V^* = \chi_+ I \otimes P_\psi$. The case $\det A < 0$ can be considered similarly. □

We denote by P_{-k} the orthogonal projection of $L_2(\mathbb{R})$ onto the space E_{-k} generated by φ_{-k}.

Theorem 2.2 *The mapping* $V\colon \mathcal{A}^2_{(n)}(\Pi) \to L_2(\mathbb{R}_+) \otimes E_{-n}$ *is an isomorphism, or equivalently,* $VB_{(n)}V^* = \chi_+ I \otimes P_{-n}$. *The projection* B_n *is unitarily equivalent to* $\chi_+ I \otimes (P_{-1} + \cdots + P_{-n})$.

Proof. Follows from Theorem 2.1 by taking $A = I$. □

Let $\mathcal{R}_{n,A} = \mathcal{R}(\mathbb{C}I; B_n, B_{n,A})$ and $\mathcal{R}_{(n),A} = \mathcal{R}(\mathbb{C}I; B_{(n)}, B_{(n),A})$.

Lemma 2.3 *If* $\det A < 0$, *then* $\mathcal{R}_{n,A} \cong \mathcal{R}_{(n),A} \cong \mathbb{C}^3$. *The isomorphism* Ψ *is defined on* $\mathcal{R}_{n,A}$ *by*

$$B_n \mapsto (0,1,0), \qquad B_{n,A} \mapsto (0,0,1).$$

If $\det A > 0$, *then* $\mathcal{R}_{(n),A}$ *can be identified with a sub-algebra of* $\mathbb{C} \times M_2(\mathbb{C})$. *The identification is given by the transformation* Ψ *defined as follows:*

$$\Psi(B_{(n)}) = \left(0, \begin{pmatrix} 1 & 0 \\ 0 & 0 \end{pmatrix}\right),$$

$$\Psi(B_{(n),A}) = \left(0, \begin{pmatrix} 1-t & \sqrt{t(1-t)} \\ \sqrt{t(1-t)} & t \end{pmatrix}\right),$$

where $t = 1 - |\langle \psi_{-n}, \varphi_{-n} \rangle|^2$.

Proof. We continue using the notation $\bar{T} = VTV^*$. If $\det A < 0$, then $\bar{B}_n \pm \bar{B}_{n,A}$ is a direct sum/difference of two orthoprojections. Thus $\operatorname{sp}(B_n + B_{n,A}) = \{0,1\}$ and $\operatorname{sp}(B_n - B_{n,A}) = \{0, \pm 1\}$. The result for $\det A < 0$ is a simple application of [14].

Assume now that $\det A > 0$. By Theorem 2.1, we have $\bar{B}_{(n)} = \chi_+ I \otimes P_{-n}$ and $\bar{B}_{(n),A} = \chi_+ I \otimes P_{\psi,-n}$.

Let H_1 be the space generated by φ_{-n}, and let H_2 be the space generated by ψ_{-n}. Introduce $H_0 = H^2_- \ominus (H_1 + H_2)$. Obviously, the restrictions to H_0 of both P_{-n} and $P_{\psi,-n}$ are equal to the zero operator. Let $\tilde{\phi}_n = \psi_n - \langle \psi_n, \varphi_n \rangle \varphi_n$ and $\phi_n = \tilde{\phi}_n / \|\tilde{\phi}_n\|$. Then $\mathcal{B} = \{\varphi_{-n}, \phi_{-n}\}$ is an orthonormal basis in $H_1 + H_2$. According to the splitting $H^2_- = H_0 \oplus (H_1 + H_2)$ we have

$$P_{-n} = 0 \oplus \begin{pmatrix} 1 & 0 \\ 0 & 0 \end{pmatrix} \text{ and } P_{\psi,-n} = 0 \oplus \begin{pmatrix} 1-t & \sqrt{t(1-t)}e^{i\theta} \\ \sqrt{t(1-t)}e^{-i\theta} & t \end{pmatrix},$$

where $t = 1 - |\langle \psi_{-n}, \varphi_{-n} \rangle|^2$ and $\langle \psi_{-n}, \varphi_{-n} \rangle = |\langle \psi_{-n}, \varphi_{-n} \rangle| e^{i\theta}$. The rest of the proof is now obvious. □

Write $\varphi(-\bar{\omega})$ and $\psi(i)$ in the polar form: $\varphi(-\bar{\omega}) = r\mu$ and $\psi(i) = r\nu$, where $r \geq 0$. Using the change of variable $\mu z = \varphi(y)$ we obtain

$$\langle \varphi_{-m}, \psi_{-n} \rangle = c \int_{|z|=1} \frac{f_n(z)}{z^m}\, dz, \tag{2.1}$$

where

$$f_n(z) = \frac{(z-r)^{n-1}}{(rz-1)^n} \quad \text{and} \quad c = \frac{\sqrt{|\operatorname{Im}\omega|}}{\pi(i-\bar{\omega})}. \tag{2.2}$$

We continue with the assumption $\det A > 0$. Thus $\operatorname{Im}\omega > 0$ and $r < 1$. The latter inequality implies that $f_n(z)$ is analytic in the unit disk for each $n \geq 1$. The integral in (2.1) is just $2\pi i$ times the residue of $f_n(z)/z^m$ at $z = 0$. Therefore $\langle \varphi_{-m-1}, \psi_{-n} \rangle = 2\pi i c\, f_n^{(m)}(0)/m!$.

The spectrum of $B_{(n)} - B_{(n),A}$ can be expressed in terms of the Legendre polynomials by using the equality $\operatorname{sp}(B_{(n)} - B_{(n),A}) = \{0, \pm\sqrt{t}\}$ and the following lemma.

Lemma 2.4 *For each $n \geq 0$ we have*

$$\langle \varphi_{-n-1}, \psi_{-n-1} \rangle = 2\pi i c\, (-1)^{n+1}\, P_n(1-2r^2) \tag{2.3}$$

and

$$r^2 = 1 - 4\pi^2|c|^2 = \frac{(\|A\|_2)^2 - 2\det A}{(\|A\|_2)^2 + 2\det A}, \tag{2.4}$$

where $\|A\|_2$ is the Euclidean norm of A and $P_n(y)$ is the Legendre polynomial defined by

$$P_n(y) = \frac{1}{2^n n!}\frac{d^n}{dy^n}(y^2-1)^n.$$

Proof. Let $g(z) = (z-r)^n$ and $h(z) = 1/(rz-1)^{n+1}$. By Leibnitz formula we have

$$f_{n+1}^{(m)}(0) = \sum_{k=\max\{0,m-n\}}^{m} \binom{m}{k} \frac{n!}{(n-m+k)!}(-r)^{n-m+k} \cdot \frac{(k+n)!}{n!}(-1)^{n+1} r^k$$

$$= (-1)^{m+1}\frac{m!}{n!}\, r^{n-m}\, \frac{d^n}{dx^n}\left[x^n \sum_{k=\max\{0,m-n\}}^{m} \binom{n}{m-k} x^k \right]\Bigg|_{-r^2}.$$

For $m = n$ the expression in the brackets is simply $x^n(1+x)^n$. Using the change of variable $x = (y-1)/2$ we obtain $f_{n+1}^{(n)}(0) = n!(-1)^{n+1}P_n(1-2r^2)$. Formula (2.3) follows from $\langle \varphi_{-m-1}, \psi_{-n-1} \rangle = 2\pi i c\, f_{n+1}^{(m)}(0)/m!$. A simple computation gives (2.4). □

Lemma 2.5 *Let $\mathcal{A}$ be a C^*-algebra with the unit e. Then*

1) *If $\mathcal{B}$ is a C^*-subalgebra of $\mathcal{A}$ with unit p, then pap is invertible in $\mathcal{B}$ if and only if $ap + q$ is invertible in $\mathcal{A}$, where $q = e - p$.*

2) *An element $a = a_0 - \sum_{k=1}^{2} a_k b_k \in \mathcal{A}$ is invertible if and only if the following matrix is invertible*

$$\begin{pmatrix} e & 0 & b_1 \\ 0 & e & b_2 \\ a_1 & a_2 & a_0 \end{pmatrix}.$$

Part 1) is very well known and has been used in the theory of singular integral operators starting probably with [10]. See [4] for part 2).

Now our aim is to characterize $\Delta = \mathrm{sp}(B_{n,A} - B_n)^2$ in case $\det A > 0$. By Theorems 2.1 and 2.2 we have that $\mathrm{sp}(B_{n,A} - B_n) = \{0\} \cup \mathrm{sp}(P_\psi - P_\varphi)$. It is easy to see that $P_\psi - P_\varphi = P_{\psi_+} - P_{\varphi_+}$, where $\psi_+ = \bar{\nu}\psi$ and $\varphi_+ = \bar{\mu}\varphi$. Let $\varphi_- = \overline{\varphi_+}$ and $\psi_- = \overline{\psi_+}$.

Lemma 2.6 *For each $\lambda \neq 0$, the operator $(P_{\psi_+} - P_{\varphi_+}) - \lambda P_-$ is invertible if and only if $Q = M_\lambda P_- + P_+$ is invertible, where*

$$M_\lambda = \begin{pmatrix} 1-\lambda^2 & \varphi_+^n \psi_-^n \\ \varphi_-^n \psi_+^n & 1 \end{pmatrix}.$$

Proof. Applying part 2) and then part 1) of Lemma 2.5 to the operator $(P_{\psi_+} - P_{\varphi_+}) - \lambda P_- = -\lambda P_- + P_- \varphi_-^n P_- \varphi_+^n P_- - P_- \psi_-^n P_- \psi_+^n P_-$ we infer that it is invertible if and only if $R = \widetilde{M}_\lambda P_- + P_+$ is invertible, where

$$\widetilde{M}_\lambda = \begin{pmatrix} 1 & 0 & -\varphi_+^n \\ 0 & 1 & \psi_+^n \\ \varphi_-^n & \psi_-^n & -\lambda \end{pmatrix}.$$

The operator R is invertible if and only if all (left) partial indices of $\widetilde{M}_\lambda$ are zero (see [2] or [8]). In the representation

$$\widetilde{M}_\lambda = \begin{pmatrix} \frac{1}{\lambda} & 0 & \varphi_+^n/\lambda \\ 0 & -1-\frac{1}{\lambda} & -\psi_+^n/\lambda \\ 0 & 0 & 1 \end{pmatrix} \begin{pmatrix} 1-\lambda^2 & \varphi_+^n \psi_-^n & 0 \\ \varphi_-^n \psi_+^n & 1 & 0 \\ 0 & 0 & 1 \end{pmatrix} \begin{pmatrix} \frac{-1}{\lambda+1} & 0 & 0 \\ 0 & -1 & 0 \\ \varphi_-^n & \psi_-^n & -\lambda \end{pmatrix}$$

the left factor is invertible in $\dot{H}_+^2 \otimes M_3(\mathbb{C})$, and the right factor in $\dot{H}_-^2 \otimes M_3(\mathbb{C})$. The middle factor is just $M_\lambda \oplus 1$. Thus, R is invertible if and only if $Q = M_\lambda P_- + P_+$ is invertible. □

Let S_l and S_r be the left and right shift operators on $\mathbb{C}^n$, respectively, and let T be the operator on $\mathbb{C}^n$ defined by $T(a_1, ..., a_n) = (a_n, ..., a_1)$. We have that $S_r^* = TS_rT = S_l$ and

$$S_l = \begin{pmatrix} 0 & 1 & & 0 \\ & \ddots & \ddots & \\ & & \ddots & 1 \\ 0 & & & 0 \end{pmatrix}, \qquad T = \begin{pmatrix} 0 & & & 1 \\ & & \iddots & \\ & \iddots & & \\ 1 & & & 0 \end{pmatrix}.$$

Lemma 2.7 *$\lambda \in \operatorname{sp}(P_\psi - P_\varphi) \setminus \{0\}$ if and only if $|\lambda|$ is a singular value of $B = (I - rS_l)^{-n}(rI - S_l)^n$, or equivalently, if and only if $\det(\lambda^2 - BB^*) = 0$.*

Proof. By Lemma 2.5, the operator $Q = M_\lambda P_- + P_+$ is invertible if and only if $X - \lambda^2 P_-$ is invertible, where $X = P_- - P_-\sigma P_-\bar{\sigma}P_-$ and $\sigma = \varphi_-^n \psi_+^n$. Let $H_\sigma \colon H_+^2 \to H_-^2$ be the Hankel operator with the symbol σ. Since $X = H_\sigma H_\sigma^*$, we have $\lambda^2 \in \operatorname{sp}(X)$ if and only if $|\lambda|$ is a singular value of H_σ. With the change of variable $z = \varphi_+(y)$ we have $\psi_+(z) = (r - z)/(1 - rz)$ and

$$\sigma(z) = \sum_{k=0}^{n-1} l_k z^{k-n} + \text{a function in } H^\infty,$$

where $l_0 = r^n$ and for $1 \le k \le n-1$

$$l_k = (-1)^n \frac{(-r)^{n-k}}{(n-1)!} \frac{d^{n-1}}{dx^{n-1}} [x^{k-1}(1+x)^n]|_{x=-r^2}.$$

Identifying $H_\pm^2$ with $l^2(\mathbb{N})$, the matrix representation of H_σ on $(\ker H_\sigma)^\perp$ is the finite Hankel matrix $M_\sigma = BT$, where B is given by

$$B = \begin{pmatrix} l_0 & l_1 & \cdots & l_{n-1} \\ & \ddots & \ddots & \vdots \\ & & \ddots & l_1 \\ & & & l_0 \end{pmatrix}.$$

Finally, $B = M_\sigma T = \sum_0^\infty l_k S_l^k = \psi_+^n(S_l)$ and $X = BB^*$ on $(\ker H_\sigma)^\perp$. □

The maximal singular value of H_σ is given by

$$\lambda_n = \|H_\sigma\| = \operatorname{dist}(\sigma, H^\infty).$$

Actually, there exists a function $f_\mu \in L^\infty$ such that $H_\sigma = H_{f_\mu}$ and $\|H_\sigma\| = \|f_\mu\|_\infty$ (see [1]). For $n = 2$ we have

$$f_\mu(z) = \frac{\lambda_2}{z^2} \frac{r^2 z - \lambda_1}{r^2 - \lambda_1 z}$$

where $\lambda_k = (-1)^k r(1 - r^2) + r\sqrt{(1 - r^2)^2 + r^2}$ $(k = 1, 2)$ are the nonzero singular values of H_σ.

Lemma 2.7 characterizes $\operatorname{sp}(B_{n,A} - B_n)^2 = \{0\} \cup \operatorname{sp}(P_\psi - P_\varphi)^2$, thus from [14] we get the following

Lemma 2.8 *If* $\det A > 0$*, then* $\mathcal{R}_{n,A} = \mathcal{R}(\mathbb{C}I; B_n, B_{n,A})$ *can be identified with a subalgebra of* $\mathbb{C} \times (M_2(\mathbb{C}) \otimes \mathbb{C}^n)$*. The identification is given by the following mapping of the generators of* $\mathcal{R}_{n,A}$*:*

$$\Psi(B_n) = \left(0, \begin{pmatrix} 1 & 0 \\ 0 & 0 \end{pmatrix}\right), \quad t \in \Delta_r,$$

$$\Psi(B_{n,A}) = \left(0, \begin{pmatrix} 1-t & \sqrt{t(1-t)} \\ \sqrt{t(1-t)} & t \end{pmatrix}\right), \quad t \in \Delta_r,$$

where $\Delta_r = \{\lambda^2 : \lambda \neq 0$ is a singular value of $(I - rS_l)^{-n}(rI - S_l)^n\}$.

3. Symbol algebra of $\mathcal{R}_n = \mathcal{R}(C(\overline{G})I; B_n, WB_nW)$

Let $G \subset \mathbb{C}$ be a $(m+1)$-connected bounded domain with smooth boundary. The poly-Bergman projection on $L_2(G)$ admits the representation [3]

$$B_n = I - P_G S_n^* P_G S_n P_G + L,$$

where L is a compact operator, $P_G = \chi_G I$, and S_n is the singular integral operator defined by

$$(S_n f)(z) = \frac{(-1)^n |n|}{\pi} \int_{\mathbb{C}} \frac{e^{2in\theta}}{|\zeta - z|^2} f(\zeta)\, d\mu(\zeta), \quad \theta = \arg(\zeta - z).$$

Actually $S_{-1} = S_{\mathbb{R}^2}$ and $S_{-n} = S_{\mathbb{R}^2}^n = S_n^*$. Recall that the equality $(F \otimes F) S_{\mathbb{R}^2} (F^* \otimes F^*) = \Phi(z) I$ holds for $\Phi(z) = \overline{z}/z$. Thus, S_n is unitarily equivalent to $[\Phi(z)]^{-n} I$.

Let α be a C^2-diffeomorphism on $\overline{G}$ satisfying the Carleman condition $\alpha \circ \alpha = \mathrm{id}_G$. Introduce the unitary self-adjoint operator

$$W\varphi = \sqrt{|\det J_\alpha|}\, \varphi \circ \alpha,$$

where J_α denotes the Jacobian of α with respect to the real variables.

Let $\mathcal{R}_n = \mathcal{R}(C(\overline{G})I; B_n, WB_nW)$, and let $\mathcal{C}$ be the ideal of all compact operators on $L_2(G)$. Let also $\pi \colon \mathcal{R}_n \to \widehat{\mathcal{R}}_n$ be the natural projection of $\mathcal{R}_n$ onto $\widehat{\mathcal{R}}_n := \mathcal{R}_n/\mathcal{C}$. Since all singular integral operators are of local type [9], $Z = \pi(C(\overline{G})I)$ is a central subalgebra of $\widehat{\mathcal{R}}_n$. Let J_{ζ_0} be the maximal ideal of Z corresponding to $\zeta_0 \in \overline{G}$, and let $\pi_{\zeta_0} \colon \widehat{\mathcal{R}}_n \to \mathcal{R}_n(\zeta_0) := \widehat{\mathcal{R}}_n / I_{\zeta_0}$ be the natural projection, where $I_{\zeta_0} = \widehat{\mathcal{R}}_n \cdot J_{\zeta_0}$.

Two operators A, B are said to be *equivalent at* $\zeta_0 \in \overline{G}$ if $\inf \|P_U(A - B)P_U + K\| = 0$, where the infimum is taken over all neighborhoods U of ζ_0 and all compact operators K [9]. This equivalence relation is denoted by $A \overset{\zeta_0}{\sim} B$. If $A, B \in \mathcal{R}_n$, then $A \overset{\zeta_0}{\sim} B$ if and only if $\pi_{\zeta_0}(\pi(A)) = \pi_{\zeta_0}(\pi(B))$.

For each $\zeta_0 \in \overline{G}$, the operator WS_nW is equivalent to $S_{n,\zeta_0} = (F^* \otimes F^*)\Phi_n(\xi, \zeta_0)I(F \otimes F)$, where $\Phi_n(\xi, \zeta_0) = [\Phi(A_{\zeta_0}\xi)]^{-n}$ and $A_{\zeta_0} = (J_\alpha(\zeta_0)^*)^{-1}$ (see [9]). Therefore

$$WB_nW \overset{\zeta_0}{\sim} I - P_G S^*_{n,\zeta_0} P_G S_{n,\zeta_0} P_G.$$

If $\zeta_0 \in G$, then $P_G S_n P_G$ is equivalent to S_n at ζ_0. Thus $B_n \overset{\zeta_0}{\sim} 0$.

Lemma 3.1 *If $\zeta_0 \in G$, then $\mathcal{R}_n(\zeta_0) \cong \mathbb{C}$.*

Now we will proceed to describe the local algebra $\mathcal{R}_n(\zeta_0)$ for each $\zeta_0 \in \partial G$. Without lost of generality, we can assume that $\zeta_0 = 0$ and that ∂G is tangent to the real line at ζ_0. After using a local change of variable at 0 if necessary, we can also assume that the boundary of G near 0 is a straight line segment. Therefore

$$B_n \overset{\zeta_0}{\sim} I - P_\Pi S^*_n P_\Pi S_n P_\Pi, \quad WB_nW \overset{\zeta_0}{\sim} I - P_\Pi S^*_{n,\zeta_0} P_\Pi S_{n,\zeta_0} P_\Pi. \tag{3.1}$$

The isomorphic description of $\mathcal{R}_n(\zeta_0)$ depends on the local spectrum of $B_n \pm WB_nW$ [14]. Relations (3.1) lead to the situation considered in Section 2.

If $\det J_\alpha < 0$, then $\mathcal{R}_n(\zeta_0) \cong \mathbb{C}^3$ (see Lemma 2.3). The local images of the generators of $\mathcal{R}_n$ in $\mathcal{R}_n(\zeta_0)$ are given by

$$\pi_{\zeta_0}(\pi(a(z)I + b(z)B_n + c(z)WB_nW)) = (a(\zeta_0), a(\zeta_0) + b(\zeta_0), a(\zeta_0) + c(\zeta_0)).$$

Following the techniques of [11] and pasting together the local algebras of $\mathcal{R}_n$ we get the following

Theorem 3.2 *If $\det J_\alpha < 0$, then the Calkin algebra of $\mathcal{R}_n$ is isomorphic to $C(\overline{G}) \times C(\partial G) \times C(\partial G)$. Every operator in $\mathcal{R}_n$ has the form*

$$R = a_1(z)(I - B_n - WB_nW) + a_2(z)B_n + a_3(z)WB_nW + K, \tag{3.2}$$

where $a_k \in C(\overline{G})$ and $K \in \mathcal{C}$. Under the identification $\widehat{\mathcal{R}}_n = C(\overline{G}) \times C(\partial G) \times C(\partial G)$, the natural projection of operator (3.2) in $\widehat{\mathcal{R}}_n$ is given by $\pi(R) = (a_1, a_2|_{\partial G}, a_3|_{\partial G})$.

Remark Theorem 3.2 remains valid if we replace $\mathcal{R}_n$ by

$$\mathcal{R}_{(n)} = \mathcal{R}(C(\overline{G})I; B_{(n)}, WB_{(n)}W).$$

Suppose that $\det J_\alpha > 0$ and let $\zeta_0 \in \partial G$. By Lemma 2.3, the local algebra $\mathcal{R}_{(n)}(\zeta_0)$ is isomorphic to a subalgebra of $\mathbb{C} \times M_2(\mathbb{C})$. The images of the generators of $\mathcal{R}_{(n)}$ in $\mathcal{R}_{(n)}(\zeta_0)$are given by

$$\pi_{\zeta_0}(\pi(B_{(n)})) = \left(0, \begin{pmatrix} 1 & 0 \\ 0 & 0 \end{pmatrix}\right),$$

$$\pi_{\zeta_0}(\pi(WB_{(n)}W)) = \left(0, \begin{pmatrix} 1 - t_0 & \sqrt{t_0(1-t_0)} \\ \sqrt{t_0(1-t_0)} & t_0 \end{pmatrix}\right),$$

where $t_0 = 1 - (1 - r^2)(P_{n-1}(1 - 2r^2))^2$ and r is defined as in (2.4) but with A_{ζ_0} in place of A. It is easy to see that $r = r(A)$ is invariant under the transformation $A \mapsto (A^*)^{-1}$. Furthermore, it is well known that all Legendre polynomials map $(-1, 1)$ into itself and that $[P_n(1)]^2 = 1$. Thus, $\mathcal{R}_{(n)}(\zeta_0)$ has a two-dimensional irreducible representation only if $r \neq 0$ and P_{n-1} does not vanish at $1 - 2r^2$.

Let $\gamma(\zeta) = 1 - (1 - \beta(\zeta))[P_{n-1}(1 - 2\beta(\zeta))]^2$, where

$$\beta(\zeta) = \frac{(\|J_\alpha(\zeta)\|_2)^2 - 2\det J_\alpha(\zeta)}{(\|J_\alpha(\zeta)\|_2)^2 + 2\det J_\alpha(\zeta)}, \quad \zeta \in \partial G.$$

Denote by $\Gamma_{(n)}$ the C^*-algebra consisting of all pairs $(a(z), \sigma(\zeta)) \in C(\overline{G}) \times M_2(C(\partial G))$ with the following properties:

1) $\sigma(\zeta)$ is diagonal and $a(\zeta) = \sigma_{22}(\zeta)$ if $\beta(\zeta) = 0$.

2) $\sigma(\zeta)$ is diagonal if $P_{n-1}(1 - 2\beta(\zeta)) = 0$.

Theorem 3.3 *If* $\det J_\alpha > 0$, *then the Calkin algebra of* $\mathcal{R}_{(n)}$ *is isomorphic to* $\Gamma_{(n)}$. *Under the identification* $\widehat{\mathcal{R}}_{(n)} = \Gamma_{(n)}$, *the natural projection* $\pi\colon \mathcal{R}_{(n)} \to \widehat{\mathcal{R}}_{(n)}$ *is defined by the following mapping of the generators of* $\mathcal{R}_{(n)}$:

$$\pi(a(z)I + b(z)B_{(n)}) = \left(a(z), \left.\begin{pmatrix} a(\zeta) + b(\zeta) & 0 \\ 0 & a(\zeta) \end{pmatrix}\right|_{\partial G}\right),$$

$$\pi(WB_{(n)}W) = \left(0, \left.\begin{pmatrix} 1 - \gamma(\zeta) & \sqrt{\gamma(\zeta)(1 - \gamma(\zeta))} \\ \sqrt{\gamma(\zeta)(1 - \gamma(\zeta))} & \gamma(\zeta) \end{pmatrix}\right|_{\partial G}\right).$$

Let $\Delta_{\beta(\zeta)} = \{\lambda^2 : \ \lambda \text{ is singular value of } (I - \beta(\zeta)S_l)^{-n}(\beta(\zeta)I - S_l)^n\}$ and let

$$G_\alpha = \{(\zeta, t) : \ \zeta \in \partial G \text{ and } t \in \Delta_{\beta(\zeta)}\}.$$

Denote by Γ_n the C^*-algebra consisting of all pairs $(a(z), \sigma(\zeta, t)) \in C(\overline{G}) \times M_2(C(G_\alpha))$ with the following property: $\sigma(\zeta, t)$ is diagonal and $a(\zeta) = \sigma_{22}(\zeta, t)$ if $\beta(\zeta) = 0$.

Theorem 3.4 *If* $\det J_\alpha > 0$, *then the Calkin algebra of* $\mathcal{R}_n$ *is isomorphic to* Γ_n. *Under the identification* $\widehat{\mathcal{R}}_n = \Gamma_n$, *the natural projection* $\pi\colon \mathcal{R}_n \to \widehat{\mathcal{R}}_n$ *is defined by the following mapping of the generators of* $\mathcal{R}_n$:

$$\pi(a(z)I + b(z)B_n) = \left(a(z), \begin{pmatrix} a(\zeta) + b(\zeta) & 0 \\ 0 & a(\zeta) \end{pmatrix}\right),$$

$$\pi(WB_nW) = \left(0, \begin{pmatrix} 1 - t & \sqrt{t(1 - t)} \\ \sqrt{t(1 - t)} & t \end{pmatrix}\right).$$

4. Symbol algebra of $\mathcal{R}_{n,W} = \mathcal{R}(C(\overline{G})I; B_n, W)$

Let $C_\alpha(\overline{G})$ be the space of all functions $a \in C(\overline{G})$ such that $a \circ \alpha = a$. Let M_α denote the quotient space induced by the following equivalence relation on $\overline{G}$: $z' \sim z$ if and only if either $z' = z$ or $z' = \alpha(z)$.

Using $Z = \pi(C_\alpha(\overline{G})I) \cong C(M_\alpha)$, the construction of the local algebra $\mathcal{R}_{n,W}(z_0)$ is similar to that of $\mathcal{R}_n(z_0)$ [12]. We denote by ν_{z_0} the natural mapping of $\mathcal{R}_{n,W}$ into $\mathcal{R}_{n,W}(z_0)$. The algebra $\mathcal{R}_{n,W}(z_0)$ is a subalgebra of $M_2(\mathcal{R}_n(z_0))$ and the natural images of the generators of $\mathcal{R}_{n,W}$ in $\mathcal{R}_{n,W}(z_0)$ are given by

$$\nu_{z_0}(aI + bB_n + dW) = \begin{pmatrix} a(z_0)e + b(z_0)p_1 & d(z_0)e \\ d(\alpha(z_0))e & a(\alpha(z_0))e + b(\alpha(z_0))p_2 \end{pmatrix},$$

where e, p_1 and p_2 are the images of I, B_n and WB_nW in $\mathcal{R}_n(z_0)$, respectively [12].

Let π_{z_0} be the natural mapping of $\mathcal{R}_n$ into $\mathcal{R}_n(z_0)$. For $C = A + BW \in \mathcal{R}$, with $A, B \in \mathcal{R}_n$, the equality

$$\nu_{z_0}(C) = \begin{pmatrix} \pi_{z_0}(A) & \pi_{z_0}(B) \\ \pi_{z_0}(WBW) & \pi_{z_0}(WAW) \end{pmatrix} \tag{4.1}$$

holds.

Let Ω_1 be the C^*-subalgebra of $M_2(C(\overline{G})) \times M_2(C(\partial G))$ consisting of all pairs of the form

$$\left(\begin{pmatrix} a_1 & b_1 \\ b_1 \circ \alpha & a_1 \circ \alpha \end{pmatrix}, \begin{pmatrix} a_2 & b_2 \\ b_3 \circ \alpha & a_3 \circ \alpha \end{pmatrix} \right), \tag{4.2}$$

where $a_1, b_1 \in C(\overline{G})$ and $a_2, b_2, a_3, b_3 \in C(\partial G)$.

Theorem 4.1 *If* $\det J_\alpha < 0$, *then the Calkin algebra* $\widehat{\mathcal{R}}_{n,W} = \mathcal{R}_{n,W}/\mathcal{C}$ *is isomorphic to* Ω_1. *Under the identification* $\widehat{\mathcal{R}} = \Omega_1$, *the element* (4.2) *is the natural image of the operator*

$$C = A + BW + T$$

in Ω_1. *Here* T *is compact,*

$$A = a_1(I - B_n - WB_nW) + a_2B_n + a_3WB_nW,$$
$$B = b_1(I - B_n - WB_nW) + b_2B_n + b_3WB_nW$$

and $a_k, b_k \in C(\overline{G})$. *If* C *is Fredholm, then* $D = a_1I + b_1W$ *is invertible and*

$$D^{-1} = \frac{1}{a_1\,(a_1 \circ \alpha) - b_1\,(b_1 \circ \alpha)}(a_1 \circ \alpha I - b_1 W).$$

Proof. From (4.1) and Theorem 3.2 we get the following symbol for C:

$$\widetilde{C} = \left(\begin{pmatrix} a_1 & b_1 \\ b_1 \circ \alpha & a_1 \circ \alpha \end{pmatrix}, \begin{pmatrix} a_2 & b_2 \\ b_3 \circ \alpha & a_3 \circ \alpha \end{pmatrix}\Big|_{\partial G}, \begin{pmatrix} a_3 & b_3 \\ b_2 \circ \alpha & a_2 \circ \alpha \end{pmatrix}\Big|_{\partial G} \right).$$

Let μ be the isomorphism on $M_2(C(\partial G))$ defined by

$$\begin{pmatrix} a & b \\ c & d \end{pmatrix} \mapsto \begin{pmatrix} d \circ \alpha & c \circ \alpha \\ b \circ \alpha & a \circ \alpha \end{pmatrix}.$$

The application of $I \oplus I \oplus \mu$ to $\widetilde{C}$ shows that $\widehat{\mathcal{R}} \cong \Omega_1$. A simple computation proves the last part of the theorem. □

Remark Theorem 4.1 remains valid if we replace $\mathcal{R}_{n,W}$ by
$\mathcal{R}_{(n),W} = \mathcal{R}(C(\overline{G})I; B_{(n)}, W)$.

Let F_α denote the set of all fixed points of α. For $\det J_\alpha > 0$, let Ω be the C^*-algebra of all pairs $\sigma = (\sigma', \sigma'') \in M_2(C(\overline{G})) \times M_4(C(G_\alpha))$ with the followings properties:

1) $\sigma'(\zeta)$ is diagonal if $\zeta \in F_\alpha$,

2) $\sigma''(x, 0)$ is diagonal if $(x, 0) \in G_\alpha$ and $x \in F_\alpha$,

3) $\sigma''(x, t)$ is 2×2 block diagonal if $(x, t) \in G_\alpha \bigcap (\partial G \times \{0, 1\})$,

4) if $\zeta = x \in \partial G$ and $(x, 0) \in G_\alpha$, then

$$\sigma'(\zeta) = \begin{pmatrix} \sigma''_{33}(x,0) & \sigma''_{34}(x,0) \\ \sigma''_{43}(x,0) & \sigma''_{44}(x,0) \end{pmatrix},$$

5) $(\sigma'(\zeta), \sigma''(x, t))$ and $(\sigma'(\alpha(\zeta)), \sigma''(\alpha(x), t))$ are related as follows

$$(\sigma'(\zeta), \sigma''(x, t)) = N(\sigma'(\alpha(\zeta)), \sigma''(\alpha(x), t))N^*,$$

where $N = (N', N'')$, $N'(\zeta) = \operatorname{diag}\{-1, 1\}$ and

$$N''(x,t) = \begin{pmatrix} \sqrt{1-t} & 0 & i\sqrt{t} & 0 \\ 0 & -\sqrt{1-t} & 0 & -i\sqrt{t} \\ -i\sqrt{t} & 0 & -\sqrt{1-t} & 0 \\ 0 & i\sqrt{t} & 0 & \sqrt{1-t} \end{pmatrix}.$$

As a simple application of [12] we get the following

Theorem 4.2 *The Calkin algebra* $\operatorname{Sym} \mathcal{R}_{n,W} = \mathcal{R}_{n,W}/\mathcal{C}$ *is isomorphic to* Ω. *Under the identification* $\operatorname{Sym} \mathcal{R}_{n,W} = \Omega$, *the natural homomorphism*

$$\operatorname{sym} : \mathcal{R}_{n,W} \to \operatorname{Sym} \mathcal{R}_{n,W}$$

is defined by the following mapping of the generators of $\mathcal{R}_{n,W}$*:*

$$\operatorname{sym} a(\zeta)I = (\sigma'_a, \sigma''_a), \qquad \operatorname{sym} K = (0_{2\times 2}, \sigma''_K), \qquad \operatorname{sym} W = (\sigma'_W, \sigma''_W),$$

where

$$\sigma'_a(\zeta) = \begin{pmatrix} a_+(\zeta) & a_-(\zeta) \\ a_-(\zeta) & a_+(\zeta) \end{pmatrix}, \ \sigma''_a(x,t) = \begin{pmatrix} a_+(x) & a_-(x) & 0 & 0 \\ a_-(x) & a_+(x) & 0 & 0 \\ 0 & 0 & a_+(x) & a_-(x) \\ 0 & 0 & a_-(x) & a_+(x) \end{pmatrix},$$

with $2a_\pm = a \pm a \circ \alpha$,

$$\sigma'_W(\zeta) = \operatorname{diag}\{1, -1\}, \qquad \sigma''_W(x,t) = \operatorname{diag}\{1, -1, 1, -1\},$$

and

$$\sigma''_K(x,t) = \frac{1}{2}\begin{pmatrix} 2-t & t & i\sqrt{t(1-t)} & -i\sqrt{t(1-t)} \\ t & 2-t & -i\sqrt{t(1-t)} & i\sqrt{t(1-t)} \\ -i\sqrt{t(1-t)} & i\sqrt{t(1-t)} & t & -t \\ i\sqrt{t(1-t)} & -i\sqrt{t(1-t)} & -t & t \end{pmatrix}.$$

Corollary 4.3 *An operator* $A \in \mathcal{R}$ *is Fredholm if and only if its symbol is invertible.*

5. Proof of Theorem 1.1

The following Fourier transforms of some simple functions are needed. For $a > 0$ and $\zeta \in \mathbb{C}$, with $\operatorname{Re} \zeta > 0$, we have

$$F^*\left(\frac{k!}{(\zeta \pm it)^{k+1}}\right) = \sqrt{2\pi}\chi_\pm(y)|y|^k e^{-\zeta|y|}. \tag{5.1}$$

For each non-negative integer k and $x \in \mathbb{R}$, consider the function $\Psi_{x,k}(y) = \Psi_k(x + iy)$, where

$$\Psi_k(z) = \frac{k!(2x)^{k+1}}{(x+iy)^{k+1}}, \qquad z = x + iy.$$

Lemma 5.1 *The operator* $Q_{x,k} = F^*[\Psi_{x,k}(t)I]F$*, acting on* $L_2(\mathbb{R})$*, admits the following integral representation:*

$$(Q_{x,k}f)(y) = |x| \int_{\mathbb{R}} f(t)e^{-|x||y-t|}(2|x||y-t|)^k[1 + \operatorname{sgn}(x)\operatorname{sgn}(y-t)]\,dt. \tag{5.2}$$

Proof. For a fixed x, let T_x be the operator defined in the right-hand side of (5.2). From (5.1) and the convolution property $F(f * g) = \sqrt{2\pi}F(f)F(g)$ it follows that $FT_xF^* = \Psi_{x,k}I$. □

Introduce $Q_k = (F^* \otimes F^*)\Psi_k(\zeta)I(F \otimes F)$. Consider the polynomial $q(y) = a_0 + a_1 y + \cdots + a_n y^n$ and the operator

$$Q = \sum_{k=0}^{n} a_k Q_k.$$

Lemma 5.2 *The operator UQU^* admits the following integral representation:*

$$(UQU^* f)(z) = \frac{1}{2}\int_{\mathbb{R}} f(x,t)e^{-|y-t|/2}q(|y-t|)[1+\operatorname{sgn}(x)\operatorname{sgn}(y-t)]\,dt. \quad (5.3)$$

Proof. It suffices to consider $Q = Q_k$. It is easy to see that $((F \otimes I)Q_k(F^* \otimes I)f)(z)$ is just the right-hand side of (5.2). The integral representation (5.3) is obtained by applying the change of variable $\tau = 2|x|t$ in the integral representation of $U_2(F \otimes I)Q_k(F^* \otimes I)U_2^*$. □

Since $(F \otimes F)S_{\mathbb{R}^2}(F^* \otimes F^*) = \Phi(z)I = (\overline{z}/z)I$, we have $S_{\mathbb{R}^2}^n = (F^* \otimes F^*)(\Phi(\zeta))^n I(F \otimes F)$. From the binomial expansion of $(\Phi(z))^n = (-1+\Psi_0(z))^n$ and Lemma 5.2 it follows that

$$\begin{aligned}&(US_{\mathbb{R}^2}^n U^* f)(y)\\ &= (-1)^n f(z) - \frac{(-1)^n}{2}\int_{\mathbb{R}} f(x,t)e^{-|y-t|/2}q_n(|y-t|)[1+\operatorname{sgn}(x)\operatorname{sgn}(y-t)]\,dt,\end{aligned}$$

where

$$q_n(y) = \sum_{k=0}^{n-1}\binom{n}{k+1}\frac{(-y)^k}{k!}.$$

Note that $US_{\mathbb{R}^2}^n U^* = (\chi_- I \otimes V_n) \oplus (\chi_+ I \otimes V_n^*)$, where

$$(V_n\varphi)(y) = (-1)^n\varphi(y) - (-1)^n\int_{\mathbb{R}}\varphi(t)e^{-|y-t|/2}q_n(|y-t|)\chi_-(y-t)\,dt.$$

The operator V_n is unitary, and $\chi_+ V_n^* \chi_+ I = V_n^* \chi_+ I$. These simple properties of V_n together with $U\chi_\Pi U^* = I \otimes \chi_+ I$ imply

$$UD_nU^* = \chi_+ I \otimes (\chi_+ I - W_n),$$

where

$$W_n = \chi_+ V_n^* \chi_+ V_n \chi_+ I \text{ and } D_n = \chi_\Pi I - \chi_\Pi S_{\mathbb{R}^2}^n \chi_\Pi (S_{\mathbb{R}^2}^*)^n \chi_\Pi I.$$

A straightforward computation shows that

$$(W_n\varphi)(y) = \varphi(y) - \int_{\mathbb{R}_+}\varphi(t)e^{-|y-t|/2}q_n(|y-t|)\,dt + \int_{\mathbb{R}_+}\varphi(t)K_n(y,t)\,dt, \quad (5.4)$$

where

$$K_n(y,t) = e^{-|y-t|/2} \int_0^{\min\{t,y\}} e^{-\tau} q_n(\tau) q_n(\tau+a)\, d\tau, \qquad a = |y-t|.$$

From $q_n(y) = -L_n'(y)$ and $L_m(y) = L_m'(y) - L_{m+1}'(y)$ it follows that $q_n(y) = \sum_{m=0}^{n-1} L_m(y)$. Therefore, $q_{n+1}(y) = q_n(y) + L_n(y)$ and

$$q_{n+1}(\tau)q_{n+1}(\tau+a) = q_n(\tau)q_n(\tau+a) + L_n(\tau)L_n(\tau+a) - \frac{d}{d\tau}[L_n(\tau)L_n(\tau+a)].$$

Thus

$$\begin{aligned} &K_{n+1}(y,t) - K_n(y,t) \\ &= e^{|y-t|/2} \int_0^{\min\{t,y\}} e^{-\tau} \left(L_n(\tau)L_n(\tau+a) - \frac{d}{d\tau}[L_n(\tau)L_n(\tau+a)] \right) d\tau \\ &= -e^{-(y+t)/2} L_n(y)L_n(t) + e^{-|y-t|/2} L_n(|y-t|). \end{aligned}$$

This relation together with (5.4) leads to $W_{n+1} - W_n = -P_n$. Consequently, $U(D_{n+1} - D_n)U^* = \chi_+ I \otimes P_n = B_{(n+1)}$. The equality $B_n = \sum_1^n B_{(k)}$ completes the proof.

References

[1] Adamjan, V. M., Arov, D. Z., and Krein, M. G., Infinite Hankel matrices and generalized problems of Carathéodory-Fejér and F. Riesz, *Funkcional. Anal. i Prilozhen.*, **2** (1968), no. 1, 1–19 (in Russian). English translation: *J. Funct. Anal. and Appl.*, **2** (1968), 1–18.

[2] Clancey, K. and Gohberg, I., *Factorization of Matrix Functions and Singular Integral Operators*, Operator Theory: Advances and Applications, **3**. Birkhäuser Verlag, Basel, 1981.

[3] Dzhuraev, A., *Methods of Singular Integral Equations*, Pitman Monographs and Surveys in Pure and Applied Mathematics, **60**. Longman Scientific & Technical, Harlow, 1992.

[4] Krupnik, N. Y., *Banach Algebras with Symbol and Singular Integral Operators*, Operator Theory: Advances and Applications, **26**. Birkhäuser, Basel, 1987.

[5] Litvinchuk, G. S., *Solvability Theory of Boundary Value Problems and Singular Integral Equations with Shift*, Mathematics and its Applications, **523**. Kluwer Academic Publishers, Dordrecht, 2000.

[6] Litvinchuk, G. S. and Spitkovsky, I. M., Exact estimates of defect numbers of the generalized Riemann boundary value problem, *Dokl. Akad. Nauk SSSR*, **255** (1980), no. 5, 1042–1046 (in Russian). English translation: *Soviet Math. Dokl.*, **22** (1980), no. 3, 781–785.

[7] Litvinchuk, G. S. and Spitkovsky, I. M., Sharp estimates of the defect numbers of a generalized Riemann boundary value problem, factorization of Hermitian matrix-functions, and some problems on approximation of meromorphic functions, *Mat. Sb. (N.S.)*, **117(159)** (1982), no. 2, 196–215 (in Russian). English translation: *Math. USSR Sbornik*, **45** (1983), no. 2, 205–224.

[8] Litvinchuk, G. S. and Spitkovskii, I. M., *Factorization of Measurable Matrix Functions*, Operator Theory: Advances and Applications, **25**. Birkhäuser Verlag, Basel, 1987.

[9] Mikhlin, S. G. and Prössdorf, S., *Singular Integral Operators.* Springer-Verlag, Berlin, 1986.

[10] Simonenko, I. B., The Riemann boundary value problem for n pairs of functions with measurable coefficients and its application to the investigation of singular integrals in the spaces L^p with weight, *Izv. Akad. Nauk SSSR, Ser. Mat.*, **28** (1964), 277–306 (in Russian).

[11] Varela, J., Duality of C^*-algebras. In: *Recent advances in the representation theory of rings and C^*-algebras by continuous sections* (Sem., Tulane Univ., New Orleans, La., 1973), Memoirs AMS, **148**, 97–108. AMS, Providence, R.I., 1974.

[12] Vasilevski, N. L., On an algebra generated by abstract singular operators and a shift operator, *Math. Nachr.*, **162** (1993), 89–108.

[13] Vasilevski, N. L., On the structure of Bergman and poly-Bergman spaces, *Integr. Equat. Oper. Theor.*, **33** (1999), no. 4, 471–488.

[14] Vasilevski, N. and Spitkovsky, I., On the algebra generated by two projections, *Dokl. Akad. Nauk Ukrain. SSR, Ser. A*, **8** (1981), no. 8, 10–13 (in Russian).

HOW TO COMPUTE THE PARTIAL INDICES OF A REGULAR AND SMOOTH MATRIX-VALUED FUNCTION?

Bernd Silbermann
Faculty of Mathematics, University of Technology Chemnitz, Germany
Bernd.Silbermann@mathematik.tu-chemnitz.de

Dedicated to G. S. Litvinchuk on the occasion of his seventieth birthday

Abstract This paper is aimed at the stable computation of the partial indices of regular and smooth matrix functions defined on the complex unit circle under special emphasis on the speed of convergence. A crucial role plays the k-splitting property of appropriately constructed block matrices, namely modified finite sections A_n of Toeplitz operators. It is proved that the singular values $s_k(A_n)$ tend with high speed to zero as $n \to \infty$ for smooth regular functions where k stands for the splitting number.

Keywords: partial indices, Toeplitz operators, singular values.

1. Introduction

Let L^∞ stand for the Banach algebra of all essentially bounded measurable functions $a : \mathbb{T} \to \mathbb{C}$ defined on the unit circle $\mathbb{T} = \{t \in \mathbb{C} : |t| = 1\}$. Given a set A, we write $A_{N \times N}$ for the collection of all $N \times N$ matrices with entries in A. We will especially deal with the algebra C of all complex valued and continuous functions defined on $\mathbb{T}$ and its subalgebras constituted by smooth functions.

A function $a \in C_{N \times N}$ is called regular if

$$\det a(t) \neq 0 \text{ for all } t \in \mathbb{T}.$$

It is well-known that a regular function $a \in C_{N \times N}$ admits a (generalized) Wiener-Hopf factorization, that is there exist integers (called the partial indices)

$$\kappa_1 \leq \kappa_2 \leq \cdots \leq \kappa_N \qquad (1.1)$$

and matrix functions a_- and a_+ such that

$$a(t) = a_-(t) \text{ diag } (t^{\kappa_1}, \dots, t^{\kappa_N})a_+(t)$$

with

$$a_+^{\pm 1} \in H^2_{N\times N}\, , a_-^{\pm 1} \in \overline{H}^2_{N\times N}\, ,$$

where H^2 means the familiar Hardy space. The integers (1.1) are uniquely determined ([7]). Our main aim is to compute them in a stable manner.

This problem is involved since the partial indices (1.1) are stable under small perturbations if and only if

$$\kappa_N - \kappa_1 \leq 1$$

(see [10]). Notice that there is an immense number of papers dealing with the factorization problem (see for instance the books [5] and [10]). But even for rational matrix functions the determination of the partial indices is highly complicated. The interesting paper [1] gives a theoretical answer for such functions by means of the state-space method.

It is well-known that the partial indices are known if the kernel dimensions of related Toeplitz operators are available (Section 2). In [12] there was proposed a method to compute the kernel dimension of a Fredholm Toeplitz operator with a smooth matrix-valued generating function. This method is based on the observation that the behaviour of the singular values of modified finite sections of the Toeplitz operator is subject to the so-called k-splitting property (see Sections 2 and 3). We shall study this property deeper, namely we show in Section 4 that the sequence $(s_k(A_n))$ of the kth singular values of the modified finite sections A_n converge with high speed to zero for n tending to infinity if the generating function is smooth. It is also shown that the proposed algorithm is numerically stable under small perturbations.

2. Toeplitz operators and their finite sections

We denote by l^2_N the space of all sequences $x : \mathbb{Z}_+ \longrightarrow \mathbb{C}^N$ such that

$$||x|| := \left(\sum_{n=0}^{\infty} ||x_n||^2 \right)^{\frac{1}{2}} < \infty \, ,$$

where $||x_n||^2 = ||(x_n^{(1)}, \dots, x_n^{(N)})||^2 = |x_n^{(1)}|^2 + \dots + |x_n^{(N)}|^2$.

If $a_k \in \mathbb{C}_{N\times N}$ denote the matrix Fourier coefficients ($k \in \mathbb{Z}$) of a function $a \in L^\infty_{N\times N}$, then the block Toeplitz matrix

$$A = \begin{pmatrix} a_0 & a_{-1} & a_{-2} & \cdots \\ a_1 & a_0 & a_{-1} & \cdots \\ a_2 & a_1 & a_0 & \cdots \\ \cdots & \cdots & \cdots & \cdots \end{pmatrix}$$

induces a bounded linear operator on l^2_N. This operator is called Toeplitz operator with generating function a and will be denoted by $T(a)$. If $a \in C_{N\times N}$ is regular, then the following theorem connects the Toeplitz operator $T(a)$ with the partial indices (1.1).

Theorem 2.1 **([6])** *For $a \in C_{N\times N}$ the Toeplitz operator $T(a) : l^2_N \longrightarrow l^2_N$ is Fredholm if and only if a is regular. If a is regular then* $\dim\ker T(a) = \sum\limits_{\kappa_j<0} |\kappa_j|$ *and*

$$Ind\ T(a) = -(\kappa_1 + \cdots + \kappa_N)\,.$$

Here $\operatorname{Ind} T(a)$ stands for the Fredholm index of $T(a)$. Clearly, $T(a)$ is invertible if and only if all its partial indices are zero. Theorem 2.1 shows that the knowledge of the partial indices allows to compute the kernel dimension of $T(a)$. Conversely, if the kernel dimensions of $T(a \operatorname{diag}(t^r,\ldots,t^r))$, $r \in \mathbb{Z}$, are known, then the partial indices can easily be computed. We shall see, that the kernel dimension of $T(a)$ with regular a can be stably computed, although the kernel dimension of a Fredholm operator is unstable in general (however it is upper semi-continuous). Recall that this kernel dimension is reflected in the singular value behaviour of the modified finite sections of the Toeplitz operator $T(a)$ [12]. For, let us introduce operators $P_n, W_n : l^2_N \longrightarrow l^2_N$ by

$$\begin{array}{rcl}(x_0,x_1,\ldots,x_{n-1},x_n,\ldots) & \mapsto & (x_0,\ldots,x_{n-1},0,0,\ldots)\,,\\ (x_0,x_1,\ldots,x_{n-1},x_n,\ldots) & \mapsto & (x_{n-1},x_{n-2},\ldots,x_0,0,0,\ldots)\,,\end{array}$$

respectively and observe that for $T_n(a) := P_nT(a)P_n$ the relation

$$W_nT_n(a)W_n = T_n(\tilde{a})$$

holds, where $\tilde{a}(t) = a\left(\frac{1}{t}\right)$. Moreover, for compact operators $K : l^2_N \longrightarrow l^2_N$ the sequence (W_nKW_n) tends strongly to zero. In what follows we identify $T_n(a)$ with the $nN \times nN$ matrix $(a_{i-j})_{i,j=1}^{n-1}$ which is also called the nth *finite section* of $T(a)$.

Introduce the collection $\mathcal{F}$ of all sequences $(A_n) = (A_n)_{n=1}^{\infty}$ of operators (matrices) $A_n :\ \operatorname{Im} P_n \longrightarrow \operatorname{Im} P_n$ ($\operatorname{Im} P_n \cong \mathbb{C}^{nN\times nN}$) for which

$$||(A_n)||_{\mathcal{F}} := \sup_{n\geq 1} ||A_n|| < \infty\,.$$

This set actually forms a C^*-algebra with the algebraic operations

$$(A_n) + (B_n) = (A_n + B_n), (A_n)(B_n) = (A_nB_n), (A_n)^* = (A_n^*)\,.$$

Given a matrix $A \in \mathbb{C}^{nN\times nN}$ we denote the (ordered) singular values of A by $s_1(A),\ldots,s_{nN}(A)$. Thus we have

$$0 \leq s_1(A) \leq \cdots \leq s_{nN}(A) = ||A||\,.$$

For the sake of convenience, let us also put $s_0(A) = 0$. The singular values $s_k(A)$, defined as square roots of the spectrum points of A^*A, can also be defined as approximation numbers: For $j \in \{0, 1, \ldots, nN\}$, let $\mathcal{F}_j^{(n)}$ denote the collection of all $nN \times nN$ matrices of rank at most j, and define the jth approximation number of a matrix $A \in \mathbb{C}^{nN \times nN}$ as

$$\text{dist}\,(A, \mathcal{F}_j^{(n)}) := \inf\left\{||A - F_j|| : F_j \in \mathcal{F}_j^{(n)}\right\}.$$

The following theorem is crucial in Section 4.

Theorem 2.2 *If $A \in \mathbb{C}^{nN \times nN}$ then*

$$s_k(A) = \;\textit{dist}\,(A, \mathcal{F}_{nN-k}^{(n)}),$$

$k = 0, 1, \ldots, nN$.

For a proof of this well-known result we refer to [8].

Definition 2.3 A sequence (A_n) of $nN \times nN$ matrices A_n is said to have the k-*splitting* property if

$$\lim_{n \to \infty} s_k(A_n) = 0 \text{ and } \lim_{n \to \infty} s_{k+1}(A_n) > 0.$$

The number k is called the *splitting number.*

For rectangular matrices the k-splitting property is defined analogously. It is known that many interesting sequences meet the k-splitting property ([11] or [9]). The following situation is of special interest.

Let $\mathcal{A} \subset \mathcal{F}$ be the smallest C^*-subalgebra generated by all sequences $(T_n(a))$, $a \in C_{N \times N}$. One can prove [4] that any sequence $(B_n) \in \mathcal{F}$ of the form

$$B_n = P_n K P_n \text{ or } B_n = W_n K W_n$$

with $K : l_N^2 \longrightarrow l_N^2$ being compact belongs to $\mathcal{A}$. Moreover, any sequence $(B_n) \in \mathcal{F}$ with $||B_n|| \longrightarrow 0$ belongs also to $\mathcal{A}$.

For each $(A_n) \in \mathcal{A}$ there exist the strong limits

$$\begin{aligned} s - \lim A_n &=: \mathcal{W}_1(A_n), \\ s - \lim W_n A W_n &=: \mathcal{W}_2(A_n). \end{aligned}$$

Using these notations we have

$$\begin{aligned} s - \lim T_n(a) &= T(a), \\ s - \lim W_n T_n(a) W_n &= T(\tilde{a}), \\ s - \lim W_n K W_n &= 0 \text{ if } K \text{ is compact}. \end{aligned}$$

Now we are ready to formulate the related result.

Theorem 2.4 *Let $(A_n) \in \mathcal{A}$ and $\mathcal{W}_1(A_n)$ be Fredholm. Then $\mathcal{W}_2(A_n)$ is also Fredholm and (A_n) is subject to the k-splitting property with*

$$k = \dim\ker \mathcal{W}_1(A_n) + \dim\ker \mathcal{W}_2(A_n)\,.$$

Especially, if $a \in C_{N\times N}$ is regular, then $(T_n(a))$ meets the k-splitting property with

$$k = \dim\ker T(a) + \dim\ker T(\tilde{a})\,.$$

A full proof can be found in [11]. Because (as a rule) the computation of dim ker $T(\tilde{a})$ is equally difficult as the computation of dim ker $T(a)$, the last relation is not very helpful. But it suggests that one has to modify the finite sections in such a way that the occurring kernel dimensions can be controlled.

3. Modified finite sections

Here we will briefly present a result which is taken from [12]. Given $a \in C_{N\times N}$ and $r \in \{0, 1, \ldots\}$, form the modified finite sections

$$T_{n,r}(a) = (a_{i-j})\ , \ \ 0 \le i \le n-1, 0 \le j \le n-r-1\,.$$

These matrices can be extended to $nN \times nN$ matrices by filling in zeros in the remaining places. The resulting matrices will be denoted by $\tilde{T}_{n,r}(a)$. Thus, $\tilde{T}_{n,0}(a) = T_n(a)$. It is easy to see that

$$\tilde{T}_{n,r}(a) = T_n(a \text{ diag } (t^{-r}, \ldots, t^{-r}))T_n(\text{diag } (t^r, \ldots, t^r))\,.$$

Thus, the sequence $(\tilde{T}_{n,r}(a))$ belongs to $\mathcal{A}$ and is subject to Theorem 2.4 for a regular function a (recall again that r is fixed).

Theorem 3.1 *For a regular function a the sequence $(\tilde{T}_{n,r}(a))$ meets the k-splitting property with*

$$k = k(r) = \dim\ker T(a) + \dim\ker C_r(a)\,,$$

where $C_r(a) := T(\tilde{a}$ diag $(t^r, \ldots, t^r))T($diag $(t^{-r}, \ldots, t^{-r}))$.

If r is large enough then the kernel dimension of the operator $C_r(a)$ equals Nr. It is easy to see that such an r is achieved if

$$\dim\ker C_{r+1}(a) - \dim\ker C_r(a) = N\,.$$

These results can be reformulated in terms of the sequence $(T_{n,r}(a))$; in what follows the singular values of $T_{n,r}(a)$ are computed via $T^*_{n,r}(a)T_{n,r}(a)$.

Corollary 3.2 *If a is regular then $(T_{n,r}(a))$ is subject to the k-splitting property with*

$$k = k(r) = \dim\ker T(a)$$

for r large enough. A correct r is found if $k(r+1) = k(r)$.

It was proved in [12] that this process is stable under small perturbations by sequences which belong to $\mathcal{A}$. The proof is mainly based on Theorems 6.27 and 7.12 in [9]. A second and more simple proof follows directly from Theorem 2.2. Computations have shown that for smooth and regular functions $a \in C_{N\times N}$ the k-splitting property can clearly be observed by help of the computer (see [12]). Moreover the computations gave raise to the conjecture that the singular value $s_k(T_{n,r}(a))$ (with k being the splitting number) tends to zero with high speed if a is sufficiently smooth. This conjecture is supported by a result of A. Böttcher [2]. He gave an alternative proof of the splitting property of the sequence $(T_n(a))$ with continuous a in the case $N = 1$. His proof also yields estimates for the of convergence speed of $s_k(T_n(a))$ to zero (compare [4], Corollary 4.9 or [3], Theorem 5.4). However, Böttcher's proof seems to be limited to the case $N = 1$. We shall show in the next section (and this is the main result of the paper) that estimates for the speed of convergence of $s_k(\tilde{T}_{n,r}(a))$ to zero can also be obtained in this much more general case.

4. Speed of convergence

We start with some preparations. Notice first of all that $\mathcal{A}$ actually forms a standard model (algebra) in the sense of [9], Section 6.1. Moreover, for regular $a \in C_{N\times N}$ the sequence $(\tilde{T}_{n,r}(a)) \in \mathcal{A}$ turns out to be a Fredholm sequence [9], Chapter 6). Thus the theory developed in [9] (or in [11]) applies. This means in our concrete situation that there is a sequence $(\Pi_n) \in \mathcal{A}$ constituted by projections such that $||\tilde{T}^*_{n,r}(a)\tilde{T}_{n,r}(a)\Pi_n|| \longrightarrow 0$ (Theorem 2.22 in [9]) and $\dim \operatorname{Im} \Pi_n = \dim\ker T(a) + \dim\ker C_r(a)$ for n large enough. (Theorem 6.11 in [9]).

Consider $B_n = \tilde{T}^*_{n,r}(a)\tilde{T}_{n,r}(a)(P_n - \Pi_n) + P_n P_{\ker T(a)} P_n + W_n P_{\ker C_r(a)} W_n \in \mathcal{A}$, where P_X stands for the orthoprojection onto the closed subspace $X \subset l^2_N$. Obviously, for regular $a \in C_{N\times N}$ the strong limits

$$\begin{aligned} \mathcal{W}_1(B_n) &= T^*(a)T(a) + P_{\ker T(a)} \\ \mathcal{W}_2(B_n) &= C^*_r(a)C_r(a) + P_{\ker C_r(a)} \end{aligned}$$

are invertible, whence it follows that the sequence B_n is stable, that is $B_n : \operatorname{im} P_n \longrightarrow \operatorname{im} P_n$ is invertible for n large enough and $||B_n^{-1}P_n|| \leq C < \infty$ (Theorem 6.14 in [4]).

Thus, we have for n large enough, say for $n \geq n_0$, that

$$\begin{aligned} \tilde{T}_{n,r}(a) = \tilde{T}_{n,r}(a)B_nB_n^{-1} = \tilde{T}_{n,r}(a)\tilde{T}^*_{n,r}(a)\tilde{T}_{n,r}(a)(P_n - \Pi_n)B_n^{-1} + \\ + \tilde{T}_{n,r}(a)(P_nP_{\ker T(a)}P_n + W_nP_{\ker C_r(a)}W_n)B_n^{-1}\,. \end{aligned}$$

The first term on the right-hand side of the last expression belongs to $\mathcal{F}^{(n)}_{nN-k}$, $k = \dim\ker T(a) + \dim\ker C_r(a)$.

Thus, by Theorem 2.2,

$$\begin{aligned} s_k(\tilde{T}_{n,r}(a)) &\leq C||\tilde{T}_{n,r}(a)(P_n P_{\ker T(a)} P_n + W_n P_{\ker C_r(a)} W_n)|| \\ &\leq C||\tilde{T}_{n,r}(a) P_n P_{\ker T(a)} P_n|| + C||P_n C_r(a) P_n P_{\ker C_r(a)} P_n||, \end{aligned} \tag{4.1}$$

for $n \geq n_0$. Here we have used that

$$\begin{aligned} W_n \tilde{T}_{n,r}(a) W_n &= P_n T(\tilde{a} \text{ diag } (t^r, \ldots, t^r)) P_n T(\text{diag } (t^{-r}, \ldots, t^{-r})) P_n \\ &= P_n C_r(a) P_n \end{aligned}$$

because

$$P_n T(\text{diag } (t^{-r}, \ldots, t^{-r})) P_n = T(\text{diag } (t^{-r}, \ldots, t^{-r})) P_n, W_n^2 = P_n,$$
and $||W_n|| = 1$.

Thus, we have to estimate the expressions in the right-hand side of (4.1). Writing $P_n T(a) P_{n-r}$ instead of $\tilde{T}_{n,r}(a)$, we get

$$\tilde{T}_{n,r}(a) P_n P_{\ker T(a)} = P_n T(a) P_{n-r} P_{\ker T(a)} P_n = -P_n T(a) Q_{n-r} P_{\ker T(a)} P_n$$

because $T(a) P_{\ker T(a)} = 0$; where Q_{n-r} denotes $I - P_{n-r}$.

Analogously, $P_n C_r(a) P_n P_{\ker C_r(a)} P_n = -P_n C_r(a) Q_n P_{\ker C_r(a)} P_n$. These relations show that we have to estimate the expressions

$$||Q_{n-r} P_{\ker T(a)}|| \text{ and } ||Q_n P_{\ker C_r(a)}||.$$

Now we can use that ker $T(a)$ and ker $C_r(a)$ are finite dimensional spaces. Let $\rho_1, \ldots, \rho_e$ be an orthonormal basis of $\ker T(a)$. Then any element $Q_{n-r} P_{\ker T(a)} x$ can be written as

$$Q_{n-r}(\lambda_1 \rho_1 + \cdots + \lambda_e \rho_e) = \lambda_1 Q_{n-r} \rho_1 + \cdots + \lambda_e Q_{n-r} \rho_e$$

with $|\lambda_1|^2 + \cdots + |\lambda_e|^2 \leq ||x||^2$ because $||Q_{n-r} P_{\ker T(a)}|| \leq 1$. Now it follows that

$$||Q_{n-r} P_{\ker T(a)} x|| \leq \max_{1 \leq j \leq e} \sqrt{N} ||Q_{n-r} \rho_j|| \, ||x||.$$

In the same fashion one proves that

$$||Q_n P_{\ker C_r(a)} x|| \leq \max_{1 \leq k \leq m} \sqrt{N} ||Q_n \psi_k|| \, ||x||,$$

where $\psi_1, \ldots, \psi_m$ is some orthonormal basis of ker $C_r(a)$. Thus, we have proved the following theorem

Theorem 4.1 *If $a \in C_{N \times N}$ is regular then the singular values $s_k(\tilde{T}_{n,r}(a))$ tend to zero for $k = \dim \ker T(a) + \dim \ker C_r(a)$ and*

$$s_k(\tilde{T}_{n,r}(a)) \leq C_1 \max\{||Q_{n-r}\rho_1||, \ldots, ||Q_{n-r}\rho_e||, ||Q_n \psi_1||, \ldots, ||Q_n \psi_m||\},$$

where C_1 is some constant not depending on n.

This inequality can be used to estimate the convergence speed. Here are two sample results.

Corollary 4.2 *Let $a \in C_{N\times N}$ be regular and*

$$\sum_{n\in\mathbb{Z}} |n|^\alpha ||a_n|| < \infty$$

for some $\alpha > 0$ (where $a_n, n \in \mathbb{Z}$, are the block Fourier coefficients of a), then $s_k(\tilde{T}_{n,r}(a)) = O(\frac{1}{(n-r)^\alpha})$ as $n \to \infty$ (with k defined in Theorem 4.1). If, in addition, a is rational, then there is a $\rho > 0$ such that

$$s_k\left(\tilde{T}_{n,r}(a)\right) = O\left(\frac{1}{e^{\rho(n-r)}}\right).$$

The proof is based on the following observation. Introduce the spaces

$$l_N^{2,\alpha} = \left\{ (x_n) : \sum_{n=0}^{\infty} (1+n)^{2\alpha} ||x_n||^2 < \infty \right\}$$

and

$$l_N^2\left(e^{\rho n}\right) = \left\{ (x_n) : \sum_{n=0}^{\infty} e^{2\rho n} ||x_n||^2 < \infty \right\}.$$

Observe that in both cases the operators $T(a)$ and $C_r(a)$ are bounded on $l_N^{2,\alpha}$ and $l_N^2(e^{\rho n})$, respectively. Moreover if ρ is small enough, then the operators $T(a)$ and $C_r(a)$ are on the unweighted and the weighted spaces simultaneously Fredholm operators with unchanged indices. Then the kernels of these operators coincide, respectively. The given estimates are now an easy consequence.

Remark 4.3 The proof of Theorem 4.1 can be slightly modified and completed in such a way that it is a completely new proof of the splitting property.

5. Collocation-based approximations

We have to know the Fourier coefficients of $a \in C_{N\times N}$ in order to apply the scheme described in the previous sections. If we known only the values of the matrix function a, then it would be valuable to have a possibility to obtain the results of the previous sections only by using these values, that is to work in the "time domain".

This can be done in the following way.

Put $t_j = \exp\frac{2\pi ij}{2n+1}, j = 0, \ldots, 2n$.

Form the $N(2n+1) \times N(2n+1)$ matrices $B_n(a)$ defined by

$$\begin{aligned} B_n(a) &= \mathcal{F}_{2n+1}^{-1} \left(a(t_j)\delta_{k,j}\right)_{k,j=0}^{n} \mathcal{F}_{2n+1} \text{ diag } \{\underbrace{I_N, \ldots, I_N}_{n+1\text{-times}}, \underbrace{0I_N, \ldots, 0I_N}_{n\text{-times}}\} \\ &+ \text{ diag } \underbrace{0I_N, \ldots, 0I_N}_{n+1\text{-times}}, \underbrace{I_N, \ldots, I_N}_{n\text{-times}}, \end{aligned}$$

where $\delta_{k,j}$ is the Kronecker delta and

$$\begin{aligned} (\mathcal{F}_{2n+1}\xi)_k : &= \frac{1}{\sqrt{2n+1}} \sum_{j=0}^{2n} e^{i2\pi kj/n}\xi_j , \\ \left(\mathcal{F}_{2n+1}^{-1}\xi\right)_k : &= \frac{1}{\sqrt{2n+1}} \sum_{j=0}^{2n} e^{-i2\pi kj/n}\xi_j ; \end{aligned}$$

the operators $\mathcal{F}_{2n+1}$ and $\mathcal{F}_{n+1}^{-1}$ acting on the space of all vectors with $2n+1$ entries each of which belongs to $\mathbb{C}^N$.

The matrices $B_{n,r}(a)$ are defined by omitting the first and last r columns (the entries of which are $N \times N$ matrices).

Theorem 5.1 *If $a \in C_{N\times N}$ is regular then the sequence $(B_{n,r}(a))$ has for r large enough the k-splitting property with*

$$k = \dim \ker T(a) .$$

Idea of the proof: The matrices $B_{n,r}(a)$ can be identified with the operators $L_n(aP+Q)P_{n-r}$, where

- L_n is the interpolation operator associated to the points t_j (that is L_n assigns to a function a its Lagrange interpolation polynomial $L_n a$ where $(L_n a)(t_j) = a(t_j)$ for $j = 0, 1, \ldots, 2n$).

- P_n is the Fourier projection $a \mapsto \sum_{k=-n}^{n} a_k t^k$, and

- P is the Riesz projection in $L^2(\mathbb{T})$, $Q = I - P$.

It can be proved that $(L_n(aP+Q)P_{n-r})$ belongs to the (smallest) C^*-algebra $\mathcal{B}$ generated by all sequences $(L_n(gP+hQ)P_n)$, $h, g \in C_{N\times N}$. This algebra can be studied completely (see for instance [9], Subsection 4.4.2). Using the abstract theory explained in Chapter 6 of [9] one obtains (in analogy to the finite section method) the results. Notice that it is also possible to obtain estimates of the convergence speed of $s_k(B_{n,r}(a))$ to zero. For instance, if a belongs to the Hölder-Zygmund space $\mathcal{H}^s_{N\times N}(s > 0)$, then $s_k(B_{n,r}(a)) = O(\frac{1}{(n-r)^{s-\delta}})$, where $\delta > 0$ is arbitrary.

References

[1] Bart, H., Gohberg, I., and Kaashoek, M., Explicit Wiener-Hopf factorization and realization. In: Operator Theory: Advances and Applications, **21**, 235–316. Birkhäuser Verlag, 1986.

[2] Böttcher, A., On the approximation numbers of large Toeplitz matrices, *Documenta Mathematica*, **2** (1997), 1–29.

[3] Böttcher, A. and Grudsky, S., *Toeplitz Matrices, Asymptotical Linear Algebra and Functional Analysis.* Hindustan Book Agency, New Delhi, 2000; Birkhäuser Verlag, Basel, 2000.

[4] Böttcher, A. and Silbermann, B., *Introduction to Large Truncated Toeplitz Matrices.* Springer-Verlag, New York, 1999.

[5] Clancey, K. and Gohberg, I., *Factorization of Matrix Functions and Singular Integral Operators*, Operator Theory: Advances and Applications, **3**. Birkhäuser Verlag, Basel, 1981.

[6] Gohberg, I., The factorization problem in normed rings, functions of isometric and symmetric operators, and singular integral equations, *Uspekhi Mat. Nauk*, **19** (1964), 71–124 (in Russian).

[7] Gohberg, I. and Feldman, I. A., *Convolution Equations and Projection Methods for Their Solution*, Nauka, Moscow, 1971 (in Russian). English translation: Translations of Mathematical Monographs, **41**. AMS, Providence, R.I., 1974.

[8] Gohberg, I. C. and Krein, M. G., *Introduction to the Theory of Linear Nonselfadjoint Operators*, Translations of Mathematical Monographs, **18**. AMS, Providence, R.I., 1969.

[9] Hagen, R., Roch, S., and Silbermann, B., *C^*-Algebras and Numerical Analysis.* Marcel Dekker, Inc., New York, Basel, 2001.

[10] Litvinchuk, G. S. and Spitkovskii, I. M., *Factorization of Measurable Matrix Functions*, Operator Theory: Advances and Applications, **25**. Birkhäuser Verlag, Basel, 1987.

[11] Roch, S. and Silbermann, B., Index calculus for approximation methods and singular value decomposition, *J. Math. Anal. Appl.*, **225** (1998), 401–426.

[12] Silbermann, B., Modified finite sections for Toeplitz operators and their singular values, *SIAM J. Matrix Analysis Appl.*, submitted.

THE MULTIPLICATIVE AND SPECTRAL STRUCTURE OF ANALYTIC OPERATOR-VALUED FUNCTIONS

Anna Tarasenko
Universidad Autonoma del Estado de Hidalgo, Pachuca, Hgo, México
anataras@uaeh.reduaeh.mx

Abstract We study the limiting values of multiplicative integrals under the following assumptions: operator functions belong to the Holder class H_μ, $0 < \mu \leq 1$, and the derivative of operator functions belongs to the Hardy class $\mathbb{H}_1$. We obtain the description of the typical multiplicative and spectral structure of analytic families of weak contractions.

Keywords: multiplicative integral, limiting values, exceptional values, kernel operator, analytic operator-function, Hausdorff's measure, Hölder class, Hardy class.

1. Introduction

In Section 2 we study the limiting values as $\zeta \to 1$ of multiplicative integrals [12] of the form

$$W(\zeta) = \int_{\alpha}^{\beta} \exp\left(\frac{\zeta + e^{i\vartheta}}{\zeta - e^{i\vartheta}} B(\vartheta) d\vartheta\right),$$

where $\beta - \alpha < 2\pi$ and an operator function $B(\vartheta)$ satisfies the Hölder condition on $[\alpha, \beta]$. The results obtained there are proved basing on the papers [14] and [13] of L.A.Sakhnovich.

Theorems of Section 2 can be used in the theory of exceptional values of analytic matrix functions. It was shown in [3], [6] that if $W(\zeta)$ is a contracting analytic matrix function for $|\zeta| < 1$, then for almost every matrix B with $\|B\| < 1$, the matrix $W_B(\zeta) = L[W(\zeta), B]$, where $L[A; B] = (I - BB^*)^{-1/2}(A - B)(I - B^*A)^{-1}(I - B^*B)^{1/2}$, can be represented as the product of Blashke-Potapov function and an exterior matrix function. The obtained results allow us, in some cases, to characterize the set of those matrices A for which such a representation does not hold.

In the papers [7],[8] of Yu.P.Ginzburg and N.A.Talyush there were obtained some generalizations of the classical O.Frostman-M.Heins's scalar theorem [1],

[11]. Let us formulate a matrix analog of this theorem. Let A, B be matrices such that $\|A\| \leq 1, \|B\| < 1$. Let C^n_{be} be the class consisting of contracting analytic $n \times n$ matrix-function $X(\varsigma), (|\varsigma| < 1)$ such that the inner part of the function $\det X(\varsigma)$ is the Blashke product. The result is as follows: if $X(\varsigma)$ is an analytical function in $|\varsigma| < 1$, $\|X(\varsigma)\| \leq 1$ and $F(z)$ is a matrix-valued polynomial with the non-degenerate leading coefficient, $|F(\varsigma|) < 1$ in $z \in G \subset \mathbb{C}$, then $X_z := L[X, F(z)] \in C^n_{be}$ for almost all $z \in G$.

There arises a natural question about an operator analog of the above theorem for contracting infinitely dimensional operator-functions and about the investigation of the corresponding operator families. Some progress in this direction was obtained in the papers [3]-[8], [4]-[5]. We formulate the result which is the closest to our study.

Let $\mathcal{C}$ be the set of analytic operator-valued functions $X(\zeta)$ in $|\zeta| < 1$, whose values are contractions $X(\zeta)$ acting in a separable Hilbert space $\mathcal{H}$ and satisfy the following conditions

1) $I - X(\zeta)$ are nuclear operators in $|\zeta| < 1$,
2) There exists ζ_0, $|\zeta_0| < 1$ such that $X(\zeta_0)$ is bounded invertible.

Let $\mathcal{C}_{be}$ be the set of functions of the class $\mathcal{C}$, such that $X(\zeta)$ can be represented in the form

$$X(\zeta) = U \overset{\frown}{\prod_j} \left(\frac{\zeta_j - \zeta}{1 - \overline{\zeta_j}\zeta} \cdot \frac{|\zeta_j|}{\zeta_j} P_j + Q_j \right) \cdot \overset{\frown}{\int_0^{2\pi}} \exp \left(\frac{\zeta + e^{i\varphi}}{\zeta - e^{i\varphi}} H(\varphi) d\varphi \right),$$

where $U^*U = 1$, P_j are orthogonal projectors, $Q_j = I - P_j$, $\sum_j (1 - |\zeta_j|) < \infty$, $j = 1, 2, ...,$ $H(\varphi) \geqslant 0$, $\|H(\varphi)\| \in L_1[0, 2\pi]$.

The result is as follows ([4]): if $X \in \mathcal{C}_{be}$ and $F(t)$ is nuclear-valued analytic function in $[0, a]$ with $F(0) = 0$, then $L[X, F(t)] \in \mathcal{C}_{be}$ for almost all $t \in [0, a]$ with respect to the Lebesgue measure.

Let $M_{X,F}$ be the set of those $z \in G$, for which the linear-fractional transformation $L[X; F(z)]$ does not belong to the class $\mathcal{C}_{be}$.

In Section 3 we continue the study of operator analogs of O.Frostman-M.Heins's theorem, multiplicative structure of the infinite dimensional families $L[X(\varsigma), F(z)]$ and analytical families of contracting operators connected with them. Descriptions of smallness of the set $M_{X,F}$ in terms of the linear Hausdorff measure are found.

2. Limiting values of multiplicative integrals

Suppose that $B(\vartheta)$ $(\alpha \leq \vartheta \leq \beta)$ is an operator function in some Hilbert space. We say that the operator function $B(\vartheta)$ $(\alpha \leq \vartheta \leq \beta)$ is in the Hölder class $H_\mu[\alpha, \beta]$ with index $\mu \in]0, 1]$, if it satisfies the inequality

$$\|B(\vartheta_1) - B(\vartheta)_2\| \leq K|\vartheta_1 - \vartheta_2|^\mu \text{ for some } K > 0 \text{ and any } \vartheta_1, \vartheta_2 \in [\alpha, \beta].$$

Let us consider a multiplicative integral of the form

$$W(\zeta) = \overset{\frown}{\int\limits_{\alpha}^{\beta}} \exp\left(\frac{\zeta + e^{i\vartheta}}{\zeta - e^{i\vartheta}} B(\vartheta) d\vartheta\right) \tag{2.1}$$

for $\zeta \in D = \{\zeta,\ |\zeta| < 1\}$, $\beta - \alpha < 2\pi$, $0 < B(\vartheta) \in\ H_\mu[\alpha, \beta]$.

We introduce the notation:

$$q(\mu) = \frac{\mu}{1+\mu}, \quad h(\rho, \varphi, \vartheta) = \frac{\rho e^{i\varphi} + e^{i\vartheta}}{\rho e^{i\varphi} - e^{i\vartheta}}, \quad h(\varphi, \vartheta) = h(1, \varphi, \vartheta) = i \cot \frac{\vartheta - \varphi}{2}.$$

Lemma 2.1 *Suppose that $B(\vartheta) \in H_\mu[\alpha, \beta]$. Then, for $\varepsilon \to +0$, $\varepsilon_1 <\ \varepsilon$, $\rho = 1 - \varepsilon^{1+\mu}$ and $\alpha < \vartheta_0 - \varepsilon < \vartheta_0 < \vartheta_0 + \varepsilon < \beta$, the following estimates are fulfilled uniformly on any segment that belongs to $]\alpha, \beta[$*

$$\left\| \exp\left\{ B(\vartheta_0) \int\limits_{\vartheta_0-\varepsilon}^{\vartheta_0+\varepsilon} h(\rho, \vartheta_0, \vartheta) d\vartheta \right\} - e^{-2\pi B(\vartheta_0)} \right\| = O(\varepsilon),$$

$$\left\| \overset{\frown}{\int\limits_{\alpha}^{\vartheta_0-\varepsilon}} \exp\{h(\rho, \theta_0, \vartheta) B(\vartheta) d\vartheta\} - \overset{\frown}{\int\limits_{\alpha}^{\vartheta_0-\varepsilon}} \exp\{h(\theta_0, \vartheta) B(\vartheta) d\vartheta\} \right\| = O(\varepsilon^\mu),$$

$$\left\| \overset{\frown}{\int\limits_{\vartheta_0+\varepsilon}^{\beta}} \exp\{h(\rho, \theta_0, \vartheta) B(\vartheta) d\vartheta\} - \overset{\frown}{\int\limits_{\vartheta_0+\varepsilon}^{\beta}} \exp\{h(\theta_0, \vartheta) B(\vartheta) d\vartheta\} \right\| = O(\varepsilon^\mu),$$

$$\left\| \overset{\frown}{\int\limits_{\vartheta_0-\varepsilon}^{\vartheta_0-\varepsilon_1}} \exp\{h(\rho, \theta_0, \vartheta) B(\vartheta) d\vartheta\} - \exp\{B(\vartheta_0) \int\limits_{\vartheta_0-\varepsilon}^{\vartheta_0-\varepsilon_1} h(\rho, \theta_0, \vartheta) d\vartheta\} \right\| = O(\varepsilon^\mu).$$

Proof. These estimates follow from the results of the paper [2]:

$$\left\| \exp\left\{ B(\theta_0) \left(2\varepsilon + 2i \ln \frac{e^{i\varepsilon} - \rho}{e^{-i\varepsilon} - \rho} \right) \right\} - \exp\{-2\pi\ B(\theta_0\} \right\|$$
$$\leq \|B(\theta_0\| \cdot |2\varepsilon - 2[\arg(e^{i\varepsilon} - \rho) - \arg(e^{-i\varepsilon} - \rho)] + 2\pi| = O(\varepsilon).$$

and the paper [14]: if the operator functions $B_k(t)(k=1,2)$ are continuous and satisfy the inequalities $-i(B_k(t) - B_k^*(t)) \leq 0\ (a \leq t \leq b)$, then the inequalities are fulfilled

$$\left\| \overset{\frown}{\int\limits_{a}^{b}} e^{iB_1(t)dt} - \overset{\frown}{\int\limits_{a}^{b}} e^{iB_2(t)dt} \right\| \leq \int\limits_{a}^{b} \|B_1(t) - B_2(t)\|\, dt, \tag{2.2}$$

$$\int\limits_{\alpha}^{\vartheta_0-\varepsilon} \|B(\vartheta)\| \cdot |h(\rho, \theta_0, \vartheta) - h(\theta_0, \vartheta)| d\vartheta \leq 4M\varepsilon^{1+\mu} \int\limits_{\alpha}^{\vartheta_0-\varepsilon} \frac{d\vartheta}{|\vartheta_0 - \vartheta|^2},$$

where $M = \sup_{\vartheta\in[\alpha,\vartheta_0]} \|B(\vartheta)\|$. □

We introduce the notations:

$$U_\varepsilon(\vartheta_0) = \exp\left\{-\int_\alpha^{\vartheta_0-\varepsilon} h(\vartheta_0,\vartheta)B(\vartheta_0)d\vartheta\right\}\cdot \overset{\frown}{\int_\alpha^{\vartheta_0-\varepsilon}} \exp\{h(\vartheta_0,\vartheta)B(\vartheta)d\vartheta\},$$

$$V_\varepsilon(\vartheta_0) = \overset{\frown}{\int_{\vartheta_0+\varepsilon}^{\beta}} \exp\{h(\vartheta_0,\vartheta)B(\vartheta)d\vartheta \cdot \exp\left\{-\int_{\vartheta_0+\varepsilon}^{\beta} h(\vartheta_0,\vartheta)B(\vartheta_0)d\vartheta\right\}.$$

If $B(\vartheta)\in\ H_\mu[\alpha,\beta]\ \ (\varepsilon_1<\varepsilon)$, then for any $\vartheta_0\in]\alpha,\beta[$ the limits

$$\lim_{\varepsilon\to+0} U_\varepsilon(\vartheta_0) = U(\vartheta_0),\ \ \lim_{\varepsilon\to+0} V_\varepsilon(\vartheta_0) = V(\vartheta_0)$$

exist and $\|U(\vartheta_0) - U_\varepsilon(\vartheta_0)\| = O(\varepsilon^\mu)$, $\|V(\vartheta_0) - V_\varepsilon(\vartheta_0)\| = O(\varepsilon^\mu)$ as $\varepsilon\to+0$.

Theorem 2.2 *Suppose that an Hermitian-valued operator function $B(\vartheta)$ is in $H_\mu[\alpha,\beta]$. Then for any $\vartheta_0\in]\alpha,\beta[$ there exists a limiting value $W(e^{i\vartheta_0})$ of the multiplicative integral (2.1) with respect to any sequence $\zeta_j\to e^{i\vartheta_0}$, $|\zeta_j|<1$, and uniformly on any segment contained in $]\alpha,\beta[$:*

$$\left\|W\left(\rho e^{i\vartheta_0}\right) - W\left(e^{i\vartheta_0}\right)\right\| = O\left((1-\rho)^{q(\mu)}\right) \tag{2.3}$$

and

$$W\left(e^{i\vartheta_0}\right) =$$

$$= \lim_{\varepsilon\to+0} \overset{\frown}{\int_{\vartheta_0+\varepsilon}^{\beta}} \exp\{h(\vartheta_0,\vartheta)B(\vartheta)d\vartheta\}\cdot e^{-2\pi B(\vartheta_0)}\cdot \overset{\frown}{\int_\alpha^{\vartheta_0-\varepsilon}} \exp\{h(\vartheta_0,\vartheta)B(\vartheta)d\vartheta\}.$$

Proof. We give the proof for the case $B(\vartheta)\ge 0$. Set

$$I_\varepsilon(\vartheta_0) = \overset{\frown}{\int_{\vartheta_0+\varepsilon}^{\beta}} \exp\left\{h\left(\vartheta_0,\vartheta\right)B\left(\vartheta\right)d\vartheta\right\}\cdot e^{-2\pi B(\vartheta_0)}\cdot \overset{\frown}{\int_\alpha^{\vartheta_0-\varepsilon}} \exp\left\{h\left(\vartheta_0,\vartheta\right)B\left(\vartheta\right)d\vartheta\right\}.$$

From Lemma 2.1 it follows that

$$\left\|W\left(\rho e^{i\vartheta_0}\right) - I_\varepsilon(\vartheta_0)\right\|$$

$$\leq \left\| \int\limits_{\alpha}^{\vartheta_0-\varepsilon} \exp\left\{\frac{\rho e^{i\vartheta_0}+e^{i\vartheta}}{\rho e^{i\vartheta_0}-e^{i\vartheta}} B(\vartheta)\, d\vartheta\right\} - \int\limits_{\alpha}^{\vartheta_0-\varepsilon} \exp\left\{h(\vartheta_0,\vartheta)\, B(\vartheta)\, d\vartheta\right\} \right\|$$

$$+ \left\| \int\limits_{\vartheta_0+\varepsilon}^{\beta} \exp\left\{\frac{\rho e^{i\vartheta_0}+e^{i\vartheta}}{\rho e^{i\vartheta_0}-e^{i\vartheta}} B(\vartheta)\, d\vartheta\right\} - \int\limits_{\vartheta_0+\varepsilon}^{\beta} \exp\left\{h(\vartheta_0,\vartheta)\, B(\vartheta)\, d\vartheta\right\} \right\|$$

$$+ \left\| \int\limits_{\vartheta_0-\varepsilon}^{\vartheta_0+\varepsilon} \exp\left\{\frac{\rho e^{i\vartheta_0}+e^{i\vartheta}}{\rho e^{i\vartheta_0}-e^{i\vartheta}} B(\vartheta)\, d\vartheta\right\} - \exp\left\{B(\vartheta_0) \int\limits_{\vartheta_0-\varepsilon}^{\vartheta_0+\varepsilon} \frac{\rho e^{i\vartheta_0}+e^{i\vartheta}}{\rho e^{i\vartheta_0}-e^{i\vartheta}} d\vartheta\right\} \right\|$$

$$+ \left\| \exp\left\{B(\vartheta_0) \int\limits_{\vartheta_0-\varepsilon}^{\vartheta_0+\varepsilon} \frac{\rho e^{i\vartheta_0}+e^{i\vartheta}}{\rho e^{i\vartheta_0}-e^{i\vartheta}} d\vartheta\right\} - e^{-2\pi B(\vartheta_0)} \right\|$$

$$= O(\varepsilon^{\mu}), \left(\rho = 1-\varepsilon^{1+\mu}\right). \tag{2.4}$$

Put

$$A_\varepsilon(\vartheta_0) = \exp\left\{\int\limits_{\vartheta_0+\varepsilon}^{\beta} h(\vartheta_0,\vartheta) B(\vartheta_0) d\vartheta\right\} \cdot e^{-2\pi B(\vartheta_0)} \cdot \exp\left\{\int\limits_{\alpha}^{\vartheta_0-\varepsilon} h(\vartheta_0,\vartheta) B(\vartheta_0) d\vartheta\right\}.$$

We have, as $\varepsilon \to +0$,

$$\begin{aligned} &\|I_\varepsilon(\vartheta_0) - V(\vartheta_0) A_\varepsilon(\vartheta_0) U(\vartheta_0)\| \\ &= \|V_\varepsilon(\vartheta_0) A_\varepsilon(\vartheta_0) U_\varepsilon(\vartheta_0) - V(\vartheta_0) A_\varepsilon(\vartheta_0) U(\vartheta_0)\| \\ &\leq \|V_\varepsilon(\vartheta_0) A_\varepsilon(\vartheta_0)\| \cdot \|U_\varepsilon(\vartheta_0) - U(\vartheta_0)\| \\ &\quad + \|V_\varepsilon(\vartheta_0) - V(\vartheta_0)\| \cdot \|A_\varepsilon(\vartheta_0) U(\vartheta_0)\| = O(\varepsilon^{\mu}), \end{aligned} \tag{2.5}$$

Here we used the estimates

$$\|U(\vartheta_0) - U_\varepsilon(\vartheta_0)\| = O(\varepsilon^{\mu}), \|V(\vartheta_0) - V_\varepsilon(\vartheta_0)\| = O(\varepsilon^{\mu});$$

$$\|V_\varepsilon(\vartheta_0) A_\varepsilon(\vartheta_0)\| \leq 1, \|A_\varepsilon(\vartheta_0) U(\vartheta_0)\| \leq 1.$$

We have thus proved the existence of

$$\lim_{\varepsilon\to+0} V_\varepsilon(\vartheta_0) A_\varepsilon(\vartheta_0) U_\varepsilon(\vartheta_0) = V(\vartheta_0) A(\vartheta_0) U(\vartheta_0),$$

with $A(\vartheta_0) = \lim\limits_{\varepsilon\to+0} A_\varepsilon(\vartheta_0)$, and (see [2]) $\|A_\varepsilon(\vartheta_0) - A(\vartheta_0)\| = O(\varepsilon)$. Then there exists the limit

$$\lim_{\varepsilon\to+0} I_\varepsilon(\vartheta),$$

which is equal to $\lim_{\rho\uparrow+1} W(\rho e^{i\vartheta_0})$ by (2.4). According to (2.4) and (2.5)

$$\|W(\rho e^{i\vartheta_0}) - V(\vartheta_0)A_\varepsilon(\vartheta_0)U(\vartheta_0)\| = O(\varepsilon^\mu).$$

Consequently, $\|W(\rho e^{i\vartheta_0}) - W(e^{i\vartheta_0})\| = O(\varepsilon^\mu)$, from which (2.3) follows. Now, we will get rid of the restriction $B(\vartheta) > 0$. The operator-function $B(\vartheta)$ is continuous. Hence there exists $m > 0$ such that

$$-mI \leq B(\vartheta) \leq mI, \ (\vartheta \in [\alpha_1, \beta_1] \subset]\alpha, \beta[).$$

For the operator-function $\tilde{B}(\vartheta) = B(\vartheta) + mI \geq 0$ the conclusion of Theorem 2.2 is fulfilled. From here the general case follows. □

Theorem 2.3 *If* $B(\vartheta) \in H_\mu[\alpha, \beta]$, $B(\vartheta) = B^*(\vartheta)$, *then* $W(e^{i\vartheta}) \in H_{q(\mu)}$ *on any segment contained in* $]\alpha, \beta[$.

Proof. We give the proof for the case $B(\vartheta) \geq 0$. The general case is reduced to this case by analogy with the proof of Lemma 2.1.

Let $\vartheta_0 > \vartheta_1 + \varepsilon$, then $\|U_\varepsilon(\vartheta_0) - U_\varepsilon(\vartheta_1)\| \leq \mathcal{I}_1 + \mathcal{I}_2$, where

$$\mathcal{I}_1 = \left\| \exp\left\{ -\int_\alpha^{\vartheta_0-\varepsilon} h(\vartheta_0, \vartheta)B(\vartheta_0)d\vartheta \right\} \cdot \overset{\curvearrowleft}{\int_\alpha^{\vartheta_0-\varepsilon}} \exp\{h(\vartheta_0, \vartheta)B(\vartheta)d\vartheta\} - \right.$$

$$\left. - \exp\left\{ -\int_\alpha^{\vartheta_1-\varepsilon} h(\vartheta_0, \vartheta)B(\vartheta_0)d\vartheta \right\} \cdot \overset{\curvearrowleft}{\int_\alpha^{\vartheta_1-\varepsilon}} \exp\{h(\vartheta_0, \vartheta)B(\vartheta)d\vartheta\} \right\|;$$

$$\mathcal{I}_2 = \left\| \exp\left\{ -\int_\alpha^{\vartheta_1-\varepsilon} h(\vartheta_0, \vartheta)B(\vartheta_0)d\vartheta \right\} \cdot \overset{\curvearrowleft}{\int_\alpha^{\vartheta_1-\varepsilon}} \exp\{h(\vartheta_0, \vartheta)B(\vartheta)d\vartheta\} - \right.$$

$$\left. - \exp\left\{ -\int_\alpha^{\vartheta_1-\varepsilon} h(\vartheta_1, \vartheta)B(\vartheta_1)d\vartheta \right\} \cdot \overset{\curvearrowleft}{\int_\alpha^{\vartheta_1-\varepsilon}} \exp\{h(\vartheta_1, \vartheta)B(\vartheta)d\vartheta\} \right\|.$$

By (2.2) we obtain $\mathcal{I}_1 \leq \frac{4K}{\mu}$.

Later on, the following inequalities will be necessary:

$$\begin{aligned} \|U_1V_1 - U_2V_2\| &\leq \|U_1 - U_2\| + \|V_1 - V_2\|, \\ \|U_1V_1 - U_2V_2\| &\leq \|U_2^{-1} - V_2\| + \|U_1 - V_1^{-1}\|, \end{aligned}$$

where U_i and V_i are unitary operators.

Let $\vartheta_2 = \vartheta_1 - (\vartheta_0 - \vartheta_1)^\gamma$ $(0 < \gamma < 1)$. Some estimates are produced

$$
\begin{aligned}
\mathcal{I}_2 \le \Bigg\| & \exp\left\{ - \int\limits_{\alpha}^{\vartheta_1-\varepsilon} h(\vartheta_0,\vartheta)B(\vartheta_0)d\vartheta \right\} \cdot \overset{\curvearrowleft}{\int\limits_{\vartheta_2}^{\vartheta_1-\varepsilon}} \exp\{h(\vartheta_0,\vartheta)B(\vartheta)d\vartheta\} \\
& - \exp\left\{ - \int\limits_{\alpha}^{\vartheta_1-\varepsilon} h(\vartheta_1,\vartheta)B(\vartheta_1)d\vartheta \right\} \cdot \overset{\curvearrowleft}{\int\limits_{\vartheta_2}^{\vartheta_1-\varepsilon}} \exp\{h(\vartheta_1,\vartheta)B(\vartheta)d\vartheta\} \Bigg\| \\
& + \left\| \overset{\curvearrowleft}{\int\limits_{\alpha}^{\vartheta_2}} \exp\{h(\vartheta_0,\vartheta)B(\vartheta)d\vartheta\} - \overset{\curvearrowleft}{\int\limits_{\alpha}^{\vartheta_2}} \exp\{h(\vartheta_1,\vartheta)B(\vartheta)d\vartheta\} \right\| \\
\le & \left\| \exp\left\{ - \int\limits_{\alpha}^{\vartheta_2} h(\vartheta_0,\vartheta)B(\vartheta_0)d\vartheta \right\} - \exp\left\{ - \int\limits_{\alpha}^{\vartheta_2} h(\vartheta_1,\vartheta)B(\vartheta_1)d\vartheta \right\} \right\| \\
& + \left\| \exp\left\{ \int\limits_{\vartheta_2}^{\vartheta_1-\varepsilon} h(\vartheta_1,\vartheta)B(\vartheta_1)d\vartheta \right\} - \overset{\curvearrowleft}{\int\limits_{\vartheta_2}^{\vartheta_1-\varepsilon}} \exp\{h(\vartheta_1,\vartheta)B(\vartheta)d\vartheta\} \right\| \\
& + \left\| \exp\left\{ \int\limits_{\vartheta_2}^{\vartheta_1-\varepsilon} h(\vartheta_0,\vartheta)B(\vartheta_0)d\vartheta \right\} - \overset{\curvearrowleft}{\int\limits_{\vartheta_2}^{\vartheta_1-\varepsilon}} \exp\{h(\vartheta_0,\vartheta)B(\vartheta)d\vartheta\} \right\| \\
& + \left\| \overset{\curvearrowleft}{\int\limits_{\alpha}^{\vartheta_2}} \exp\{h(\vartheta_0,\vartheta)B(\vartheta)d\vartheta\} - \overset{\curvearrowleft}{\int\limits_{\alpha}^{\vartheta_2}} \exp\{h(\vartheta_1,\vartheta)B(\vartheta)d\vartheta\} \right\| \le \\
\le & \, T_1 + T_2 + T_3 + T_4,
\end{aligned}
$$

Here

$$
T_1 = \int\limits_{\alpha}^{\vartheta_2} \|B(\vartheta_0)h(\vartheta_0,\vartheta) - B(\vartheta_1)h(\vartheta_1,\vartheta)\| d\vartheta,
$$

$$
T_2 = \int\limits_{\vartheta_2}^{\vartheta_1-\varepsilon} \|B(\vartheta_1) - B(\vartheta)\| \cdot |h(\vartheta_1,\vartheta)| d\vartheta,
$$

$$
T_3 = \int\limits_{\vartheta_2}^{\vartheta_1-\varepsilon} \|B(\vartheta_0) - B(\vartheta)\| \cdot |h(\vartheta_0,\vartheta)| d\vartheta,
$$

$$T_4 = \int_{\alpha}^{\vartheta_2} \|B(\vartheta)\| \cdot |h(\vartheta_0, \vartheta) - h(\vartheta_1, \vartheta)| d\vartheta.$$

We estimate separately each term

$$T_2 \leq 4K \int_{\vartheta_2}^{\vartheta_1 - \varepsilon} (\vartheta_1 - \vartheta)^{\mu - 1} d\vartheta \leq \frac{4K}{\mu}[(\vartheta_0 - \vartheta_1)^{\gamma\mu} + (\vartheta_0 - \vartheta_1)^{\mu}].$$

Analogously,

$$T_3 \leq \frac{4K}{\mu}[(\vartheta_0 - \vartheta_1)^{\gamma\mu} + (\vartheta_0 - \vartheta_1)^{\mu}], T_4 \leq 4M \int_{\alpha}^{\vartheta_2} \frac{\vartheta_0 - \vartheta_1}{(\vartheta_0 - \vartheta)(\vartheta_1 - \vartheta)} d\vartheta$$

$$\leq 4M(\vartheta_0 - \vartheta_1) \int_{\alpha}^{\vartheta_2} \frac{d\vartheta}{(\vartheta_1 - \vartheta)^2} \leq 4M \left[(\vartheta_0 - \vartheta_1)^{1-\gamma} + \frac{\vartheta_0 - \vartheta_1}{-\alpha + \vartheta_1} \right],$$

with $M = \sup_{[\alpha,\beta]} \|B(\vartheta)\|$,

$$T_1 \leq \int_{\alpha}^{\vartheta_2} \|B(\vartheta_0)\| \cdot |h(\vartheta_0, \vartheta) - h(\vartheta_1, \vartheta)| d\vartheta + \int_{\alpha}^{\vartheta_2} \|B(\vartheta_0) - B(\vartheta_1)\| \cdot |h(\vartheta_1, \vartheta)| d\vartheta$$

$$\leq 4M \left[(\vartheta_0 - \vartheta_1)^{1-\gamma} + \frac{\vartheta_0 - \vartheta_1}{\vartheta_1 - \alpha} \right] + \frac{4K}{\mu} \left[(\vartheta_0 - \vartheta_1)^{\gamma\mu} + (\vartheta_0 - \vartheta_1)^{m} \right].$$

So, we proved that

$$\mathcal{I}_2 \leq C \left\{ (\vartheta_0 - \vartheta_1)^{1-\gamma} + \frac{\vartheta_0 - \vartheta_1}{\vartheta_1 - \alpha} + (\vartheta_0 - \vartheta_1)^{\gamma\mu} + (\vartheta_0 - \vartheta_1)^{m} \right\},$$

$$C = \max\left(\frac{12K}{\mu}, 8M \right).$$

It follows that $\|U(\vartheta_0) - U(\vartheta_1)\| \leq L|\vartheta_0 - \vartheta_1|^{\alpha_0}$, where L is a constant and $\alpha_0 = \min\{\mu\gamma, 1 - \gamma\} = \frac{\mu}{1+\mu}$.
Similar arguments show that $\|V(\vartheta_0) - V(\vartheta_1)\| \leq L|\vartheta_0 - \vartheta_1|^{\alpha}$.

Since $W(e^{i\vartheta}) = V(\vartheta)\widetilde{W}_{\vartheta}(e^{i\vartheta})U(\vartheta)$, where

$$\widetilde{W}_{\vartheta}(\zeta) = \exp\left\{ \int_{\alpha}^{\beta} \frac{\zeta + e^{i\varphi}}{\zeta - e^{i\varphi}} B(\vartheta) d\varphi \right\},$$

$$\widetilde{W}_{\vartheta}(e^{i\vartheta}) = \lim_{\zeta \to e^{i\vartheta}} \widetilde{W}_{\vartheta}(\zeta) = \exp\left\{ \left[i\pi + \beta - \alpha + 2i \ln \frac{e^{i\beta} - e^{i\vartheta}}{e^{i\vartheta} - e^{i\alpha}} \right] B(\vartheta) \right\},$$

and $V(\vartheta), \widetilde{W}_\vartheta(e^{i\vartheta})$ and $U(\vartheta)$ belong to the class $H_{q(\mu)}$, we obtain that the operator-function $W(e^{i\vartheta})$ also belongs to $H_{q(\mu)}$. □

Theorem 2.4 *Suppose that* $B(\vartheta) \in H_\mu[\alpha, \beta]$, $\beta - \alpha < 2\pi$ *and*

$$W_\alpha(\zeta) = W(\zeta) \cdot (\zeta - e^{i\alpha})^{2iB(\alpha)}, \tag{2.6}$$

$$W_\beta(\zeta) = (\zeta - e^{i\beta})^{-2iB(\beta)} \cdot W(\zeta). \tag{2.7}$$

Then the following limits exist

$$\lim_{\zeta \to e^{i\alpha}} W_\alpha(\zeta) =: M_\alpha; \quad \lim_{\zeta \to e^{i\beta}} W_\beta(\zeta) =: M_\beta;$$

where

$$M_\alpha = \lim_{\varepsilon \to +0} \int\limits_{\alpha+\varepsilon}^{\overset{\frown}{\beta}} \exp\{h(\alpha, \vartheta)B(\vartheta)d\vartheta\} \cdot \varepsilon^{2iB(\alpha)} e^{-(\pi+2\alpha)B(\alpha)}, \tag{2.8}$$

$$M_\beta = \lim_{\varepsilon \to +0} e^{(-\pi+2\beta)B(\beta)} \cdot \varepsilon^{-2iB(\beta)} \cdot \int\limits_{\alpha}^{\overset{\frown}{\beta-\varepsilon}} \exp\{h(\beta, \vartheta)B(\vartheta)d\vartheta\}. \tag{2.9}$$

(Here and later on, z^A, where A is an operator, is understood as $\exp\{A \ln z\}$, where $\ln z$ is the branch of the logarithmic function which takes real values for positive z).

Proof. We introduce the operator-function

$$\tilde{B}(\vartheta) = \begin{cases} B(\alpha), \alpha - l \le \vartheta < \alpha \\ B(\vartheta), \alpha \le \vartheta \le \beta \end{cases},$$

where $l < 2\pi + (\alpha - \beta)$.

Then $W(\zeta) = P(\zeta)F^{-1}(\zeta)$, where

$$W(\zeta) = \int\limits_{\alpha}^{\overset{\frown}{\beta}} \exp\left\{\frac{\zeta + e^{i\vartheta}}{\zeta - e^{i\vartheta}} B(\vartheta)d\vartheta\right\}, P(\zeta) = \int\limits_{\alpha-l}^{\overset{\frown}{\beta}} \exp\left\{\frac{\zeta + e^{i\vartheta}}{\zeta - e^{i\vartheta}} \widetilde{B}(\vartheta)d\vartheta\right\},$$

$$F(\zeta) = \exp \int\limits_{\alpha-l}^{\alpha} \frac{\zeta + e^{i\vartheta}}{\zeta - e^{i\vartheta}} B(\alpha)d\vartheta$$

$$= \exp\left\{B(\alpha) \cdot l + 2iB(\alpha) \left[\ln\left(\zeta - e^{i\alpha}\right) - \ln\left(\zeta - e^{i(\alpha-l)}\right)\right]\right\}.$$

According to Theorem 2.2, the limit of $P(\zeta)$ as $\zeta \to e^{i\alpha}$ exists and

$$\lim_{\zeta \to e^{i\alpha}} P(\zeta) =$$

$$= \lim_{\varepsilon\to+0} \int\limits_{\alpha+\varepsilon}^{\beta} \exp\{h(\alpha,\vartheta)B(\vartheta)d\vartheta\} \cdot e^{-2\pi B(\alpha)} \int\limits_{\alpha-l}^{\alpha-\varepsilon} \exp\{h(\alpha,\vartheta)B(\alpha)d\vartheta\}$$

$$= \lim_{\varepsilon\to+0} \int\limits_{\alpha+\varepsilon}^{\beta} \exp\{h(\alpha,\vartheta)B(\vartheta)d\vartheta\} \cdot e^{-2\pi B(\alpha)} \cdot$$

$$\cdot \exp\left\{B(\alpha)\left[l + 2i \ln \frac{e^{i\alpha} - e^{i(\alpha-\varepsilon)}}{e^{i\alpha} - e^{i(\alpha-l)}}\right]\right\}.$$

Thus $W(\zeta) = W_\alpha(\zeta)(\zeta - e^{i\alpha})^{-2iB(\alpha)}$,

$$W_\alpha(\zeta) = \int\limits_{\alpha-l}^{\beta} \exp\left\{\frac{\zeta+e^{i\vartheta}}{\zeta-e^{i\vartheta}}\tilde{B}(\vartheta)d\vartheta\right\} \cdot \exp\left\{-B(\alpha)l - 2iB(\alpha)\ln\left(\zeta - e^{i(\alpha-l)}\right)\right\}.$$

Moreover, there exists also the following limit

$$\lim_{\zeta\to e^{i\alpha}} W_\alpha(\zeta) = \lim_{\varepsilon\to+0} \int\limits_{\alpha+\varepsilon}^{\beta} \exp\{h(\alpha,\vartheta)B(\vartheta)d\vartheta\} \cdot e^{-2\pi B(\alpha)} \exp\{(\pi+2\alpha)B(\alpha)\}\varepsilon^{2iB(\alpha)}.$$

Representation (2.9) is proved in a similar way. □

Theorem 2.5 *Suppose that* $B(\vartheta) \in H_\mu[0,2\pi]$, $B^*(\vartheta) = B(\vartheta)$,

$$W(\zeta) = \int\limits_{0}^{2\pi} \exp\left\{\frac{\zeta+e^{i\vartheta}}{\zeta-e^{i\vartheta}}B(\vartheta)d\vartheta\right\} \tag{2.10}$$

and

$$W_0(\zeta) = (\zeta-1)^{-2iB(2\pi)} \cdot W(\zeta) \cdot (\zeta-1)^{2iB(0)}. \tag{2.11}$$

Then there exists the limit $\lim\limits_{\zeta\to1} W_0(\zeta) =: U_0$,

$$U_0 = e^{\pi B(2\pi)} \cdot \lim_{\varepsilon\downarrow0}\left(\varepsilon^{-2iB(2\pi)} \int\limits_{\varepsilon}^{2\pi-\varepsilon} \exp\{i\cot\frac{\vartheta}{2}B(\vartheta)d\vartheta\} \cdot \varepsilon^{2iB(0)}\right)e^{-\pi B(0)}. \tag{2.12}$$

Proof. Suppose that $\lambda \in]0,2\pi[$. We split the multiplicative integral (2.10) into the product of integrals $W(\zeta) = F_1 \cdot F_2$, where

$$F_1 = \int\limits_{0}^{\lambda} \exp\left\{\frac{\zeta+e^{i\vartheta}}{\zeta-e^{i\vartheta}}B(\vartheta)d\vartheta\right\}, \quad F_2 = \int\limits_{\lambda}^{2\pi} \exp\left\{\frac{\zeta+e^{i\vartheta}}{\zeta-e^{i\vartheta}}B(\vartheta)d\vartheta\right\}.$$

We introduce the notation:

$$W_1(\zeta) = F_1(\zeta) \cdot (\zeta - 1)^{2iB(0)}; \; W_2(\zeta) = (\zeta - 1)^{-2iB(2\pi)} \cdot F_2(\zeta).$$

In virtue of Theorem 2.4, there exist the limits

$$\lim_{\zeta \to 1} W_1(\zeta) =: U_1 = \lim_{\varepsilon \to +0} \overset{\curvearrowleft}{\int_{\varepsilon}^{\lambda}} \exp\{h(0,\vartheta)B(\vartheta)d\vartheta\} \cdot e^{-\pi B(0)} \cdot \varepsilon^{2iB(0)};$$

$$\lim_{\zeta \to 1} W_2(\zeta) =: U_2 = \lim_{\varepsilon \to +0} \varepsilon^{-2iB(2\pi)} \cdot e^{\pi B(2\pi)} \cdot \overset{\curvearrowleft}{\int_{\lambda}^{2\pi-\varepsilon}} \exp\{h(2\pi,\vartheta)B(\vartheta)d\vartheta\}.$$

We set

$$W_0 = W_2 \cdot W_1 = (\zeta - 1)^{-2iB(2\pi)} \cdot W(\zeta) \cdot (\zeta - 1)^{2iB(0)},$$

and obtain

$$\lim_{\zeta \to 1} W_0(\zeta) = \lim_{\zeta \to 1} W_2(\zeta) \cdot W_1(\zeta) = U_2 \cdot U_1$$

$$= e^{\pi B(2\pi)} \cdot \lim_{\varepsilon \downarrow 0} \left(\varepsilon^{-2iB(2\pi)} \overset{\curvearrowleft}{\int_{\varepsilon}^{2\pi-\varepsilon}} \exp\{i \cot \frac{\vartheta}{2} B(\vartheta)d\vartheta\} \cdot \varepsilon^{2iB(0)} \right) e^{-\pi B(0)} =: U_0.$$

□

With every bounded self-adjoint operator H we associate the number $\delta(H) = \max_{\|x\|=1} (Hx, x) - \min_{\|x\|=1} (Hx, x)$ (the "measure of non-scalarity" of H).

Corollary 2.6 *Suppose that* 1) $B(\vartheta) \in H_\mu[0, 2\pi]$, $B(\vartheta) = B^*(\vartheta)$; 2) $B(0) = U_0^{-1} B(2\pi) U_0$, *where* U_0 *is the unitary operator defined by* $B(\vartheta)$ *via* (2.12); 3) $\delta(B(0)) < q(\mu)$. *Then operator function* (2.10) *can be extended by continuity to a closed unit disc.*

Let us restate Theorem 2.2 for the case of an operator function of the form

$$W(\lambda) = \overset{\curvearrowleft}{\int_{a}^{b}} \exp\left\{-i \frac{B(t)}{t - \lambda} dt\right\}.$$

Corollary 2.7 *Suppose that the Hermitian-valued operator function* $B(\vartheta)$ *is in* $H_\mu[0, 2\pi]$; *then for any* $\sigma \in]a, b[$ *there exists a limiting value* $W(\sigma) = \lim_{\lambda \to \sigma} W(\lambda)$ *and we have*

$$W(\sigma) = \lim_{\varepsilon \to +0} \overset{\curvearrowleft}{\int_{\sigma+\varepsilon}^{b}} \exp\left\{-i \frac{B(t)}{t - \varepsilon} dt\right\} e^{\pi B(\sigma)} \overset{\curvearrowleft}{\int_{a}^{\sigma-\varepsilon}} \exp\left\{-i \frac{B(t)}{t - \sigma} dt\right\}, \qquad (2.13)$$

where $W(\sigma) \in H_{q(\mu)}]a, b[$.

3. The case of the derivative in the Hardy class

The results of the section are presented without proof, they are given with complete demonstrations in the paper accepted for publication [15].

Let γ_1 be the class of nuclear operators. The nuclear Hardy class $\mathbb{H}_1$ is introduced as the class of analytic operator-valued functions in $|\zeta| < 1$ with values in γ_1 such that

$$\sup_r \int_0^{2\pi} \left\|F\left(re^{i\vartheta}\right)\right\|_1 d\vartheta < \infty,$$

where $||\cdot||_1$ denotes the norm of the nuclear operator of the class γ_1.

Let $\mathcal{E}_G$ be the set of γ_1 - analytic, strongly contracting operator-valued functions in the domain $G \subset \mathbb{C}$, which take nuclear values.

Let $X(\zeta) \in \mathcal{C}$. For those ζ $(|\zeta| < 1)$, for which the operator $X(\zeta)$ is invertible, from the formula

$$\frac{d}{d\zeta} \ln \det X(\zeta) = \operatorname{tr} \left[X^{-1}(\zeta)\frac{d}{d\zeta}X(\zeta)\right],$$

we get

$$\frac{d}{d\zeta} \det X(\zeta) = \operatorname{tr} \left[X^{-1}(\zeta) \det X(\zeta)\frac{d}{d\zeta}X(\zeta)\right].$$

For such ζ, the inequality

$$\left|\frac{d}{d\zeta} \det X(\zeta)\right| \leqslant \left\| X^{-1}(\zeta) \cdot \det X(\zeta) \cdot \frac{d}{d\zeta}X(\zeta)\right\|_1$$
$$\leqslant \| X^{-1}(\zeta) \cdot \det X(\zeta)\| \cdot \left\|\frac{d}{d\zeta} \cdot X(\zeta)\right\|_1$$

follows. As $\| X^{-1}(\zeta) \cdot \det X(\zeta)\| \leqslant 1$, we have

$$\left|\frac{d}{d\zeta} \det X(\zeta)\right| \leqslant \left\| \frac{d}{d\zeta}X(\zeta)\right\|_1 \tag{3.1}$$

at those points ζ, where $\det X(\zeta) \neq 0$.

Since functions in the class $\mathcal{C}$ are analytic in the nuclear norm [10], the inequality (3.1) is extended by continuity to the whole unit circle.

Let $\widetilde{\mathcal{E}}_G$ be the class of γ_1 -analytic strongly contracting operator-valued functions $F(z)$ on a set $G \subset \mathbb{C}$, which have nuclear values.

Lemma 3.1 *Let* $F(z) \in \widetilde{\mathcal{E}}_G$, $X'(\zeta) \in \mathbb{H}_1$ $(|\zeta| < 1)$.
Then $\det(X(\xi) - F(z))$ *is a continuous function of bounded variation in* ξ $(|\xi| = 1)$ *uniformly by* z *on any closed set* $G_1 \subset G$.

Lemma 3.2 *Let* $\frac{d}{d\zeta}X(\zeta) \in \mathbb{H}_1$, *and* $F(z)$ *be a* γ_1*-analytic function in* G. *Then* $\det(X(\xi) - F(z))$ *is a continuous function on* $\mathbb{T} \times G$.

We need the following three auxiliary statements .

Statement 3.3 [4] *Let* $X \in \mathcal{C}$, *and* $K(\in \gamma_1)$ *be a strictly contracting operator. If* $L[X, K] \in \mathcal{C}_{be}$, *then there exists a point* $\xi \in \mathbb{T}$, *such that*

$$\lim_{\zeta \to \xi} \det(X(\zeta) - K) = 0.$$

Statement 3.4 [9] *Let* $X \in \mathcal{C}_{be}, F \in \mathcal{E}_{\mathcal{G}}$. *Then for any rectifiable curve* $\mathcal{L} \subset G$, *the set* $M_{X,F} \bigcap \mathcal{L}$ *can be covered by a set of circles with an arbitrarily small sum of diameters, that is, its linear Hausdorff measure is equal to 0.*

Statement 3.5 [9] *Let* $G \subset \mathbb{C}$ *be an arbitrary domain, let* $R(\zeta, z)$ *be a scalar function* $(\zeta \in \mathbb{T}, z \in G)$ *satisfying the following conditions:*
1. *There does not exist any* ζ_0 *such that* $R(\zeta_0, z) \equiv 0$.
2. *The function* $R(\zeta, z)$ *has bounded variation in* ζ *uniformly with respect to* $z \in G_0$ $(\forall G_0 = \overline{G_0}, G_0 \subset G)$.
3. *For an arbitrary fixed* $\zeta \in \mathbb{T}$ *the function* $R(\zeta, z)$ *is analytic in* z *in* G.
4. *The function* $R(\zeta, z)$ *is continuous as a function of two complex variables* $\mathbb{T} \times G$.

Then there exists at most a countable set of rectifiable curves $\mathcal{L}_n$ $(n = 1, 2, 3, ...)$ *such that* $\mathcal{M}(R) \subset \bigcup_n \mathcal{L}_n$ *where* $\mathcal{M}(R)$ *is a union, with respect to* $\zeta \in \mathbb{T}$, *of the set of roots of the equation* $R(\zeta, z) = 0$ *in* z *under fixed* ζ.

The following theorem in fact follows immediately from the above statements.

Theorem 3.6 *Let* $X(\zeta) \in \mathcal{C}_{be}$, $\frac{d}{d\zeta}X(\zeta) \in \mathbb{H}_1$, $F(z) \in \mathcal{E}_G$, *and* $F(z_0) = 0, z_0 \in G$ *and suppose that there does not exist any* $\xi \in \mathbb{T}$ *such that* $\det(X(\xi) - F(z)) \equiv 0$. *Then the set* $M_{X,F}$ *has a linear Hausdorff measure zero.*

Theorem 3.7 *Let* $X(\zeta) \in \mathcal{C}$, $\frac{d}{d\zeta}X^{-1}(\zeta) \in \mathbb{H}_1, F(z) \in \mathcal{E}_G$, $F(z_0) = 0$, $z_0 \in G$. *Then the set* $M_{X,F}$ *has a linear Hausdorff measure zero.*

References

[1] Frostman, O., Potentiel d'equilibre et capacite des ensembles avec quelques applications a la theorie des fonctions, *Meddel. Lunds Univ., Mat. Sem.*, **3** (1935), 35–54.

[2] Gakhov, F. D., *Boundary Value Problems.* Pergamon Press, Oxford, New York, Paris, 1966. Extended Russian edition: Nauka, Moscow, 1977.

[3] Ginzburg, Yu. P., The almost invariant spectral properties of contractions and multiplicative properties of analytic operator-functions, *Funkts. Anal. Prilozhen.*, **5** (1971), no. 3, 32–41 (in Russian).

[4] Ginzburg, Yu. P., Almost invariant properties of analytic operator families, Funktsional. Anal. Prilozhen., **16** (1982), no. 1, 68–69 (in Russian).

[5] Ginzburg, Yu. P., Analogues of a theorem of Frostman on linear fractional transformations of inner functions and the typical spectral structure of analytic families of weak contractions. In: Operator Theory: Advances and Applications, **123**, 323–336. Birkhäuser Verlag, 2001.

[6] Ginzburg, Yu. P. and Talyush, N. A., A matrix analogue of a theorem of Heins and the typical spectral structure of contractions, *Funkts. Anal. Prilozhen.*, **7** (1973), no. 1, 66–67 (in Russian).

[7] Ginzburg, Yu. P. and Talyush, N. A., On polynomial pencils of analytic operator-functions and families of contracting extensions of isometric operators, *Izv. Vyssh. Uchebn. Zaved., Mat.*, **4** (1982), 19–27 (in Russian).

[8] Ginzburg, Yu. P. and Talyush, N. A., Exceptional sets of analytic matrix-functions, contracting and dissipative operators, *Izv. Vyssh. Uchebn. Zaved., Mat.*, **8** (1984), 9–14 (in Russian).

[9] Ginzburg, Yu. P. and Tarasenko, A. A., Analytic families of linear-fractional transformations of analytic operator-valued functions, In: *Functional Analysis*, **20**, 51–59. Ul'yanovsk. Gos. Ped. Inst., Ul'yanovsk, 1983 (in Russian).

[10] Gohberg, I. C. and Krein, M. G., *Introduction to the Theory of Linear Nonselfadjoint Operators*, Translations of Mathematical Monographs, **18**. AMS, Providence, R.I., 1969.

[11] Heins, M., On the Lindelöf principle, *Ann. Math.*, **61** (1955), no. 3, 440–473.

[12] Potapov, V. P., The multiplicative structure of J-contractive matrix functions, *Trudy Moskov. Mat. Obsch.*, **4** (1955), 125–236 (in Russian).

[13] Sakhnovich, L. A., On limit values of multiplicative integrals, *Uspekhi Mat. Nauk*, **12** (1957), no. 3, 205–210 (in Russian).

[14] Sakhnovich, L. A., Dissipative operators with absolutely continuous spectrum, *Trudy Moskov. Mat. Obsch.*, **19** (1968), 211–270 (in Russian).

[15] Tarasenko, A. A., On the exceptional values of operator-valued functions with the derivative in the Hardy class, *Proc. of A. Razmadze Math. Institute*, **128** (2002), 113–119.

TOEPLITZ OPERATORS ON THE BERGMAN SPACE

Nikolai Vasilevski
Departamento de Matemáticas, CINVESTAV del I.P.N., Mexico City, México
nvasilev@math.cinvestav.mx

To my teacher, Professor G. Litvinchuk, in occasion of his 70th birthday

Abstract The paper is a survey of recent results grouped around a central (quite unexpected) fact, the existence of a rich family of *commutative* C^*-algebras of Toeplitz operators.

Keywords: Bergman space, Toeplitz operator, radial symbol, commutator and semicommutator, boundedness, compactness, spectrum, hyperbolic geometry.

1. Introduction

Let $\mathbb{D}$ be the unit disk in $\mathbb{C}$. Introduce the space $L_2(\mathbb{D})$ with the usual Lebesgue plane measure $d\mu(z) = dxdy$, and the Bergman space $\mathcal{A}^2(\mathbb{D})$, the closed subspace of $L_2(\mathbb{D})$ which consists of analytic functions in $\mathbb{D}$.The Bergman orthogonal projection $B_{\mathbb{D}}$ of $L_2(\mathbb{D})$ onto $\mathcal{A}^2(\mathbb{D})$ has the form

$$(B_{\mathbb{D}}\varphi)(z) = \frac{1}{\pi}\int_{\mathbb{D}} \frac{\varphi(\zeta)\, d\mu(\zeta)}{(1-z\overline{\zeta})^2},$$

Given a function $a = a(z) \in L_\infty(\mathbb{D})$, the Toeplitz operator T_a with symbol a is defined as follows

$$T_a : \varphi \in \mathcal{A}^2(\mathbb{D}) \longmapsto B_{\mathbb{D}}\, a\varphi \in \mathcal{A}^2(\mathbb{D}).$$

The paper is a survey of a number of recent results grouped around a central (quite unexpected) fact, the existence of a rich family of *commutative* C^*-algebras of Toeplitz operators.

The exposition is based mainly on papers [11, 12, 23, 22, 25, 27], where proofs, comments, additional results and examples can be found. In Section 2 we present the classification of commutative C^*-algebras of Toeplitz operators in terms of pencils of geodesics in the hyperbolic geometry of the unit disk. The proof of the commutativity for a model case of parabolic pencils is sketched in Section 3. Section 3 contains as well the description of decomposition of L_2 in the upper half-plane onto the direct sum of Bergman and poly-Bergman type spaces. In Section 4 we deal with unbounded symbols generating bounded and compact Toeplitz operators, illustrating the peculiarities caused by unboundedness of symbols. In Section 5 we analyze the difference between compactness of commutators and semicomuutators of Toeplitz operators and its impact on the representation theory of algebras connected with Toeplitz operators. In the last Section 6 we study Toeplitz operators with radial symbols on weighted Bergman spaces analyzing the behavior of their different properties (boundedness, compactness, spectral properties, etc.) in dependence on a parameter of weight.

For reader benefits we extend the bibliography adding related papers even not being sited below.

2. Commutative algebras of Toeplitz operators

Surprisingly all known commutative C^*-algebras of Toeplitz operators can be perfectly classified in terms of pencils of geodesics in the hyperbolic geometry of $\mathbb{D}$. Note in this connection, that the commutativity properties do not depend at all of smoothness properties of symbols, the symbols can be merely measurable. The prime cause here is a geometric behavior of the level lines of symbols.

A *geodesic*, or a hyperbolic straight line, in $\mathbb{D}$ is (a part of) an Euclidean circle or a straight line orthogonal to $S^1 = \partial\mathbb{D}$. Each pair of geodesics, say L_1 and L_2, lie in a geometrically defined object, one-parameter family $\mathcal{P}$ of geodesics, which is called the *pencil* determined by L_1 and L_2. Each pencil has an associated family of lines, called *cycles*, the orthogonal trajectories to geodesics forming the pencil. The pencil $\mathcal{P}$ determined by L_1 and L_2 is called

- *parabolic* if L_1 and L_2 are parallel, in this case $\mathcal{P}$ is a set of all geodesics parallel to L_1 (or L_2)

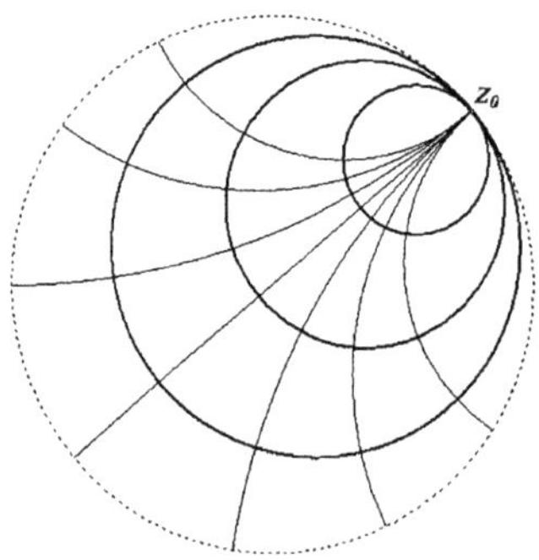

Parabolic pencil.

- *elliptic* if L_1 and L_2 are intersecting, in this case $\mathcal{P}$ is a set of all geodesics passing through the common point of L_1 and L_2

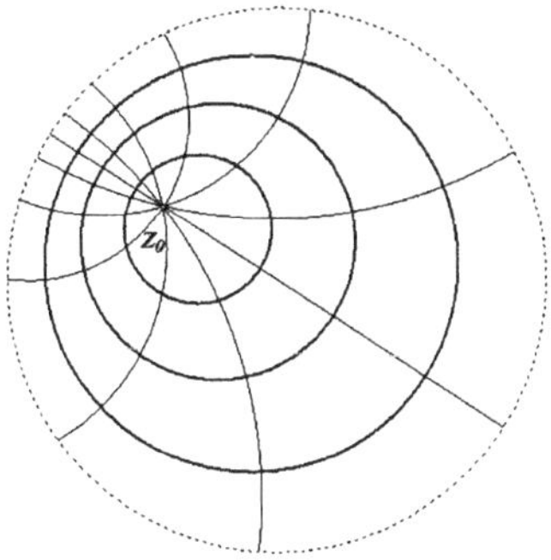

Elliptic pencil.

- *hyperbolic* if L_1 and L_2 are disjoint, in this case $\mathcal{P}$ is a set of all geodesics orthogonal to the common orthogonal of L_1 and L_2

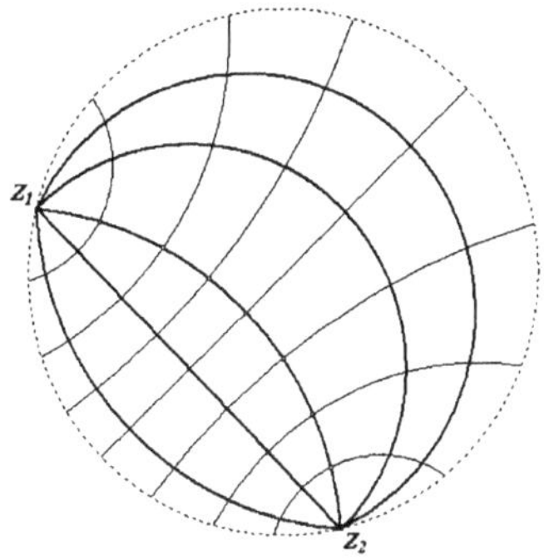

Hyperbolic pencil.

Theorem 2.1 *Given a pencil $\mathcal{P}$ of geodesics, consider the set of L_∞-symbols which are* constant *on the corresponding cycles. The C^*-algebra generated by Toeplitz operators with such symbols is commutative.*

That is, each pencil of geodesics generates a commutative C^*-algebra of Toeplitz operators.

The theorem statement for model cases of each type of pencils follows from Theorems 3.6, 3.7, and 3.8, respectively. For general pencils the result extends via Möbius transformations. The model case for a parabolic pencil is the upper half-plane and symbols depending only on y, imaginary part of $z = x + iy$. The model case for an elliptic pencil is the unit disk and radial symbols, i.e., symbols depending only on $r = |z|$. The model case for a hyperbolic pencil is again the upper half-plane but homogeneous of order zero symbols, i.e., symbols depending only on $\frac{z}{|z|}$, or on $\theta = \arg z$.

3. Bergman space structure and spectral form of special classes of Toeplitz operators

We begin with the upper half-plane Π in $\mathbb{C}$. Introduce the space $L_2(\Pi)$ with the usual Lebesgue plane measure $d\mu(w) = dudv$, $w = u + iv$, and its Bergman subspace $\mathcal{A}^2(\Pi)$, consisting of all functions analytic in Π. The Bergman orthogonal projection B_Π of $L_2(\Pi)$ onto $\mathcal{A}^2(\Pi)$ is given by

$$(B_\Pi\varphi)(w) = -\frac{1}{\pi}\int_\Pi \frac{\varphi(\omega)}{(w-\overline{\omega})^2}\, d\mu(\omega).$$

Let $H^2_\pm(\mathbb{R})$ be the Hardy space on the upper (lover) half-plane in $\mathbb{C}$. Recall one old and well known result: *the Fourier transform F gives an isometrical isomorphism of the space $L_2(\mathbb{R})$, under which*

$$\begin{aligned} F &: H^2_+(\mathbb{R}) \longrightarrow L_2(\mathbb{R}_+), \\ F &: H^2_-(\mathbb{R}) \longrightarrow L_2(\mathbb{R}_-). \end{aligned}$$

Furthermore, $L_2(\mathbb{R}) = H^2_+(\mathbb{R}) \oplus H^2_-(\mathbb{R})$.

In this context a number of natural questions arises for the Bergman space $\mathcal{A}^2(\Pi)$:

- how much room does the Bergman space $\mathcal{A}^2(\Pi)$ occupy inside $L_2(\Pi)$,
- does there exist a complete decomposition of $L_2(\Pi)$ onto analytic and analytic like spaces,
- does there exist an unitary operator on $L_2(\Pi)$, which *simultaneously* reduces all the pieces of decomposition to some simple transparent form,

Answering to these questions we introduce the unitary operators

$$U_1 = F \otimes I : L_2(\Pi) = L_2(\mathbb{R}) \otimes L_2(\mathbb{R}_+) \longrightarrow L_2(\mathbb{R}) \otimes L_2(\mathbb{R}_+),$$

where $F : L_2(\mathbb{R}) \to L_2(\mathbb{R})$ is the Fourier transform

$$(Ff)(u) = \frac{1}{\sqrt{2\pi}}\int_{\mathbb{R}} e^{-iu\xi} f(\xi)\, d\xi,$$

and

$$U_2 : L_2(\Pi) = L_2(\mathbb{R}) \otimes L_2(\mathbb{R}_+) \longrightarrow L_2(\mathbb{R}) \otimes L_2(\mathbb{R}_+)$$

which is defined by the rule

$$U_2 : \varphi(u,v) \longmapsto \frac{1}{\sqrt{2|x|}}\, \varphi(x, \frac{y}{2|x|}).$$

Letting $\ell_0(y) = e^{-y/2}$, we have $\ell_0(y) \in L_2(\mathbb{R}_+)$ and $\|\ell_0(y)\| = 1$. Denote by L_0 the one-dimensional subspace of $L_2(\mathbb{R}_+)$ generated by $\ell_0(y)$.

Theorem 3.1 *The unitary operator $U = U_2U_1$ is an isometric isomorphism of the space $L_2(\Pi) = L_2(\mathbb{R}) \otimes L_2(\mathbb{R}_+)$ under which the Bergman space $\mathcal{A}^2(\Pi)$ is mapped onto $L_2(\mathbb{R}_+) \otimes L_0$,*

$$U : \mathcal{A}^2(\Pi) \longrightarrow L_2(\mathbb{R}_+) \otimes L_0.$$

Besides of Bergman space $\mathcal{A}^2(\Pi)$ we introduce the space $\widetilde{\mathcal{A}}^2(\Pi)$ as the (closed) subspace of $L_2(\Pi)$ consisting of all anti-analytic functions. Furthermore introduce the spaces of poly-analytic and poly-anti-analytic functions (see, for example, [2, 6, 7]), the poly-Bergman spaces.

We define the space $\mathcal{A}_n^2(\Pi)$ of n-analytic functions as the (closed) subspace of $L_2(\Pi)$ of all functions $\varphi = \varphi(w, \overline{w}) = \varphi(u, v)$, which satisfy the equation

$$\left(\frac{\partial}{\partial \overline{w}}\right)^n \varphi = \frac{1}{2^n}\left(\frac{\partial}{\partial u} + i\frac{\partial}{\partial v}\right)^n \varphi = 0.$$

Similarly, define the space $\widetilde{\mathcal{A}}_n^2(\Pi)$ of n-anti-analytic functions as the (closed) subspace of $L_2(\Pi)$ of all functions $\varphi = \varphi(w, \overline{w}) = \varphi(u, v)$, which satisfy the equation

$$\left(\frac{\partial}{\partial w}\right)^n \varphi = \frac{1}{2^n}\left(\frac{\partial}{\partial u} - i\frac{\partial}{\partial v}\right)^n \varphi = 0.$$

Of course, we have $\mathcal{A}_1^2(\Pi) = \mathcal{A}^2(\Pi)$ and $\widetilde{\mathcal{A}}_1^2(\Pi) = \widetilde{\mathcal{A}}^2(\Pi)$, for $n = 1$, as well as $\mathcal{A}_n^2(\Pi) \subset \mathcal{A}_{n+1}^2(\Pi)$ and $\widetilde{\mathcal{A}}_n^2(\Pi) \subset \widetilde{\mathcal{A}}_{n+1}^2(\Pi)$, for each $n \in \mathbb{N}$.

Introduce now the spaces $\mathcal{A}_{(n)}^2$ of true-n-analytic functions and $\widetilde{\mathcal{A}}_{(n)}^2$ of true-n-anti-analytic functions by

$$\mathcal{A}_{(n)}^2 = \mathcal{A}_n^2 \ominus \mathcal{A}_{n-1}^2, \qquad \widetilde{\mathcal{A}}_{(n)}^2 = \widetilde{\mathcal{A}}_n^2 \ominus \widetilde{\mathcal{A}}_{n-1}^2,$$

for $n > 1$, and by $\mathcal{A}_{(1)}^2 = \mathcal{A}_1^2$, $\widetilde{\mathcal{A}}_{(1)}^2 = \widetilde{\mathcal{A}}_1^2$, for $n = 1$. Then, of course,

$$\mathcal{A}_n^2 = \bigoplus_{k=1}^n \mathcal{A}_{(k)}^2, \qquad \widetilde{\mathcal{A}}_n^2 = \bigoplus_{k=1}^n \widetilde{\mathcal{A}}_{(k)}^2.$$

Recall (see, for example, [8]), that the Laguerre polinomial $L_n(y)$ of degree n, $n = 0, 1, 2, \ldots$, and type 0 is defined by

$$L_n(y) = \frac{e^y}{n!}\frac{d^n}{dy^n}(e^{-y}\, y^n) = \sum_{k=0}^n \frac{n!}{k!(n-k)!}\frac{(-y)^k}{k!}, \qquad y \in \mathbb{R}_+,$$

and that the system of functions $\ell_n(y) = e^{-y/2}L_n(y)$, $n = 0, 1, 2, \ldots$ forms an ortonormal base in the space $L_2(\mathbb{R}_+)$.

Denote by L_n, $n = 0, 1, 2, \ldots$, the one-dimensional subspace of $L_2(\mathbb{R}_+)$ generated by the function $\ell_n(y)$. Note that for $n = 0$ this definition gives exactly the previously defined space L_0.

Theorem 3.2 *We have the following isometric isomorphisms and decompositions of the spaces:*

1 Isomorphic images of poly-analytic spaces

$$U \; : \; \mathcal{A}^2_{(n)}(\Pi) \longrightarrow L_2(\mathbb{R}_+) \otimes L_{n-1},$$

$$U \; : \; \mathcal{A}^2_{n}(\Pi) \longrightarrow L_2(\mathbb{R}_+) \otimes \bigoplus_{k=0}^{n-1} L_k,$$

$$U \; : \; \bigoplus_{k=1}^{\infty} \mathcal{A}^2_{(k)}(\Pi) \longrightarrow L_2(\mathbb{R}_+) \otimes L_2(\mathbb{R}_+).$$

2 Isomorphic images of poly-anti-analytic spaces

$$U \; : \; \widetilde{\mathcal{A}}^2_{(n)}(\Pi) \longrightarrow L_2(\mathbb{R}_-) \otimes L_{n-1},$$

$$U \; : \; \widetilde{\mathcal{A}}^2_{n}(\Pi) \longrightarrow L_2(\mathbb{R}_-) \otimes \bigoplus_{k=0}^{n-1} L_k,$$

$$U \; : \; \bigoplus_{k=1}^{\infty} \widetilde{\mathcal{A}}^2_{(k)}(\Pi) \longrightarrow L_2(\mathbb{R}_-) \otimes L_2(\mathbb{R}_+).$$

3 Decomposition of the space $L_2(\Pi)$

$$L_2(\Pi) = \bigoplus_{k=1}^{\infty} (\mathcal{A}^2_{(k)}(\Pi) \oplus \widetilde{\mathcal{A}}^2_{(k)}(\Pi)) = \bigoplus_{k=1}^{\infty} \mathcal{A}^2_{(k)}(\Pi) \oplus \bigoplus_{k=1}^{\infty} \widetilde{\mathcal{A}}^2_{(k)}(\Pi).$$

Introduce the isometric imbedding $R_0 : L_2(\mathbb{R}_+) \to L_2(\mathbb{R}) \otimes L_2(\mathbb{R}_+)$ by the rule

$$(R_0 f)(x, y) = \chi_+(x)\, f(x)\, \ell_0(y),$$

where $\chi_+(x)$ is the characteristic function of the positive half-line. The adjoint operator $R_0^* : L_2(\Pi) \to L_2(\mathbb{R}_+)$ is given by

$$(R_0^* \varphi)(x) = \chi_+(x) \int_{\mathbb{R}_+} \varphi(x, \eta)\, \ell_0(\eta)\, d\eta.$$

The operator $R = R_0^* U$ maps the space $L_2(\Pi)$ onto $L_2(\mathbb{R}_+)$, and the restriction

$$R|_{\mathcal{A}^2(\Pi)} : \mathcal{A}^2(\Pi) \longrightarrow L_2(\mathbb{R}_+)$$

is an isometric isomorphism. The adjoint operator

$$R^* = U^* R_0 : L_2(\mathbb{R}_+) \longrightarrow \mathcal{A}^2(\Pi) \subset L_2(\Pi)$$

is an isometric isomorphism of $L_2(\mathbb{R}_+)$ onto the subspace $\mathcal{A}^2(\Pi)$ of the space $L_2(\Pi)$.

Remark 3.3 We have

$$\begin{aligned} R R^* = I \quad &: \quad L_2(\mathbb{R}_+) \longrightarrow L_2(\mathbb{R}_+), \\ R^* R = B_\Pi \quad &: \quad L_2(\Pi) \longrightarrow \mathcal{A}^2(\Pi). \end{aligned}$$

Theorem 3.4 *The isomorphism $R^* : L_2(\mathbb{R}_+) \longrightarrow \mathcal{A}^2(\Pi)$ is given by*

$$(R^* f)(z) = \frac{1}{\sqrt{\pi}} \int_{\mathbb{R}_+} \sqrt{\xi}\, f(\xi)\, e^{iz\cdot\xi}\, d\xi.$$

Corollary 3.5 *The isomorphism $R : \mathcal{A}^2(\Pi) \longrightarrow L_2(\mathbb{R}_+)$ is given by*

$$(R\varphi)(x) = \sqrt{x}\, \frac{1}{\sqrt{\pi}} \int_\Pi \varphi(w)\, e^{-i\,\overline{w}\cdot x}\, d\mu(w).$$

Introduce the C^*-algebra of bounded measurable symbols which depend only on v (the imaginary part of a variable $w = u + iv$) $\mathcal{A}(\infty) = \mathbb{C} \otimes L_\infty(\mathbb{R}_+)$ $(\subset L_\infty(\mathbb{R} \times \mathbb{R}_+))$.

Theorem 3.6 *Let $a = a(v) \in \mathcal{A}(\infty)$. Then the Toeplitz operator T_a acting on $\mathcal{A}^2(\Pi)$ is unitary equivalent to the multiplication operator $\gamma_a I = R T_a R^*$, acting on $L_2(\mathbb{R}_+)$. The function $\gamma_a(x)$ is given by*

$$\gamma_a(x) = \int_{\mathbb{R}_+} a(\frac{y}{2x})\, e^{-y}\, dy, \quad x \in \mathbb{R}_+.$$

In the case of the unit disk and radial symbols as well as for the upper half-pane and homogeneous of zero order functions we have analogous results.

Theorem 3.7 *Let $a = a(r) \in L_\infty$. Then for an appropriate unitary operator R the Toeplitz operator T_a is unitary equivalent to the multiplication operator $\gamma_a I = R T_a R^*$, acting on l_2. The sequence $\gamma_a = \{\gamma_a(n)\}_{n\in\mathbb{Z}_+}$ is given by*

$$\gamma_a(n) = (n+1) \int_0^1 a(\sqrt{r})\, r^n\, dr, \quad n \in \mathbb{Z}_+ = \mathbb{N} \cup \{0\}.$$

Theorem 3.8 *Let $a = a(\theta) \in L_\infty$, $\theta = \arg z$. Then for an appropriate unitary operator R the Toeplitz operator T_a unitary equivalent to the multiplication operator $\gamma_a' I = R T_a R^*$, acting on $L_2(\mathbb{R})$. The function $\gamma_a'(\lambda)$, $\lambda \in \mathbb{R}$, is given by*

$$\gamma_a'(\lambda) = \frac{2\lambda}{1 - e^{-2\pi\lambda}} \int_0^\pi a(\theta)\, e^{-2\lambda\theta}\, d\theta, \qquad \lambda \in \mathbb{R}.$$

4. Unbounded symbols

Theorems 3.6-3.8 suggest considering not only bounded but unbounded symbols as well. For example, as a direct corollary from Theorem 3.7 we have.

Corollary 4.1 *The Toeplitz operator T_a with a radial symbol $a(r)$, where $a(\sqrt{r}) \in L_1(0,1)$, is bounded on $\mathcal{A}^2(\mathbb{D})$ if and only if*

$$\gamma_a = \{\gamma_a(n)\}_{n\in\mathbb{Z}_+} \in l_\infty, \qquad \text{and} \qquad \|T_a\| = \sup_{n\in\mathbb{Z}_+} |\gamma_a(n)|.$$

The Toeplitz operator T_a is compact if and only if $\gamma_a \in c_0$, that is

$$\lim_{n\to\infty} \gamma_a(n) = 0.$$

The spectrum of bounded Toeplitz operator T_a is given by

$$\operatorname{sp} T_a = \overline{\{\gamma_a(n) : n \in \mathbb{Z}_+\}}.$$

The behavior of a sequence $\gamma_a(n)$, when $n \to \infty$, is completely determined by the behavior of a function $a(r)$ in a neighborhood of the point $r = 1$. Given a radial symbol, introduce the function

$$B(s) = \int_s^1 a(\sqrt{r})dr.$$

Theorem 4.2 *If the function $B(s)$ when $s \to 1$ has the form*

$$|B(s)| = O(1-s), \tag{4.1}$$

then

$$\sup_{n\in\mathbb{Z}_+} |\gamma_a(n)| < \infty.$$

If

$$|B(s)| = o(1-s), \tag{4.2}$$

then

$$\lim_{n\to\infty} \gamma_a(n) = 0.$$

In fact Theorem 4.2 says that the behavior near the boundary of a certain average of symbols, rather than the behavior of the symbols themselves, is responsible for the boundedness and compactness properties of corresponding Toeplitz operators. That is, in spite of bad behavior of a symbol, which can be even unbounded near the boundary, the corresponding Toeplitz operator can be bounded and even compact.

Example 4.3 Let

$$a(r) = (1 - r^2)^{-\beta} \sin(1 - r^2)^{-\alpha}, \quad \alpha > 0, \quad \beta < 1.$$

Then the corresponding Toeplitz operator T_a

- is bounded on $\mathcal{A}^2(\mathbb{D})$ for $\alpha \geq \beta$;
- is compact on $\mathcal{A}^2(\mathbb{D})$ for $\alpha > \beta$.

For $\beta \leq 0$ the symbol is bounded, while for $\beta > 0$ the symbol is *unbounded near the boundary* $\partial\mathbb{D}$.

Note that the last case ($\alpha > \beta > 0$) provides the most surprising example, a *compact* Toeplitz operator whose symbol is *unbounded near each point* of the boundary $\partial\mathbb{D}$.

Conditions (4.1) and (4.2) are sufficient for boundedness and compactness of an operator T_a in general. It is known [15], that for bounded symbols $a(r) \in L_\infty(0,1)$ the condition (4.2) is necessary and sufficient for compactness of T_a on $\mathcal{A}^2(\mathbb{D})$.

A case when conditions (4.1) and (4.2) are necessary for L_1 symbols is described in the next theorem. Note that for general L_1 symbols the conditions (4.1) and (4.2) fail to be necessary.

Theorem 4.4 *Let $a(\sqrt{r}) \in L_1(0,1)$, and $a(r) \geq 0$ almost everywhere. Then the conditions* (4.1) *and* (4.2) *are necessary and sufficient for $\gamma_a \in l_\infty$ and $\gamma_a \in c_0$, respectively.*

Example 4.5 Consider the following family of radial symbols

$$a_\alpha(r) = (1 - r)^{\alpha - 1}, \quad \text{where} \quad \alpha > 0,$$

which scales the (polynomial) growth of symbols near the boundary. We have

$$B_\alpha(s) = \int_s^1 a_\alpha(\sqrt{r})\, dr = \frac{2}{\alpha} s(1-s)^\alpha + \frac{2}{\alpha(\alpha+1)}(1-s)^{\alpha+1}.$$

By Theorem 4.4 the operator T_{a_α} is bounded if and only if $\alpha \geq 1$, and compact if and only if $\alpha > 1$. That is, in this scale unbounded symbols generate unbounded Toeplitz operators. Moreover, as it will follow from Corollary 4.6, to generate bounded or compact Toeplitz operator its unbounded symbol must necessarily have sufficiently sophisticated oscillating behavior near the unit circle $\partial\mathbb{D}$.

For a non negative symbol $a(r)$ introduce the function

$$m_a(u) = \inf_{r \in [u,1)} a(r),$$

which is obviously always monotone.

Corollary 4.6 *If* $\lim_{u\to 1} m_a(u) = +\infty$ *(which is equivalent to* $\lim_{r\to 1} a(r) = +\infty$*), then the Toeplitz operator* T_a *is unbounded.*

The C^*-algebra generated by bounded Toeplitz operators with (unbounded) radial symbols has certain peculiarities. In particular, contrary to commonly known and studied cases (see, for example [30]), the Toeplitz operator algebra is commutative and, at the same time, the semicommutators $[T_{a_1}, T_{a_2}) = T_{a_1} \cdot T_{a_2} - T_{a_1\cdot a_2}$ are not compact in general. Moreover, the symbols under study do not form an algebra under pointwise multiplication. That is, having two radial symbols $a_1(r)$ and $a_2(r)$, for which the corresponding Toeplitz operators $T_{a_1(r)}$ and $T_{a_2(r)}$ are bounded, the Toeplitz operator $T_{a_1\cdot a_2}$ which corresponds to the product of these symbols is not necessarily bounded. The natural structure on the set of symbols under consideration is a linear space (in the algebraic sense, i.e., no norm structure assumed).

Example 4.7 Let

$$a_1(r) = \sin(1-r^2)^{-\alpha}, \quad a_2(r) = (1-r^2)^{-\beta}\sin(1-r^2)^{-\alpha},$$

where $0 \le \beta < 1$, $\beta < \alpha$. Both T_{a_1} and T_{a_2} are bounded and compact. At the same time

- for $\beta = 0$ the operator $T_{a_1\cdot a_2}$ is bounded but *not compact*; i.e., the semicommutator $[T_{a_1}, T_{a_2}) = T_{a_1} \cdot T_{a_2} - T_{a_1\cdot a_2}$ is *not compact*;
- for $\beta > 0$ the operator $T_{a_1\cdot a_2}$ is *not bounded* at all.

5. Commutator properties and representations of C^*-algebras

Let $\mathcal{A}(\mathbb{D})$ be a C^*-subalgebra of $L_\infty(\mathbb{D})$. Denote by $\mathcal{T}(\mathcal{A}(\mathbb{D}))$ the C^*-algebra (on $\mathcal{A}^2(\mathbb{D})$) generated by the Toeplitz operators T_a, with $a \in \mathcal{A}(\mathbb{D})$, and denote by $\mathcal{R}(\mathcal{A}(\mathbb{D}), B_{\mathbb{D}})$ by the C^*-algebra (on $L_2(\mathbb{D})$) generated by the operators

$$A = aI + bB_{\mathbb{D}}, \qquad a,\, b \in \mathcal{A}(\mathbb{D}).$$

Given $\mathcal{A}(\mathbb{D})$, consider the following statements (which may be false or true depending on $\mathcal{A}(\mathbb{D})$) :

1) For each $a \in \mathcal{A}(\mathbb{D})$ the commutator $[B_{\mathbb{D}}, aI] = B_{\mathbb{D}}aI - aB_{\mathbb{D}}$ is compact.
2) For each pair $a,\, b \in \mathcal{A}(\mathbb{D})$ the semi-commutator $[T_a, T_b) = T_aT_b - T_{ab}$ is compact.
3) For each pair $a,\, b \in \mathcal{A}(\mathbb{D})$ the commutator $[T_a, T_b] = T_aT_b - T_bT_a$ is compact.

We have obviously 1) $\Rightarrow$ 2) $\Rightarrow$ 3). In fact 1) $\Leftrightarrow$ 2), but 3) is certainly stronger then both 1) and 2). A simpler and better known example here is the algebra

$\mathcal{A}(\mathbb{D}) = PC(\overline{\mathbb{D}}, L)$ of piece-wise continuous functions, i.e., the functions continuous in $\overline{\mathbb{D}} \setminus L$, and having one-sided limit values at each point of a simple (smooth) line L, which connects two points of the boundary $\partial\mathbb{D}$ of the unit disk (see, for example, [21]).

The very important question of description of the maximal algebra satisfying the statement 2) was solved in [28] (see also [30]). One of the main features of algebras having property 2) is that the corresponding Toeplitz operator algebras $\mathcal{T}(\mathcal{A}(\mathbb{D}))$ admit the commutative symbolic calculus, i.e., the symbol (Calkin) algebra $\operatorname{Sym}\mathcal{T}(\mathcal{A}(\mathbb{D})) = \mathcal{T}(\mathcal{A}(\mathbb{D}))/\mathcal{K}$, where $\mathcal{K}$ is the ideal of compact operator, is commutative.

Note, that under the condition 2) the symbol algebra $\operatorname{Sym}\mathcal{R}(\mathcal{A}(\mathbb{D}), B_{\mathbb{D}}) = \mathcal{R}(\mathcal{A}(\mathbb{D}), B_{\mathbb{D}})/\mathcal{K}$ of the algebra $\mathcal{R}(\mathcal{A}(\mathbb{D}), B_{\mathbb{D}})$ is commutative as well.

At the same time, under the (only) condition 3) the symbol algebra $\operatorname{Sym}\mathcal{R}(\mathcal{A}(\mathbb{D}), B_{\mathbb{D}}) = \mathcal{R}(\mathcal{A}(\mathbb{D}), B_{\mathbb{D}})/\mathcal{K}$ of the algebra $\mathcal{R}(\mathcal{A}(\mathbb{D}), B_{\mathbb{D}})$ is *non-commutative*, while the symbol algebra $\operatorname{Sym}\mathcal{T}(\mathcal{A}(\mathbb{D})) = \mathcal{T}(\mathcal{A}(\mathbb{D}))/\mathcal{K}$ still remains *commutative.*

The natural questions appear:

- describe functional algebras $\mathcal{A}(\mathbb{D})$ having the (only) property 3),
- how complicated the symbol algebras of the corresponding algebras $\mathcal{R}(\mathcal{A}(\mathbb{D}), B_{\mathbb{D}})$ can be, while Toeplitz operator algebras $\mathcal{T}(\mathcal{A}(\mathbb{D}))$ still admit the commutative symbolic calculus.

Note that, in spite of the sufficiently old complete description of algebras with the property 2), there was no essential progress in understanding of a difference between properties 2) and 3) during last 30 years. Basically only a single example of an algebra having the (only) property 3) was known in both Hardy and Bergman Toeplitz operator settings. Namely, Toeplitz operators with piecewise continuous functions commute modulo compact operators, while the symbol (Calkin) algebra of the algebra generated by piecewise continuous functions and by the projection has one- and two-dimensional irreducible representations.

Given a pencil of hyperbolic geodesics, consider the algebra of L_∞-functions constant on the corresponding cycles. Such algebras provide a series of new examples of algebras with even more strong property: *the corresponding Toeplitz operators just commute* (more strong that 3)), *and at the same time property 2) does not hold.*

The complete in a sense answer to the second question for the Bergman space setting is given in the next theorem. At the same time the above questions still remain open for the case of Toeplitz operators on the Hardy space.

Theorem 5.1 *For each finite set* $\Lambda = \langle n_0, n_1, ..., n_m \rangle$, *where* $1 = n_0 < n_1 < ... < n_m \leq \infty$, *and* $n_k \in \mathbb{N} \cup \{\infty\}$, *there are algebras* $\mathcal{A}_\Lambda \subset L_\infty(\mathbb{D})$, *such that the symbol algebras* $\operatorname{Sym}\mathcal{T}(\mathcal{A}_\Lambda)$ *of the algebras* $\mathcal{T}(\mathcal{A}_\Lambda)$ *are commutative, while* *the symbol algebras* Sym $\mathcal{R}(\mathcal{A}_\Lambda, B_\mathbb{D})$ *of the algebras* $\mathcal{R}(\mathcal{A}_\Lambda, B_\mathbb{D})$ *have irreducible representations exactly of predefined dimensions* $n_0, n_1, \ldots, n_m$.

6. Dynamics of properties of Toeplitz operators

Given a smooth symbol $a = a(z)$, the family of Toeplitz operators $T_a = \{T_a^{(h)}\}$, with $h \in (0,1)$, is considered under the Berezin quantization procedure [3, 4]. For a fixed h the Toeplitz operator $T_a^{(h)}$ acts on the weighted Bergman space $\mathcal{A}_h^2(\mathbb{D})$. In the special quantization procedure each Toeplitz operator $T_a^{(h)}$ is represented by its Wick symbol $\widetilde{a}_h$, and the correspondence principle says that for smooth symbols one has

$$\lim_{h\to 0} \widetilde{a}_h = a.$$

Moreover by [14] the above limit remains valid in the L_1-sense for a wider class of symbols.

Thus it is very important and interesting to study the behavior of different properties (boundedness, compactness, spectral properties, etc.) of $T_a^{(h)}$ in dependence on h, and to compare their limit behavior under $h \to 0$ with corresponding properties of the initial symbol a. We consider here Toeplitz operators only with radial symbols $a = a(r)$. In this case each Toeplitz operator is unitary equivalent to a multiplication operator acting on the one-sided space l_2. This permits us to get more explicit information than can be obtained studying general Wick symbols.

Introduce the weighted Bergman space $\mathcal{A}_\lambda^2(\mathbb{D})$ which consists of analytic in $\mathbb{D}$ functions having the finite norm

$$\|\varphi\|_{L_2(\mathbb{D},d\mu_\lambda)} = \left(\int_\mathbb{D} |\varphi(z)|^2 d\mu_\lambda(z)\right)^{1/2},$$

where

$$d\mu_\lambda(z) = (\lambda+1)(1-|z|^2)^\lambda \frac{1}{\pi} d\mu(z), \qquad \lambda \geq 0.$$

Note that our weighted Bergman spaces depend on a real parameter λ which is connected with h by the rule $\lambda + 2 = \frac{1}{h}$.

The next theorem extends the result of Theorem 3.7 to the case of weighted Bergman spaces $\mathcal{A}_\lambda^2(\mathbb{D})$.

Theorem 6.1 *Given a radial symbol* $a = a(r)$, *for a certain unitary operator* R_λ *the Toeplitz operator* $T_a^{(\lambda)}$ *acting on the weighted Bergman space* $\mathcal{A}_\lambda^2(\mathbb{D})$ *is unitary equivalent to the multiplication operator* $R_\lambda T_a^{(\lambda)} R_\lambda^{-1} = \gamma_{a,\lambda} I$ *acting on* l_2. *The sequence* $\gamma_{a,\lambda} = \{\gamma_{a,\lambda}(n)\}$ *is given by*

$$\gamma_{a,\lambda}(n) = \frac{1}{\mathrm{B}(n+1,\lambda+1)} \int_0^1 a(\sqrt{r})\,(1-r)^\lambda\, r^n dr. \tag{6.1}$$

6.1 Boundedness and compactness

For each bounded radial symbols $a(r)$ the corresponding Toeplitz operators $T_a^{(\lambda)}$ is bounded simultaneously on all weighted spaces $\mathcal{A}_\lambda^2(\mathbb{D})$. At the same time Toeplitz operators with unbounded symbols can be bounded (compact) on $\mathcal{A}_\lambda^2(\mathbb{D})$ for some values of λ and unbounded (non compact) for another.

Wishing to analyze the regions (on λ) of boundedness and compactness, for a given *unbounded* symbol $a(r)$ introduce the sets

$$B(a) = \{\lambda : T_a^{(\lambda)} \text{ is bounded}\}, \qquad K(a) = \{\lambda : T_a^{(\lambda)} \text{ is compact}\}.$$

Theorem 6.2 *Being non empty, the sets $B(a)$ and $K(a)$ can have only one of the following forms*

$$(i)\ \ [0,\infty), \qquad (ii)\ \ [0,\lambda_0), \qquad (iii)\ \ [0,\lambda_0].$$

The next theorem gives an example of a different behavior of the sets $B(a)$ and $K(a)$.

Theorem 6.3 *There is a symbol $a_{\beta,\nu}(r)$, where $\beta \in \mathbb{R}$, $\nu \in (0,1)$, for which*

$$\begin{array}{llll} a) & B(a_{\nu,\beta}) = [0,\nu), & K(a_{\nu,\beta}) = [0,\nu), & \beta > 0, \\ b) & B(a_{\nu,0}) = [0,\nu], & K(a_{\nu,0}) = [0,\nu), & \beta = 0, \\ c) & B(a_{\nu,\beta}) = [0,\nu], & K(a_{\nu,\beta}) = [0,\nu], & \beta < 0. \end{array}$$

6.2 Spectra

Let E be a subset of $\mathbb{R}$ having $+\infty$ as a limit point (normally $E = [0,+\infty)$), and let for each $\lambda \in E$ there is a set $M_\lambda \subset \mathbb{C}$. Define the set M_∞ as the set of all $z \in \mathbb{C}$ for which there exists a sequence of complex numbers $\{z_n\}_{n\in\mathbb{N}}$ such that

(i) for each $n \in \mathbb{N}$ there exists $\lambda_n \in E$ such that $z_n \in M_{\lambda_n}$,

(ii) $\lim_{n\to\infty} \lambda_n = +\infty$,

(iii) $z = \lim_{n\to\infty} z_n$.

We will write

$$M_\infty = \lim_{\lambda\to+\infty} M_\lambda,$$

and call M_∞ the (partial) limit set of a family $\{M_\lambda\}_{\lambda\in E}$ when $\lambda \to +\infty$.

The *a priori* spectral information for Toeplitz operators with L_∞-symbols is given, for example, in [5], and says that for each $a \in L_\infty(\mathbb{D})$ and each λ

$$\operatorname{sp} T_a^{(\lambda)} \subset \operatorname{conv}(\text{ess-Range}\, a).$$

For a radial symbol $a = a(r)$ the Toeplitz operator $T_a^{(\lambda)}$ is unitary equivalent to the multiplication operator $\gamma_{a,\lambda} I$, where the sequence $\gamma_{a,\lambda} = \{\gamma_{a,\lambda}(n)\}$ is given by (6.1). Thus we have obviously

$$\operatorname{sp} T_a^{(\lambda)} = \overline{M_\lambda(a)},$$

where $M_\lambda(a) = \operatorname{Range} \gamma_{a,\lambda}$.

An interesting question here is to study the limit behavior of $\operatorname{sp} T_a^{(\lambda)}$ when $\lambda \to \infty$.

Theorem 6.4 *Let* $a = a(r) \in C[0,1]$. *Then*

$$\lim_{\lambda \to +\infty} \operatorname{sp} T_a^{(\lambda)} = \lim_{\lambda \to +\infty} M_\lambda(a) = \operatorname{Range} a. \tag{6.2}$$

Obviously the set $\operatorname{Range} a$ coincides with the spectrum $\operatorname{sp} aI$ of the operator of multiplication by $a = a(r)$ acting, say, on any of $L_2(\mathbb{D}, d\mu_\lambda)$, thus another form of (6.2) is

$$\lim_{\lambda \to +\infty} \operatorname{sp} T_a^{(\lambda)} = \operatorname{sp} aI.$$

Recall that for a continuous symbol $a(r)$ and for each fixed λ the spectrum $\operatorname{sp} T_a^{(\lambda)}$ coincides with the closure of the set $\gamma_{a,\lambda} = \{\gamma_{a,\lambda}(n)\}$, i.e., is a discrete set with the unique limit point $a(1)$, and, in general, seems to have no any strict connection with the range of $a(r)$. The definite tendency in the behavior of elements of the sequence $\gamma_{a,\lambda}$ starts appearing as λ tends to ∞, and the limit set $M_\infty(a)$ of those sequences coincides with the range of the initial symbol $a(r)$.

We illustrate this effect on the symbol $a(r) = (1 + (0.9i - 1)r^2)^4$ for three values of λ: $\lambda = 0$, $\lambda = 5$, and $\lambda = 100$.

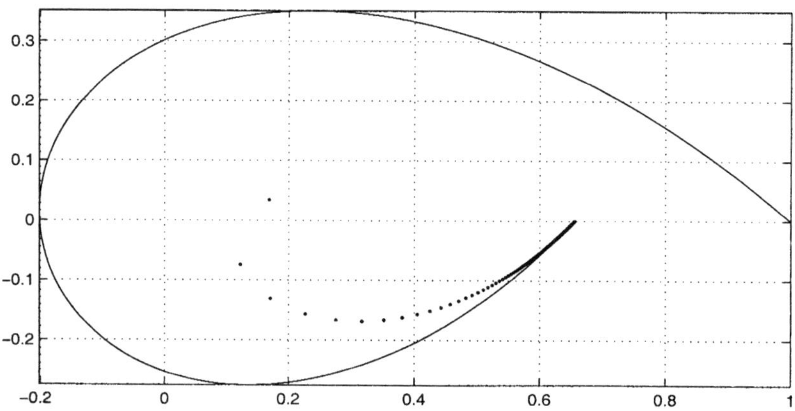

The sequence $\gamma_{a,\lambda} = \{\gamma_{a,\lambda}(n)\}$ for $\lambda = 0$.

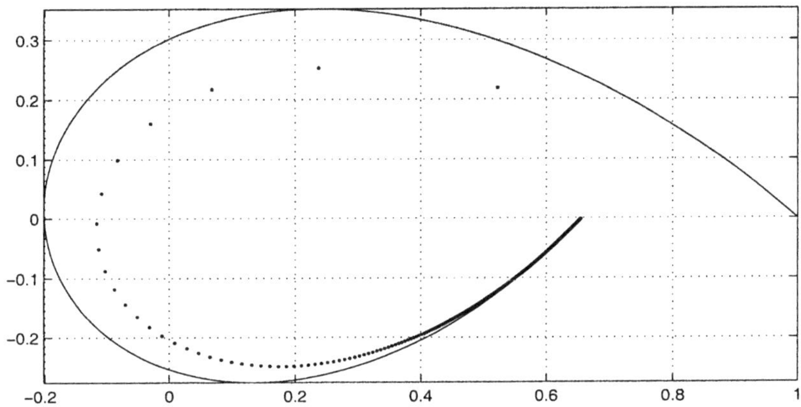

The sequence $\gamma_{a,\lambda} = \{\gamma_{a,\lambda}(n)\}$ for $\lambda = 5$.

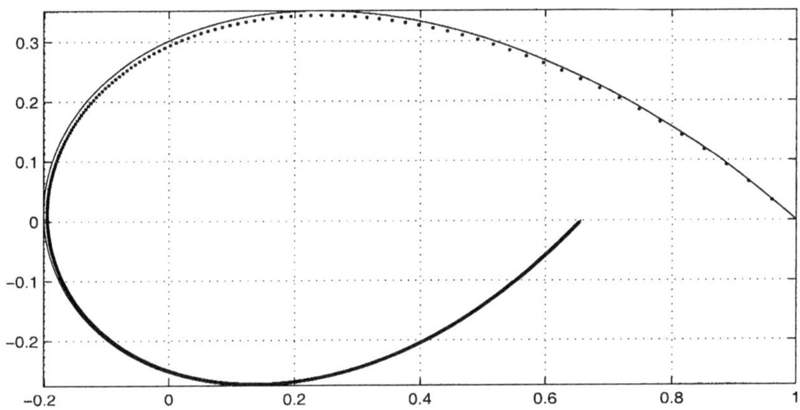

The sequence $\gamma_{a,\lambda} = \{\gamma_{a,\lambda}(n)\}$ for $\lambda = 100$.

For the piecewise continuous symbols the situation becomes a bit more complicated.

Let $b(r) = a(\sqrt{r})$ be a piecewise continuous function having jumps at a finite set of points,

$$0 < r_1 < r_2 < \ldots < r_m < 1.$$

Introduce the sets

$$J_j(a) = \{z \in \mathbb{C} : z = a(\sqrt{r}), r \in [r_j + 0, r_{j+1} - 0]\},$$

where $j = 0, 1, \ldots, m$, $r_0 = 0$ and $r_{m+1} = 1$, and $I_j(a)$ being the straight line segment with the endpoints $a(\sqrt{r_j - 0})$ and $a(\sqrt{r_j + 0})$, for all $j = 0, 1, \ldots, m$.

Introduce now

$$\widetilde{R}(a) = \left(\bigcup_{j=0}^{m} J_j(a)\right) \bigcup \left(\bigcup_{j=1}^{m} I_j(a)\right) = \operatorname{Range} a \cup \left(\bigcup_{j=1}^{m} I_j(a)\right),$$

Theorem 6.5 *Let $b(r) = a(\sqrt{r})$ be a piecewise continuous function as above. Then*

$$\lim_{\lambda \to +\infty} \operatorname{sp} T_a^{(\lambda)} = M_\infty(a) = \widetilde{R}(a).$$

Note, that the appearance of line segments which connect the one-sided limit values at the points of discontinuity of symbols, is quite typical in the theory of Toeplitz operators with piecewise continuous symbols acting either on the Hardy, or on the Bergman space (see, for example, [9, 21]). However we stress the principal difference between our case and the cases just mentioned. In the mentioned case of Toeplitz operators with piecewise continuous symbols the line segments appear in the *essential spectrum* of the Toeplitz operator. In our case any Toeplitz operator is a compact perturbation of a multiple of the identity, i.e., $T_{a(r)} = a(1)I + K$, and its essential spectrum consists of a single point $a(1)$ for all λ. For each fixed λ the spectrum of a Toeplitz operator coincides with the union of the *discrete set* (the sequence $\gamma_{a,\lambda}(n)$) with its limit point $a(1)$. The tendency of the line segment forming stars appearing for large values of λ, and the line segments themselves appear only in the *limit set of spectra.*

We illustrate this effect on the piecewise continuous symbol

$$a(r) = \begin{cases} e^{-i\pi r^2}, & r \in [0, 1/\sqrt{2}], \\ e^{i\pi r^2}, & r \in (1/\sqrt{2}, 1], \end{cases}$$

for λ taking the values 0, 4, 40, and 200.

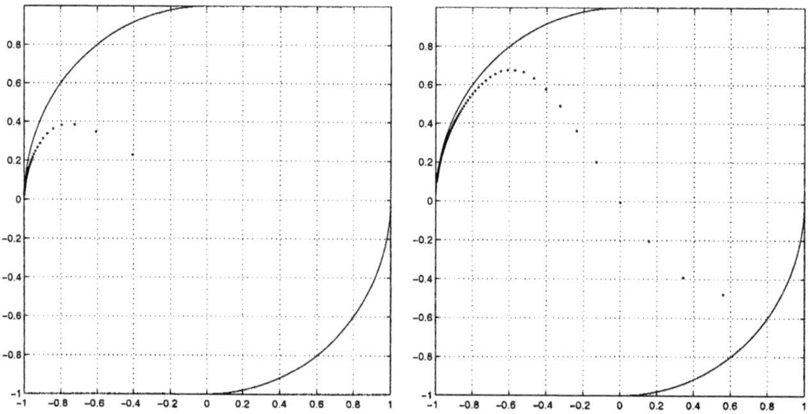

The sequence $\gamma_{a,\lambda} = \{\gamma_{a,\lambda}(n)\}$ for $\lambda = 0$ and $\lambda = 4$.

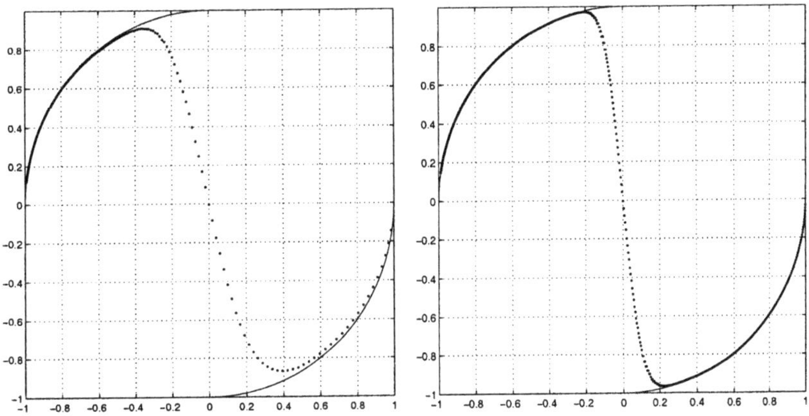

The sequence $\gamma_{a,\lambda} = \{\gamma_{a,\lambda}(n)\}$ for $\lambda = 40$ and $\lambda = 200$.

Note that the additional line segments, which connect the one-sided limit values of symbols, may essentially enlarge the limit set $M_\infty(a)$ comparing to the range of a symbol. For a measurable and, in general, unbounded symbol one always has

$$\text{Range}\, a \subset M_\infty(a) \subset \text{conv}(\text{Range}\, a),$$

and the gap between these extreme sets can be substantial. In particular, in [11] we give two examples of radial symbols with $\text{clos}(\text{Range}\, a) = S^1 = \partial\mathbb{D}$ and with $\text{Range}\, a = S^1$, such that in the first case $M_\infty(a) = \mathbb{D}$, while $M_\infty(a)$ is a countable union of circles whose radii tend to 1, in the second case. For unbounded symbols the limit set $M_\infty(a)$ may even coincide with the whole complex plane $\mathbb{C}$.

Acknowledgements

This paper has been written during the author's sabbatical stay at Chemnitz Technical University, Chemnitz, Germany. The author acknowledges CONACYT, Mexico and DFG Grant 444 MEX-112/2/01, Germany supporting in part this stay.

References

[1] Axler, S. and Zheng, D., Compact operators via the Berezin transform, *Indiana Univ. Math. J.*, **47** (1998), no. 2, 387–400.

[2] Balk, M. B. and Zuev, M. F., On polyanalytic functions, *Russ. Math. Surveys*, **25** (1970), no. 5, 201–223.

[3] Berezin, F. A., Quantization, *Math. USSR Izvestija*, **8** (1974), 1109–1165.

[4] Berezin, F. A., General concept of quantization, *Commun. Math. Phys.*, **40** (1975), 135–174.

[5] Berezin, F. A., *The Method of Second Quantization.* Nauka, Moscow, 1986 (in Russian).

[6] Dzhuraev, A., Multikernel functions of a domain, kernel operators, singular integral operators, *Soviet Math. Dokl.*, **32** (1985), no. 1, 251–253.

[7] Dzhuraev, A., *Methods of Singular Integral Equations*, Pitman Monographs and Surveys in Pure and Applied Mathematics, **60**. Longman Scientific & Technical, Harlow, 1992.

[8] Erdélyi, A., Magnus, W., Oberhettinger, F., and Tricomi, F. G., *Higher Transcendental Functions*, Vol. II. McGraw-Hill, New York, Toronto, London, 1954.

[9] Gohberg, I. and Krupnik, N., On the algebra generated by one–dimensional singular integral operators with piece–wise continuous coefficients, *Funct. Analysis Appl.*, **4** (1970), no. 3, 193–201.

[10] Grudsky, S., Karapetyants, A., and Vasilevski, N., Toeplits operators on the unit ball in $\mathbb{C}^n$ with radial symbols. Reporte Interno #292, Departamento de Matemáticas, CINVESTAV del I.P.N., Mexico City, 2001. *J. Operator Theory*, to appear.

[11] Grudsky, S., Karapetyants, A., and Vasilevski, N., Dynamics of properties of Toeplitz operators with radial symbols. Reporte Interno #317, Departamento de Matemáticas, CINVESTAV del I.P.N., Mexico City, 2002.

[12] Grudsky, S. and Vasilevski, N., Bergman-Toeplitz operators: radial component influence, *Integr. Equat. Oper. Theor.*, **40** (2001), no. 1, 16–33.

[13] Grudsky, S. and Vasilevski, N., Toeplitz operators on the Fock space: radial component effects, *Integr. Equat. Oper. Theor.*, **44** (2002), no. 1, 10–37.

[14] Hedenmalm, H., Korenblum, B., and Zhu, K., *Theory of Bergman Spaces*, Graduate Texts in Mathematics, **199**. Springer-Verlag, New York, 2000.

[15] Korenblum, B. and Zhu, K., An application of Tauberian theorems to Toeplitz operators, *J. Operator Theory*, **39** (1995), 353–361.

[16] Miao, J., Toeplitz operators with bounded radial symbols on the harmonic Bergman space of the unit ball, *Acta Sci. Math. (Szeged)*, **63**, (1997), 639–645.

[17] Peng, L., Rochberg, R., and Wu, Z., Orthogonal polynomials and middle Hankel operators on Bergman spaces, *Studia Math.*, **102** (1992), no. 1, 57–75.

[18] Sangadji and Stroethoff, K., Compact Toeplitz operators on generalized Fock spaces, *Acta Sci. Math. (Szeged)*, **64** (1998), 657–669.

[19] Stroethoff, K., Compact Toeplitz operators on Bergman spaces, *Math. Poc. Camb. Phil. Soc.*, **124** (1998), 151–160.

[20] Stroethoff, K., Compact Toeplitz operators on weighted harmonic Bergman spaces, *J. Austal. Math. Soc. (Series A)*, **64** (1998), 136–148.

[21] Vasilevski, N. L., Banach algebras generated by two-dimensional integral operators with Bergman kernel and piece-wise continuous coefficients, I, *Soviet Math. (Izv. VUZ)*, **30** (1986), no. 3, 14–24.

[22] Vasilevski, N. L., On the structure of Bergman and poly–Bergman spaces, *Integr. Equat. Oper. Theor.*, **33** (1999), 471–488.

[23] Vasilevski, N. L., On Bergman-Toeplitz operators with commutative symbol algebras, *Integr. Equat. Oper. Theor.*, **34** (1999), 107–126.

[24] Vasilevski, N. L., The Bergman space in tube domains, and commuting Toeplitz operators, *Doklady, Mathematics*, **61** (2000), no. 3, 9–12.

[25] Vasilevski, N. L., Bergman space structure, commutative algebras of Toeplitz operators and hyperbolic geometry. Reporte Interno #284, Departamento de Matemáticas, CINVESTAV del I.P.N., Mexico City, 20 p., 2000. *Integr. Equat. Oper. Theor.*, to appear.

[26] Vasilevski, N. L., Poly-Fock spaces. In: Operator Theory: Advances and Applications, **117**, 371–386. Birkhäuser Verlag, 2000.

[27] Vasilevski, N. L., Toeplitz operators on the Bergman spaces: Inside-the-domain effects, *Contemp. Math.*, **289** (2001), 79–146.

[28] Zhu, K., VMO, ESV, and Toeplitz operators on the Bergman space, *Trans. Amer. Math. Soc.*, **302** (1987), 617–646.

[29] Zhu, K., Positive Toeplitz operators on weighted Bergman space, *J. Operator Theory*, **20** (1988), 329–357.

[30] Zhu, K., *Operator Theory in Function Spaces*, Monographs and Textbooks in Pure and Applied Mathematics, **139**. Marcel Dekker Inc., New York, 1990.

MIX
Papier aus verantwortungsvollen Quellen
Paper from responsible sources
FSC® C105338

If you have any concerns about our products,
you can contact us on
ProductSafety@springernature.com

In case Publisher is established outside the EU,
the EU authorized representative is:
Springer Nature Customer Service Center GmbH
Europaplatz 3, 69115 Heidelberg, Germany

Printed by Libri Plureos GmbH
in Hamburg, Germany